Lecture Notes in Artificial Intel

T0238261

Subseries of Lecture Notes in Computer Scien...

Edited by J. G. Carbonell and J. Siekmann

Lecture Notes in Computer Science

Edited by G. Goos, J. Hartmanis and J. van Leeuwen

Springer

Berlin
Heidelberg
New York
Barcelona
Hong Kong
London
Milan
Paris
Singapore
Tokyo

Henrik I. Christensen   Horst Bunke
Hartmut Noltemeier (Eds.)

Sensor Based Intelligent Robots

International Workshop
Dagstuhl Castle, Germany
September 28 – October 2, 1998
Selected Papers

 Springer

Series Editors

Jaime G. Carbonell, Carnegie Mellon University, Pittsburgh, PA, USA
Jörg Siekmann, University of Saarland, Saarbrücken, Germany

Volume Editors

Henrik I. Christensen
Numerical Analysis and Computing Science, Royal Institute of Technology
10044 Stockholm, Sweden
E-mail: hic@nada.kth.se

Horst Bunke
Institute of Computer Science and Applied Mathematics (IAM)
University of Bern
Neubrückstr. 10, 3012 Bern, Switzerland
E-mail: bunke@iam.unibe.ch

Hartmut Noltemeier
Department of Mathematics and Computer Science
University of Würzburg
Am Hubland, 97074 Würzburg, Germany
E-mail: noltemei@informatik.uni-wuerzburg.de

Cataloging-in-Publication data applied for

Die Deutsche Bibliothek - CIP-Einheitsaufnahme

Sensor based intelligent robots : international workshop, Dagstuhl
Castle, Germany, September 28 - October 2, 1998 ; selected papers /
Henrik I. Christensen ... (ed.). - Berlin ; Heidelberg ; New York ;
Barcelona ; Hong Kong ; London ; Milan ; Paris ; Singapore ; Tokyo :
Springer, 1999
 (Lecture notes in computer science ; 1724 : Lecture notes in
 artificial intelligence)
 ISBN 3-540-66933-7

CR Subject Classification (1998): I.2, I.4, I.3, D.2, I.6

ISBN 3-540-66933-7 Springer-Verlag Berlin Heidelberg New York

© Springer-Verlag Berlin Heidelberg 1999
Printed in Germany

Typesetting: Camera-ready by author
SPIN 10705474   06/3142 – 5 4 3 2 1 0   Printed on acid-free paper

Preface

Robotics is a highly interdisciplinary research topic, that requires integration of methods for mechanics, control engineering, signal processing, planning, graphics, human-computer interaction, real-time systems, applied mathematics, and software engineering to enable construction of fully operational systems. The diversity of topics needed to design, implement, and deploy such systems implies that it is almost impossible for individual teams to provide the critical mass required for such endeavours. To facilitate interaction and progress on sensor based intelligent robotics the organisation of inter-disciplinary workshops is necessary through which in-depth discussion can be used for cross dissemination between different disciplines.

The Dagstuhl foundation has organised a number of workshops on Modelling and Integration of Sensor Based Intelligent Robot Systems. The Dagstuhl seminars are all organised over a full week in a beautiful setting in the Saarland in Germany. The setting provides an ideal environment for in-depth presentations and rich interactions between the participants. This volume contains the papers presented during the third workshop held over the period September 28 - October 2, 1998.

All papers have been reviewed by one-three reviewers over a relativly short period. We wish to thank all the reviewers for their invaluable help in making this a high quality selection of papers.

We gratefully acknowledge the support of the Schloss Dagstuhl Foundation and the staff at Springer-Verlag. Without their support the production of this volume would not have been possible.

September 1999

H. I. Christensen
H. Bunke
H. Noltemeier

Table of Contents

Markov Localization for Reliable Robot Navigation and People Detection

Dieter Fox[1,2], Wolfram Burgard[1], and Sebastian Thrun[2]

[1] University of Bonn, Department of Computer Science III, Bonn, Germany
[2] Carnegie Mellon University, School of Computer Science, Pittsburgh, PA

Abstract Localization is one of the fundamental problems in mobile robotics. Without knowledge about their position mobile robots cannot efficiently carry out their tasks. In this paper we present Markov localization as a technique for estimating the position of a mobile robot. The key idea of this technique is to maintain a probability density over the whole state space of the robot within its environment. This way our technique is able to globally localize the robot from scratch and even to recover from localization failures, a property which is essential for truly autonomous robots. The probabilistic framework makes this approach robust against approximate models of the environment as well as noisy sensors. Based on a fine-grained, metric discretization of the state space, Markov localization is able to incorporate raw sensor readings and does not require predefined landmarks. It also includes a filtering technique which allows to reliably estimate the position of a mobile robot even in densely populated environments. We furthermore describe, how the explicit representation of the density can be exploited in a reactive collision avoidance system to increase the robustness and reliability of the robot even in situations in which it is uncertain about its position. The method described here has been implemented and tested in several real-world applications of mobile robots including the deployments of two mobile robots as interactive museum tour-guides.

1 Introduction

The problem of estimating the position of a mobile robot within its environment belongs to the fundamental problems of mobile robotics [10,2]. The knowledge about its position enables a mobile robot to carry out its tasks efficiently and reliably. In general, the problem is to estimate the location of the robot, i.e. its current state in its three-dimensional (x, y, θ) configuration space within its environment given a map and incoming sensory information. Methods of this type are regarded as so-called map-matching techniques, since they match measurements of standard sensors with the given model of the environment. The advantage of map-matching techniques is that they do not require any modifications of the environment and expensive special purpose sensors. However, there are several problems, these methods have to deal with. First, they must be able to deal with

Christensen et al. (Eds.): Sensor Based Intelligent Robots, LNAI 1724, pp. 1–20, 1999.

uncertain information coming from the inherent noise of sensory data. Second, models of the environment are generally approximative especially if the environment is populated. Finally, the methods have to deal with ambiguous situations which frequently arise for example in office environments with long corridors.

The position estimation techniques developed so far can be distinguished according to the type of problem they attack. *Tracking* or *local* techniques aim at compensating odometric errors occurring during robot navigation. They require, however, that the initial location of the robot is known and occasionally fail if they lost track of the robot's position. On the opposite side are the so called *global* techniques which are designed to estimate the position of the robot even under global uncertainty. Techniques of this type solve the so-called wake-up resp. initialization and kidnapped robot problems. They can estimate the position of the robot without any prior knowledge about it, and they can recover from situations in which the robot is exposed serious positioning errors coming for example from bumping into an object.

In this paper we present Markov localization as a technique for globally estimating the position of the robot given a model of the environment. It uses a probabilistic framework and estimates a position probability density over the whole state space of the robot. This allows the technique to globally estimate the robot's position. In the beginning this technique starts with a uniform distribution over the whole three-dimensional state-space of the robot. By integrating sensory input this method keeps track of multiple hypotheses and incrementally refines the density until it ends up with a uni-modal distribution. It furthermore can re-localize the robot in the case of localization failures. Both properties are a basic precondition for truly autonomous robots which are designed to operate autonomously over longer periods of time. Furthermore, Markov localization can deal with uncertain information. This is important because sensors such as ultrasound sensors as well as models of the environment such as occupancy grid maps are generally imperfect. Our method uses a fine-grained and metric discretization of the state space. This approach has several advantages. First, it provides accurate position estimates which allow a mobile robot to efficiently perform tasks like office delivery. Second, the method can integrate raw sensory input such as a single beam of an ultrasound sensor. Most approaches for global position estimation, in contrast to that, rely on assumptions about the nature of the environment such as the orthogonality or the types of landmarks found in that environment. Therefore, such techniques are prone to fail if the environment does not align well with these assumptions. Typical examples of such environments can be found in the experimental results section of this paper. Furthermore, our Markov localization technique includes a method for filtering sensory input, which is designed to increase the robustness of the position estimation process especially in densely populated environments such as museums or exhibitions. This way, our technique can even be applied in dynamic environments in which most of the robot's sensor readings do not correspond to the expected measurements, since the senor beams are reflected by people surrounding the robot or by other un-modelled objects. The technique has been implemented and proven robust in several long-term and real-world applications of mobile robots in populated environments.

The paper is organized as follows. In the next section we will describe the mathematical framework of Markov localization. Section 3 introduces the grid-based representation of the position probability density. It furthermore presents techniques for efficiently updating these densities in real-time and also introduces two filtering schemes to deal with obstacles in the robot's environment that are not contained in the map. Section 4 presents successful applications of Markov localization in real-world deployments of mobile robots. Finally, we relate our approach to previous work in Section 5.

2 Markov Localization

The basic idea of Markov localization is to maintain a probability density over the whole state space of the robot within its environment [27,30,7,20]. Markov localization assigns to each possible pose in the (x, y, θ)-space of the robot in the given environment the probability that the robot is at that particular position and has the corresponding orientation. Let L_t denote the random variable representing the state in the (x, y, θ) space of the robot at time t. Thus, $P(L_t = l)$ denotes the robot's belief that it was at location l at time t. The state L_t is updated whenever new sensory input is received or the robot moves. Without loss of generality we assume that at every discrete point t in time first a measurement s_t is perceived and then a movement action a_t is performed. Then, given a sensory input s_t Markov localization updates the belief for each location l in the following way:

$$P(L_t = l|s_t) \leftarrow \alpha_t \cdot P(s_t \mid l) \cdot P(L_t = l) \tag{1}$$

In this equation the term $P(s_t \mid l)$ is denoted as the *perception model* since it describes the probability of measuring s_t at location l. The constant α_t simply is a normalizer ensuring that the left-hand side sums up to one over all l. Upon executing action a_t Markov localization applies the following formula coming from the domain of Markov chains to update the belief:

$$P(L_{t+1} = l) \leftarrow \sum_{l'} P(L_{t+1} = l \mid L_t = l', a_t) \cdot P(L_t = l' \mid s_t) \tag{2}$$

The term $P(L_{t+1} = l \mid L_t = l', a_t)$ is also called *action model* since it describes the probability that the robot is at location l upon executing action a_t at a position l'.

The belief $P(L_0)$ at time $t = 0$ reflects the knowledge about the starting position of the robot. If the position of the robot relative to its map is entirely unknown, $P(L_0)$ corresponds to a uniform distribution. If the initial position of the robot is known with absolute certainty, then $P(L_0)$ is a Dirac distribution centered at this position.

3   Grid-Based Markov Localization

Grid-based Markov localization — in contrast to other variants of Markov localization which use a topological discretization — uses a fine-grained geometric
discretization to represent the position of the robot (see [7]). More specifically, L
is represented by a three-dimensional fine-grained, regularly spaced grid, where
the spatial resolution is usually between 10 and 40 cm and the angular resolution
is usually 2 or 5 degrees. This approach has several desirable advantages. First,
it provides very precise estimates for the position of a mobile robot. Second, the
high resolution grids allow the integration of raw (proximity) sensor readings.
Thus, the grid-based technique is able to exploit arbitrary geometric features of
the environment such as the size of objects or the width and length of corridors
resp. rooms. It furthermore does not require the definition of abstract features
such as openings, doorways or other types of landmarks as it has to be done for
the techniques based on topological discretizations.

A disadvantage of the fine-grained discretization, however, lies in the huge
state space which has to be maintained. For a mid-size environment of size
$30 \times 30\text{m}^2$, an angular grid resolution of $2°$, and a cell size of $15 \times 15\text{cm}^2$ the
state space consists of $7,200,000$ states. The basic Markov localization algorithm
updates each of these states for each sensory input and each movement operation
of the robot. To efficiently update such large state spaces our system includes
two techniques which are described in the remainder of this section. The first
optimization is the selective update strategy which focuses the computation on
the relevant part of the state space. The second method is our sensor model
which has been designed especially for proximity sensors and allows the computation of the likelihood $P(s \mid l)$ by two look-up operations. Based on these
two techniques, grid-based Markov localization can be applied in real-time to
estimate the position of a mobile robot during its operation.

Global localization techniques, which are based on matching sensor readings
with a fixed model of the environment assume that the map reflects the true state
of the world and therefore are prone to fail in highly dynamic environments in
which crowds of people cover the robot's sensors for longer periods of time. To
deal with such situations, our system also includes a filtering technique that
analyzes sensor readings according to whether or not they come from dynamic
and un-modelled aspects of the environment. Based on this technique the robot
can robustly operate even in populated environments such as a museum.

3.1   Selective Update

An obvious disadvantage of the grid-based discretization comes from the size of
the state space. To integrate a single measurement s_t into the belief state all
cells have to be updated. In this section we therefore describe a technique which
allows a *selective* update of the belief state. The key idea of this approach is to
exclude unlikely positions from being updated. For this purpose, we introduce

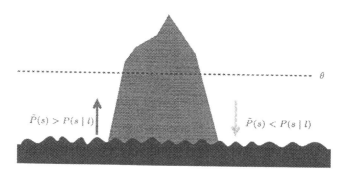

Figure 1. Basic idea of the selective update technique

a threshold θ and approximate $P(s_t \mid l)$ for cells with $P(L_{t-1} = l) \leq \theta$ by the probability $\tilde{P}(s_t)$. The quantity $\tilde{P}(s_t)$ is given by the average probability of measuring the feature s_t given a uniform distribution over all possible locations. This leads us to the following update rule for a sensor measurement s_t:

$$P(L_t = l|s_t) \longleftarrow \begin{cases} \alpha_t \cdot P(s_t \mid l) \cdot P(L_t = l) \text{ if } P(L_t = l) > \theta \\ \alpha_t \cdot \tilde{P}(s_t) \cdot P(L_t = l) \quad \text{otherwise} \end{cases} \tag{3}$$

Since α_t is a normalizing constant ensuring that $P(L_t = l \mid s_t)$ sums up to one over all l, this is equivalent to

$$P(L_t = l \mid s_t) \longleftarrow \begin{cases} \tilde{\alpha}_t \cdot \frac{P(s_t|l)}{\tilde{P}(s_t)} \cdot P(L_t = l) \text{ if } P(L_t = l) > \theta \\ \tilde{\alpha}_t \cdot P(L_t = l) \quad \text{otherwise} \end{cases} \tag{4}$$

Thus, all positions with a probability less or equal θ only have to be updated with the normalizing constant $\tilde{\alpha}_t$ which can simply be computed by summing up over all likely cells and adding to that $n \cdot \tilde{P}(s_t) \cdot \max \{P(L_t)|P(L_t) \leq \theta\}$ which is an approximation of the probability mass contained in the unlikely cells. In this term n is the number of states with $P(L_t) \leq \theta$. Since all these states are multiplied with the same value, it suffices to update only one variable instead of all of unlikely cells (see [5]).

At this point it should be noted that the approximation of $P(s \mid l)$ by $\tilde{P}(s)$ for a measurement s is a conservative approximation, since $P(s \mid l)$ is usually below the average $\tilde{P}(s)$ at unlikely positions l.

Figure 1 illustrates the key idea of the selective update scheme. In this example the belief is concentrated on a single peak which sticks out of the sea. The water level represents the maximum probability of all unlikely states. Whenever the estimated position corresponds to the true location of the robot, the probability $P(s \mid l)$ exceeds $\tilde{P}(s)$ and the belief is confirmed so that the level of the sea goes down. However, if the system loses track of the robot's position, then the obtained measurements no longer match the expected measurements and $P(s \mid l) < \tilde{P}(s)$. Thus the robot's certainty decreases and the level of the sea increases. As soon as the sea level exceeds θ, the so far unlikely states are updated again.

In extensive experimental tests we did not observe evidence that the selective update scheme impacts the robot's behavior in any noticeable way. In general, the probabilities of the active locations sum up to at least 0.99. During global localization, the certainty of the robot permanently increases so that the density quickly concentrates on the true position of the robot. As soon as the position of the robot has been determined with high certainty, a scan comprising 180 laser measurements is typically processed in less than 0.1 seconds using a 200MHz Intel Pentium Pro.

The advantage of the selective update scheme is that the computation time required to update the belief state adapts automatically to the certainty of the robot. This way, our system is able to efficiently track the position of a robot once its position has been determined. Thereby, Markov localization keeps the ability to detect localization failures and to re-localize the robot. The only disadvantage lies in the fixed representation of the grid which has the undesirable effect that the space requirement in our current implementation stays constant even if only a minor part of the state space is updated. In this context we would like to mention that recently promising techniques have been presented to overcome this disadvantage by applying alternative and dynamic representations of the state space [6,5,11,13].

3.2   The Model of Proximity Sensors

As mentioned above, the likelihood $P(s \mid l)$ that a sensor reading s is measured at position l has to be computed for all positions l in each update cycle. Therefore, it is crucial for on-line position estimation that this quantity can be computed very efficiently. In [26] Moravec proposed a method to compute a generally non-Gaussian probability density function $P(s \mid l)$ over a discrete set of possible distances measured by the sensor for each location l. In a first implementation of our approach [7] we used a similar method, which unfortunately turned out to be computationally too expensive for on-line position estimation. Furthermore, the pre-computation of all these densities is not feasible since it requires an enormous amount of memory.

To overcome these disadvantages, we developed a perception model which allows us to compute $P(s \mid l)$ by two look-up operations. The key idea is to store for each location l in the (x, y, θ) space the distance o_l to the next obstacle in the map. This distance can be extracted by ray-tracing from occupancy grid maps or CAD-models of the environment. In order to store these distances compactly, we use a discretization $d_0, \ldots, d_{n-1}$ of possible distances measured by a proximity sensor and store for each location only the index of the expected distance. Thus, to compute the probability $P(s_t \mid l)$ it remains to determine the probability $P(s_t \mid o_l)$ of measuring the value s_t given the expected distance o_l.

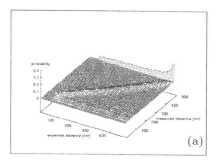
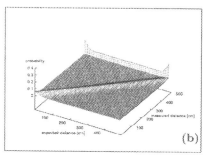

Figure 2. Measured (a) and approximated (b) densities for an ultrasound sensor.

In our approach, the density $P(s \mid o_l)$ is defined as a mixture of a Gaussian density centered around the expected distance o_l and a geometric distribution [18,12]. The geometric distribution is designed to allow the system to deal with a certain amount of un-modelled objects such as people walking by. In order to determine the parameters of this model we collected several million data pairs consisting of the expected distance o_l and the measured distance d_i during the typical operation of the robot. Based on these data we adopted the parameters of the sensor model so as to best fit the measured data. The measured data and resulting densities for ultrasound sensors are depicted in Figure 2. The similarity between the measured and the approximated distributions shows that our sensor model yields a good approximation of the data.

Based on these densities we now can compute $P(s \mid l)$ simply by two nested look-up operations. After retrieving the expected distance o_l for the currently considered location l, we compute $P(s \mid o_l)$ by a further look-up operation in the corresponding table.

Figure 3. The mobile robots Rhino (a) and Minerva (b)
acting as interactive museum tour-guides.

3.3   Filtering Techniques for Dynamic Environments

The perception model described in the previous section assigns a fixed probability to every pair of measured and expected distances. Although it also incorporates a certain amount of un-modelled objects, it is only capable to model such noise *on average*. While this approach showed to reliably deal with occasional sensor blockage, it is not sufficient in situations where a large amount of all sensor readings are corrupted. In general, localization techniques based on map-matching are prone to fail if the assumption that the map represents the state of the world is severely violated [15]. Consider, for example, a mobile robot which operates in an environment in which large groups of people permanently cover the robots sensors and thus lead to unexpected measurements. The mobile robots Rhino and Minerva which were deployed as interactive museum tour-guides in the *Deutsches Museum Bonn*, Germany and in the *National Museum of American History, Washington DC* respectively [4,32] were permanently faced with such a situation. Figure 3 shows cases in which the robots were surrounded by many visitors while giving a tour through the museum. To deal with such situations, we developed two filtering techniques which are designed to detect whether a certain sensor reading is corrupted or not. The basic idea of these filters is to sort all measurements into two buckets, one that contains all readings believed to reflect the true state of the world and another one containing all readings believed to be caused by un-modelled obstacles surrounding the robot. Compared to a fixed integration in the perception model, these filters have the advantage that they select the maximum amount of information from the obtained readings and automatically adopt their behaviour according to whether the environment is empty or highly populated.

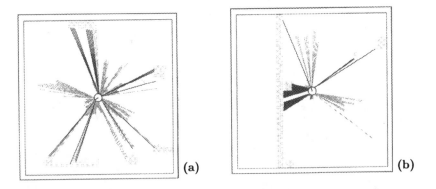

Figure 4. Typical laser scans obtained when Rhino is surrounded by visitors.

Figure 4 contains two typical laser scans obtained in such situations. Obviously, the readings are to a large extend corrupted by the people in the museum which are not contained in the static world model. The different shading of the beams indicates the two classes they belong to: the black lines correspond to static obstacles that are part of the map, whereas the grey-shaded lines are those beams reflected by visitors in the Museum (readings above 5m are excluded for the sake of clarity). Since people standing close usually increase the robot's belief of being close to modelled obstacles, the robot quickly loses track of its position when taking all incoming sensor data seriously.

In the remainder of this section we introduce two different kinds of filters. The first one is called *entropy filter* which can be applied to arbitrary sensors, since it filters a reading according to its effect on the belief $P(L)$. The second filter is the *distance filter* which selects the readings according to how much shorter they are than the expected value. It therefore is especially designed for proximity sensors.

The Entropy Filter The entropy $H(L)$ of the belief over L is defined as

$$H(L) = -\sum_l P(L = l) \, \log P(L = l) \tag{5}$$

and is a measure of uncertainty about the outcome of the random variable L [8]: The higher the entropy, the higher the robot's uncertainty as to where it is. The *entropy filter* measures the relative change of entropy upon incorporating a sensor reading into the belief $P(L)$. More specifically, let s denote the measurement of a sensor (in our case a single range measurement). The change of the entropy of $P(L)$ given s is defined as:

$$\Delta H(L \mid s) := H(L) - H(L \mid s) \tag{6}$$

Here, $H(L \mid s)$ is the entropy of the belief $P(L \mid s)$. While a negative change of entropy indicates that after incorporating s, the robot is less certain about its position, a positive change indicates an increase in certainty. The selection scheme of the entropy filter is:

Exclude all sensor measurements s with $\Delta H(L \mid s) < 0$.

Thus, the entropy filter makes robot perception highly selective, in that it considers only sensor readings confirming the robot's current belief.

The Distance Filter The advantage of the entropy filter is that it makes no assumptions about the nature of the sensor data and the kind of disturbances occurring in dynamic environments. In the case of proximity sensors, however, un-modelled obstacles produce readings that are shorter than the distance that can be expected given the map. The *distance filter* which selects sensor readings based on their distance relative to the distance to the closest obstacle in the map removes those sensor measurements s which with probability higher than θ

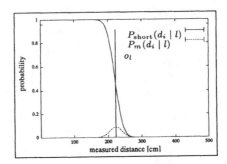

Figure 5. Probability $P_m(d_i \mid l)$ of expected measurement and probability $P_{short}(d_i \mid l)$ that a distance d_i is shorter than the expected measurement.

(which we generally set to 0.99) are shorter than expected, i.e. caused by an un-modelled object.

Consider a discrete set of possible distances $d_1, \ldots, d_n$ measured by a proximity sensor. Let $P_m(d_i \mid l) = P(d_i \mid o_l)$ denote the probability of measuring distance d_i if the robot is at position l and the sensor detects the next obstacle in the map. The distribution P_m describes the sensor measurement *expected* from the map. This distribution is assumed to be Gaussian with mean at the distance o_l to the next obstacle. The dashed line in Figure 5 represents P_m for a laser-range finder and a distance o_l of 230cm. Given P_m we now can define the probability $P_{short}(d_i \mid l)$ that a measured distance d_i is *shorter* than the expected one given the robot is at position l. This probability is obviously equivalent to the probability that the expected measurement is longer than the o_l and can be computed as:

$$P_{short}(d_i \mid l) = \sum_{j>i} P_m(d_j \mid l). \qquad (7)$$

$P_{short}(d_i \mid l)$ is the probability that a measurement d_i is shorter than expected given the robot is at location l. In practice, however, we are interested in the probability $P_{short}(d_i)$ that d_i is shorter than expected given the current belief of the robot. Thus, we have to average over all possible positions of the robot:

$$P_{short}(d_i) = \sum_l P_{short}(d_i \mid l) P(L = l) \qquad (8)$$

Based on $P_{short}(d_i)$ we now can define the distance filter as:

Exclude all sensor measurements d_i with $P_{short}(d_i) > \theta$.

4   Applications of Markov Localization in Real-World Environments

The grid-based Markov localization technique including the filter extension has been implemented and proven robust in various environments. In this section we illustrate the application of Markov localization in the context of the deployments of the mobile robots Rhino [4] and Minerva [32] as interactive museum tour-guide robots (see Figure 3). Rhino [3,33] has a ring of 24 ultrasound sensors each with an opening angle of 15 degrees. Both, Rhino and Minerva are equipped with two laser-range finders covering 360 degrees of their surrounding.

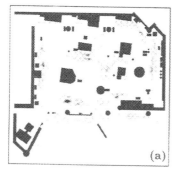 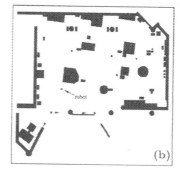

Figure 6. Global localization in the Deutsches Museum Bonn. The left image (a) shows the belief state after incorporating one laser scan. After incorporating two more scans, the robot uniquely determined its position (b).

4.1   Reliable Position Estimation in Populated Environments

During the deployment of the mobile robots Rhino and Minerva in the Deutsches Museum Bonn resp. the National Museum of American History (NMAH) in Washington DC the reliability of the localization system was a precondition for the success of the entire mission. Although both robots used the laser-range finder sensors for (global) localization, they had to deal with situations in which more than 50% of all sensor readings were corrupted because large crowds of people surrounded the robot for longer periods of time [15].

Rhino used the entropy filter to identify sensor readings that were corrupted by the presence of people. Since this filter integrates only readings which do not increase the uncertainty, it is restricted to situations in which the robot knows its approximate location. Unfortunately, when using the entropy filter, the robot is not able to recover from situations in which the robot loses its position entirely. To prevent this problem, Rhino additionally incorporated a small number of randomly chosen sensor readings into the belief state. Based on this technique,

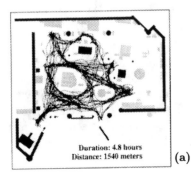 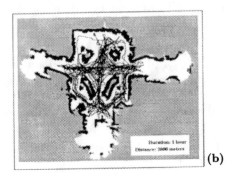

Figure 7. Typical trajectories of the robot Rhino in the Deutsches Museum Bonn (and) Minerva in the National Museum of American History (b).

Rhino's localization module was able to (1) globally localize the robot in the morning when the robot was switched on and (2) to reliably and accurately keep track of the robot's position. Figure 6 shows the process of global localization in the Deutsches Museum Bonn. RHINO is started with a uniform distribution over its belief state. The probability distribution given after integrating the first sensor scan is shown in Figure 6(a). After incorporating two more sensor scans, the robot knows its position with high certainty (see Figure 6(b)). Based on the selective update mechanism, the robot could efficiently keep track of the robot's position after determining it uniquely. Figure 7(a) shows a typical trajectory of the robot Rhino in the museum in Bonn. In the entire six-day deployment period Rhino travelled over 18km. Its maximum speed was over 80cm/sec and the average speed was 37cm/sec. Rhino completed 2394 out of 2400 requests which corresponds to a success rate of 99.75%.

Figure 7(b) shows a trajectory of 2 km of the robot Minerva in the National Museum of American History. Minerva used the distance filter to identify readings reflected by un-modelled objects. Based on this technique, Minerva was able to operate reliably over a period of 13 days. During that time over 50.000 people were in the museum and watched or interacted with the robot. At all Minerva travelled 44km with a maximum speed of 1.63m/sec.

In an extensive experimental comparison [15,12] it has been demonstrated that both filters significantly improve the robustness of the localization process especially in the presence of large amounts of sensor noise.

4.2 Probabilistic Integration of Map Information into a Reactive Collision Avoidance System

During the deployments of Rhino and Minerva as interactive museum tour-guides, safe and collision-free navigation was of uttermost importance. In the Deutsches Museum Bonn, the collision avoidance was complicated by the fact

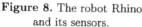
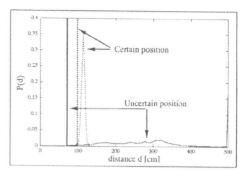

Figure 8. The robot Rhino
and its sensors.

Figure 9. Probability densities
$P(S_m = s)$ and selected measurement s^*.

that many of the obstacles and exhibits could not be sensed by the robot's sensors although Rhino possessed five state-of-the-art sensor systems (vision, laser, sonar, infrared, and tactile). Invisible obstacles included glass cages put up to protect exhibits, metal bars at various heights, and small podiums or metal plates on which exhibits were placed (c.f., the glass cage labelled "o1" and the control panel label "o2" in Figure 8). Furthermore, the area in which the robots were intended to operate was limited due to un-modelled regions or dangerous areas such as staircases. Therefore, both robots needed a technique which forces the robots not to leave their area of operation.

The collision avoidance system of the robots is based on the dynamic window algorithm (DWA) [14], a purely sensor-based approach designed to quickly react to obstacles blocking the robot's path. The basic DWA, just like any other sensor-based collision avoidance approach, does not allow for preventing collisions with obstacles which are "invisible" to the robot's sensors. To avoid collisions with those, the robot had to consult its map. Of course, obstacles in the map are specified in world coordinates, whereas reactive collision avoidance techniques such as DWA require the location of those obstacles in robo-centric coordinates.

At first glance, one might be inclined to use the maximum likelihood position estimate produced by Rhino's localization module, to convert world coordinates to robo-centric coordinates, thereby determining the location of "invisible" obstacles relative to the robot. However, such a methodology would be too brittle in situations where the localization module assigns high probability to multiple poses, which we observed quite frequently especially when the robot moves at high speeds. Our approach is more conservative. Based on the map, it generates "virtual" sensor readings that underestimate the true distance to the next obstacle in the map with probability 0.99. More specifically, let S_m be the proximity measurement that one would expect if all invisible obstacles were actually detectable. Then

$$P(S_m = s) = \sum_l P(S_m = s \mid L_t = l) \, P(L_t = l) \qquad (9)$$

is the probability distribution that s is the distance to the next obstacle in the map given the current belief state of the robot. Now the measurement s^* which underestimates the true distance to the next obstacle in the map with probability θ, is obtained as

$$s^* := \max\{s \mid P(S_m > s) \geq \theta\} \qquad (10)$$

where

$$P(S_m > s) := \sum_{s' > s} P(S_m = s'). \qquad (11)$$

Figure 9 depicts two different densities $P(S_m = s)$ based on two different belief states. Whereas the solid line corresponds to a situation, in which the robot is highly uncertain about its position, the dashed line comes from a typical belief state representing high certainty. This figure illustrates, that the robot conservatively chooses a very small distance s^* if it is uncertain about its position and a distance close to the true distance in the case of high certainty.

Our extension of the DWA worked well in the museum. Figure 7 shows a 1.6 km-long trajectory of the robot Rhino in the Deutsches Museum Bonn. The location of the "invisible" obstacles are indicated by the gray-shaded areas, which the robot safely avoided. By adding appropriate obstacles to the map, the same mechanism was used to limit the operational range of the robots. With it, Rhino and Minerva successfully avoided to enter any terrain that was out of their intended operational range.

4.3   Reacting to People Blocking the Path of the Robot

In the museum, people were not always cooperative. A typical behaviour of visitors was to intentionally block the robot's path for longer periods of time. One example of such a situation with Rhino in the Deutsches Museum Bonn is shown in Figure 10. Here a group of visitors tries to challenge the robot by forming a u-shaped obstacle. To deal with such situations, the robot needed the capability to detect that un-modelled obstacles in the environment are blocking its path. Rhino and Minerva applied the distance filter to detect such situations, since people standing close to the robot lead to readings that are shorter than expected. If the path of the robot was blocked for a certain amount of time, then Rhino blew its horn to ask for clearance [4]. Minerva used a more sophisticated way of interaction [32]. It possessed a face (see Figure 11) allowing it to express different moods according to the duration of the blockage. This face was mounted on a pan-tilt head and pointed to the person standing in the direction the robot was supposed to move. Minerva changed the mimic of the face according to the duration of the blocking. The moods ranged from "happy" (Figure 11(a)) to "angry" (Figure 11(b)). Minerva additionally used different phrases of pre-recorded texts to express the desire for free space.

Figure 10. Typical situation in which visitors try to challenge the robot by intentionally blocking its path.

Figure 11. The different moods of the robot Minerva according to its progress ranged from happy (a) to angry (b).

Rhino's and Minerva's ability to react directly to people was among the most entertaining aspects, which contributed enormously to its popularity and success. Many visitors were amazed by the fact that the robots acknowledged their presence by blowing its horn, changing the mimic or explicitely asking to stay behind the robot. One effect was that they repeatedly stepped in its way to get the acoustic "reward." Nevertheless, the ability to detect such situations and to interactively express their intentions allowed the robots to reliably reach their goals.

5   Related Work

Most of the techniques for position estimation of mobile robots developed so far belong to the class of local approaches resp. tracking techniques which are designed to compensate odometric error occurring during navigation given that the initial position of the robot is known (see [2] for a comprehensive overview). Weiß et. al. [34] store angle histograms derived from laser-range finder scans taken at different locations in the environment. The position and orientation of the robot is calculated by maximizing the correlation between the stored histograms and laser-range scans obtained while the robot moves through the environment. The estimated position together with the odometry information is then used to predict the position of the robot and to select the histogram used for the next match. Yamauchi [35] applies a similar technique, but uses hill-climbing to match local maps built from ultrasonic sensors against a given occupancy grid map. As in [34], the location of the robot is represented by the position yielding the best match.

A very popular mathematical framework for position tracking are *Kalman filters* [25], a technique that was introduced by Kalman in 1960 [21]. Kalman filter based methods represent their belief about the position of the robot by a unimodal Gaussian distribution over the three-dimensional state-space of the robot. The existing applications of Kalman filtering to position estimation for mobile robots are similar in how they model the motion of the robot. They differ mostly in how they update the Gaussian according to new sensory input. Leonard and Whyte [23] match beacons extracted from sonar scans with beacons predicted from a geometric map of the environment. These beacons consist of planes, cylinders, and corners. To update the current estimate of the robot's position, Cox [9] matches distances measured by infrared range finders against a line segment description of the environment. Schiele and Crowley [28] compare different strategies to track the robots position based on occupancy grid maps and ultrasonic sensors. They show that matching local occupancy grid maps against a global grid map results in a similar localization performance as if the matching is based on features that are extracted from both maps. Shaffer et. al. [29] compare the robustness of two different matching techniques against different sources of noise. They suggest a combination of map-matching and feature-based techniques in order to inherit the benefits of both. Gutmann, Lu, and Milios [17,24] use a scan-matching technique to precisely estimate the position of the robot based on laser-range finder scans and learned models of the environment. Arras and Vestli [1] use a similar technique to compute the position of the robot with a very high accuracy. All these variants, however, rest on the assumption that the position of the robot can be represented by a single Gaussian distribution. The advantage of this approach lies in the high accuracy that can be obtained. The assumption of a unimodal Gaussian distribution, however, is not justified if the position of a robot has to be estimated from scratch, i.e. without knowledge about the starting position of the robot.

To overcome these disadvantages, recently different variants of Markov localization have been developed and employed successfully [27,30,20,7,19,31]. The basic idea of Markov localization is to maintain an arbitrary and not necessarily Gaussian position probability density over the whole three-dimensional (x, y, θ) state space of the robot in its environment. The different variants of this technique can be roughly distinguished by the type of discretization used for the representation of the state space. In [27,30,20,19,31] Markov localization is used for landmark-based corridor navigation and the state space is organized according to the topological structure of the environment. Based on an orthogonality assumption [27,30,20] consider only four possible headings of the robot. Our fine-grained and grid-based approach of Markov localization has the advantage that it provides accurate position estimates and that it can be applied in arbitrary unstructured environments. The disadvantage of this technique, however, lies in its computational complexity and space requirements. In order to represent the whole state space of the robot within its environment, usually several million states have to be represented and updated. In this paper we therefore present different techniques to overcome these disadvantages. The result is an efficient and accurate position estimation technique for mobile robots. Our implementation of Markov localization is able to globally estimate the position of the robot from scratch and to efficiently keep track of the robots position once it has been determined. Simultaneously, it allows the robot to recover from localization failures. In a recent experimental comparison it has been demonstrated that Kalman filter based tracking techniques provide highly accurate position estimates but are less robust than Markov localization since they lack the ability to globally localize the robot and to recover from localization errors [16]. Our technique furthermore includes an approach to filter out readings coming from objects that are not contained in the map. This way our approach, in contrast to other Markov localization techniques, is able to reliably estimate the position of the robot even in densely populated and highly dynamic environments [4,12,15,32].

6   Discussion

In this paper we presented Markov localization as a robust technique for estimating the position of a mobile robot. The key idea of Markov localization is to maintain a position probability density over a fine-grained discretization of the robot's state space within the environment. This density is updated whenever new sensory input is received and whenever the robot moves.

We introduced two techniques for efficiently updating the density. First, we use a selective update scheme which focuses the computation on the relevant parts of the state space. Second, we apply a sensor model that allows to efficiently compute the necessary quantities by two look-up operations. Based on these two approaches, the belief state can be updated in real-time. Our approach to Markov localization furthermore includes filtering techniques significantly increasing the robustness of the position estimation process even in densely populated or highly dynamic environments. We presented two kinds of filters. The entropy filter is a general filter, that can be applied to arbitrary sensors including cameras. The distance filter has been developed especially for proximity sensors and therefore provides better results with sensors like a laser-range finder.

Our technique has been implemented and evaluated in several experiments at various sites. During the deployments of the mobile robots Rhino in the Deutsches Museum Bonn, Germany, and Minerva in the National Museum of American History, Washington DC, our system was successfully applied. The accuracy obtained by the grid-based and fine-grained discretization of the state space turned out to be high enough to avoid even obstacles that could not be sensed by the robot's sensors. It furthermore provided the basis for detecting situations, in which people intentionally block the path of the robot.

There are several aspects which are objectives of future research. First, the current implementation of Markov localization uses a fixed discretization of the whole state space which is always kept in memory. To overcome this disadvantage, recently different alternative representations of the density have been developed and suggested [6,5,16]. Whereas [6] uses a local grid for efficient position tracking, we introduced an Octree-based representation of the robot's state space in [5] which allows to dynamically adopt the required memory as well as the resolution of the discretization. [16] suggest a combination of Markov localization with Kalman filtering which should lead to a system inheriting the advantages of both approaches. Recently, [11,13] introduced a promising approach based on Monte Carlo methods. This approach uses a set of samples to represent the state space, thereby significantly reducing the space and the computation time required to maintain and update the density. A further restriction of the current system is that the model of the environment is assumed to be static. In this context the problem to combine map building and localization on-line is an interesting topic for future research.

References

1. K.O. Arras and S.J. Vestli. Hybrid, high-precision localization for the mail distributing mobile robot system MOPS. In *Proc. of the IEEE International Conference on Robotics and Automation*, 1998.
2. J. Borenstein, B. Everett, and L. Feng. *Navigating Mobile Robots: Systems and Techniques*. A. K. Peters, Ltd., Wellesley, MA, 1996.
3. J. Buhmann, W. Burgard, A.B. Cremers, D. Fox, T. Hofmann, F. Schneider, J. Strikos, and S. Thrun. The mobile robot Rhino. *AI Magazine*, 16(2), 1995.
4. W. Burgard, A.B. Cremers, D. Fox, D. Hähnel, G. Lakemeyer, D. Schulz, W. Steiner, and S. Thrun. The interactive museum tour-guide robot. In *Proc.of the Fifteenth National Conference on Artificial Intelligence*, 1998.
5. W. Burgard, A. Derr, D. Fox, and A.B. Cremers. Integrating global position estimation and position tracking for mobile robots: The dynamic markov localization approach. In *Proc. of the IEEE/RSJ International Conference on Intelligent Robots and Systems*, 1998.
6. W. Burgard, D. Fox, and D. Hennig. Fast grid-based position tracking for mobile robots. In *Proc. of the 21st German Conference on Artificial Intelligence, Germany*. Springer Verlag, 1997.
7. W. Burgard, D. Fox, D. Hennig, and T. Schmidt. Estimating the absolute position of a mobile robot using position probability grids. In *Proc. of the Thirteenth National Conference on Artificial Intelligence*, 1996.
8. T.M. Cover and J.A. Thomas. *Elements of Information Theory*. Wiley, 1991.
9. I.J. Cox. Blanche – an experiment in guidance and navigation of an autonomous robot vehicle. *IEEE Transactions on Robotics and Automation*, 7(2):193–204, 4 1991.

10. I.J. Cox and G.T. Wilfong, editors. *Autonomous Robot Vehicles*. Springer Verlag, 1990.
11. F. Dellaert, D. Fox, W. Burgard, and S. Thrun. Monte Carlo localization for mobile robots. In *Proceedings of the International Conference on Robotics and Automation (ICRA '99)*, 1999. To appear.
12. D. Fox. *Markov Localization: A Probabilistic Framework for Mobile Robot Localization and Navigation*. PhD thesis, Dept of Computer Science, Univ. of Bonn, Germany, 1998.
13. D. Fox, W. Burgard, F. Dellaert, and S. Thrun. Monte Carlo localization—efficient position estimation for mobile robots. In *Proc. of the Sixteenth National Conference on Artificial Intelligence*, 1999. To appear.
14. D. Fox, W. Burgard, and S. Thrun. The dynamic window approach to collision avoidance. *IEEE Robotics and Automation*, 4(1), 1997.
15. D. Fox, W. Burgard, S. Thrun, and A.B. Cremers. Position estimation for mobile robots in dynamic environments. In *Proc. of the Fifteenth National Conference on Artificial Intelligence*, 1998.
16. J.-S. Gutmann, W. Burgard, D. Fox, and K. Konolige. An experimental comparison of localization methods. In *Proc. of the IEEE/RSJ International Conference on Intelligent Robots and Systems*, 1998.
17. J.-S. Gutmann and C. Schlegel. AMOS: Comparison of scan matching approaches for self-localization in indoor environments. In *Proceedings of the 1st Euromicro Workshop on Advanced Mobile Robots*. IEEE Computer Society Press, 1996.
18. D. Hennig. Globale und lokale Positionierung mobiler Roboter mittels Wahrscheinlichkeitsgittern. Master's thesis, Department of Computer Science, University of Bonn, Germany, 1997. In German.
19. J. Hertzberg and F. Kirchner. Landmark-based autonomous navigation in sewerage pipes. In *Proc. of the First Euromicro Workshop on Advanced Mobile Robots*. IEEE Computer Society Press, 1996.
20. L.P. Kaelbling, A.R. Cassandra, and J.A. Kurien. Acting under uncertainty: Discrete bayesian models for mobile-robot navigation. In *Proc. of the IEEE/RSJ International Conference on Intelligent Robots and Systems*, 1996.
21. R.E. Kalman. A new approach to linear filtering and prediction problems. *Tansaction of the ASME – Journal of basic engineering*, pages 35–45, March 1960.
22. D. Kortenkamp, R.P. Bonasso, and R. Murphy, editors. *Artificial Intelligence and Mobile Robots*. MIT/AAAI Press, Cambridge, MA, 1998.
23. J.J. Leonard and H.F. Durrant-Whyte. Mobile robot localization by tracking geometric beacons. *IEEE Transactions on Robotics and Automation*, 7(3):376–382, 1991.
24. F. Lu and E. Milios. Robot pose estimation in unknown environments by matching 2d range scans. In *IEEE Computer Vision and Pattern Recognition Conference (CVPR)*, 1994.
25. P.S. Maybeck. Autonomous robot vehicles. In Cox and Wilfong [10].
26. H.P. Moravec. Sensor fusion in certainty grids for mobile robots. *AI Magazine*, Summer 1988.
27. I. Nourbakhsh, R. Powers, and S. Birchfield. DERVISH an office-navigating robot. *AI Magazine*, 16(2), Summer 1995.
28. B. Schiele and J.L. Crowley. A comparison of position estimation techniques using occupancy grids. In *Proc. of the IEEE International Conference on Robotics and Automation*, 1994.
29. G. Shaffer, J. Gonzalez, and A. Stentz. Comparison of two range-based estimators for a mobile robot. In *SPIE Conf. on Mobile Robots VII*, pages 661–667, 1992.

30. R. Simmons and S. Koenig. Probabilistic robot navigation in partially observable environments. In *Proc. of the International Joint Conference on Artificial Intelligence*, 1995.
31. S. Thrun. Bayesian landmark learning for mobile robot localization. *Machine Learning*, 33(1), 1998.
32. S. Thrun, M. Bennewitz, W. Burgard, A.B. Cremers, F. Dellaert, D. Fox, D. Hähnel, C. Rosenberg, J. Schulte, and D. Schulz. MINERVA: A second-generation museum tour-guide robot. In *Proceedings of the International Conference on Robotics and Automation (ICRA '99)*, 1999. To appear.
33. S. Thrun, A. Bücken, W. Burgard, D. Fox, T. Fröhlinghaus, D. Hennig, T. Hofmann, M. Krell, and T. Schimdt. Map learning and high-speed navigation in RHINO. In Kortenkamp et al. [22].
34. G. Weiß, C. Wetzler, and E. von Puttkamer. Keeping track of position and orientation of moving indoor systems by correlation of range-finder scans. In *Proc. of the IEEE/RSJ International Conference on Intelligent Robots and Systems*, 1994.
35. B. Yamauchi. Mobile robot localization in dynamic environments using dead reckoning and evidence grids. In *Proc. of the 1996 IEEE International Conference on Robotics and Automation*, 1996.

Relocalisation by Partial Map Matching

Wolfgang D. Rencken[1], Wendlin Feiten[1], and Raoul Zöllner[2]

[1] Siemens AG, Coporate Technology, 81730 München, Germany
{Wolfgang.Rencken,Wendelin.Feiten}@mchp.siemens.de
[2] Institute for Real-Time Computer Control Systems, Dept. of Computer Science,
University of Karlsruhe 76128 Karlsruhe, Germany

Abstract. The autonomous operation of an intelligent service robot in practical applications requires that the robot builds up a map of the environment by itself. A prerequisite for building large scale consistent maps is that the robot is able to recognise previously mapped areas and relocalise within these areas.

The recognition is based on constructing partial maps of geometric landmarks which are then compared to yield the optimal correspondence between these landmarks. For each landmark a signature is constructed which contains additional information about its immediate environment and its non-geometric properties. It is ensured that the signatures are robust with respect to missing landmarks, rotation and translation of landmarks and varying landmark lengths.

Both simulation and experiments on real robots have shown that the approach is capable of recognising previously mapped areas robustly in real-time[1].

1 Introduction

A service robot performing useful tasks needs to navigate within its working environment. One aspect of navigation deals with determining the position of the robot within its environment, usually known as the localisation problem. For this purpose, our robot uses a map of 2D geometric landmarks (planes, corners and cylinders). The map can be acquired in several ways. It can be generated from CAD plans, measured by hand or built up automatically. Measuring maps by hand is very time consuming and CAD plans are usually incomplete and not accurate enough. Therefore the robot should build up the map by itself.

Concurrent map building and localisation however poses a difficult problem. If the map is very large, small errors due to sensor inaccuracies (e.g. wheel slippage, range sensor noise) accumulate over time to yield an inconsistent map. This is also known as the loop closing problem. This becomes visible because several instances of the same real world objects are built up in the map (see figure 1).

---

[1] This work forms part of the INSERVUM project, partially funded by the German Ministry of Education, Science, Research and Technology under the number 01 IN 601 A2.

Christensen et al. (Eds.): Sensor Based Intelligent Robots, LNAI 1724, pp. 21–35, 1999.

Fig. 1. Loop closing problem. The environment is $120m \times 60m$

One way of reducing the inconsistencies is to use more accurate sensors. However, if the map is made large enough, the loop closing problem will reoccur. Therefore it seems more appropriate that the robot is able to recognise previously visited areas and correct the errors accordingly, thus maintaining the consistency of the map.

This paper concentrates on the aspect of recognising previously visited areas. The correction of the inconsistent map is beyond the scope of this publication.

1.1   Related Work

Most work on robot map building concentrates on extracting landmark data from sensor measurements and aggregating them over time and space (eg. [2] [4] [6] [8] [10]). More recently work has also been done on map recognition and map correction. Since this paper deals primarily with map recognition, only related work in this area will be looked at in more detail.

There are different ways of recognising that the robot has returned to a previously visited area.

One way of doing so, is to let the user tell the robot that it has returned to such an area [11]. This approach reduces the degree of the autonomy of the map building process.

If the robot's map consists of an occupancy grid, the correlation of two grid maps, from different points in time, can tell the robot if it has returned [5]. This approach is very time consuming since a large correlation space has to be searched through.

On the other hand, if the robot's map consists of a set of stored sensor data (e.g. laser range scans), the recognition problem can be solved by directly matching the sensor data [9] or by correlating the angle and distance histograms deduced from the sensor data [12]. Both approaches are computationally very expensive and require a lot of memory, since a large set of scans has to be stored and correlated.

If the robot's map consists of geometric features which can be connected into star shaped polygons, the polygon of the current sensor data can be mapped to a set of precomputed polygons which are stored in the visibility graphs of the robot's environment [7]. This method is not very robust with respect to changes in the environment and sensor errors.

Since our robot uses a map of geometric landmarks, the recognition problem is defined as matching two different sets of landmarks originating from different points in time. Such a set of landmarks will be referred to as a partial map.

2 Map

2.1 Landmark Extraction

In this paper, a landmark will be considered as a finite line segment which is extracted from range sensor data originating from a laser scanner or sonar sensors. Line segments are extracted from laser scans by first segmenting the range data and then applying a least squares fit algorithm to the individual segments. See [10] for details on extracting line segments from sonar data. It should be noted, that due to the extraction processes, the length of the line segments can vary considerably. Additional information such as colour or texture of the landmarks can be obtained from cameras.

2.2 Partial Map

The map of the environment L can be partitioned into partial maps L_i. The partitioning function p can be geometric (for example all features within a specific area, (see figure 2), temporal (for example all features extracted in a certain time interval) or a combination of both.

$$p : L \rightarrow \{L_i, i \in I\} \tag{1}$$

3 Partial Map Match

A partial map match $\mu \subset L_a \times L_b$ is a relation which is one-to-one over a suitable subset $D_a \times D_b, D_a \subset L_a, D_b \subset L_b$. On the set M of partial map matches, the match disparity $| \cdot |$ is given by ν (to be defined later on). The optimal partial map match μ_0 is then given by

$$|\mu_0| = min(|\mu|, \mu \in M) \tag{2}$$

Since the robot can approach the same region in different ways, landmarks which are present in one partial map are not necessary part of another partial map of the same environment. Furthermore, the parameters of the same

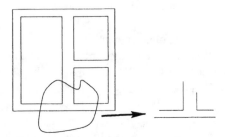

Fig. 2. A geometric partitioning of a landmark map.

landmark can differ considerably from one partial map to another. This means that when comparing two partial maps factors such as map incompleteness, map rotation and translation have to be specifically dealt with.

The straight forward approach of selecting and matching two subsets of landmarks is as follows:

- select n landmarks from the N landmarks of L_a
- select n landmarks from the M landmarks of L_b
- for each pairwise assignment of landmarks compute the disparity of the match.

The computational complexity of this approach is characterised by the $\binom{N}{n}$ and $\binom{M}{n}$ possibilities for subset selection and the up to $n!$ possible assignments of these subsets. The advantage of this approach is that it encaptures the local structure of the environment and therefore yields more robust results in the face of large uncertainties in the parameters of the single landmarks.

At the other end of the spectrum, a comparison of single landmarks yields a complexity of only $N \times M$. However, single landmarks do not have enough structure to be comparable. The large uncertainty of the landmark parameters makes the comparison even more difficult.

Therefore it seems advantageous to combine both approaches to obtain a matching algorithm which on the one hand uses the local structure of the environment, and on the other hand approaches the computational complexity of single landmark comparison.

3.1   Landmark Signature

Our approach to using local structure, while at the same time limiting the complexity of the landmark comparison, is based on the concept of landmark signatures. The signature Σ_i of landmark l_i is a set of features which makes a single landmark more distinguishable within a set of landmarks. The signature can contain information about the landmark itself, such as colour, texture or other non-geometric properties Γ. On the other hand, information about the environment of the landmark, such as relative angles and distances w.r.t. other non-parallel landmarks Ω, distances to other parallel landmarks Π and other geometric features can be stored in the signature.

$$\Sigma_i = \{\Omega_i, \Pi_i\, \Gamma_i, \ldots\} \tag{3}$$

Since the loop closing problem introduces a translational and rotational error into the map, care should be taken that the geometric properties of the signature are invariant w.r.t. these errors. Otherwise, the comparison of landmarks corresponding to the same environmental feature may fail. Examples of such geometric invariant constructs are

- distance between "parallel" landmarks $\pi \in \Pi$ (see figure 3)
- distance between and relative angles w.r.t. pairs of "orthogonal" landmarks $\omega \in \Omega$ (see figure 4)

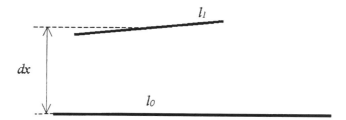

Fig. 3. A *paral* π. The distance between two parallel landmarks is invariant w.r.t. rotation and translation and the landmark lengths. The dependency on landmark lengths is negligible.

The above constructs are not only invariant w.r.t. the rotation and translation errors of the map, but also w.r.t. the individual landmark lengths. Depending on the path traversed during map acquisition, the type of range sensor used and the landmark extraction process, the landmark lengths are subject to a large uncertainty. Therefore landmark lengths cannot be reliably used when comparing signatures.

4   Comparison of Signed Landmarks

For landmarks l_i and l_j, the respective signatures are calculated. The disparity of two signatures ν is defined as

$$\nu : (\Sigma_i \times \Sigma_j) \to R \tag{4}$$

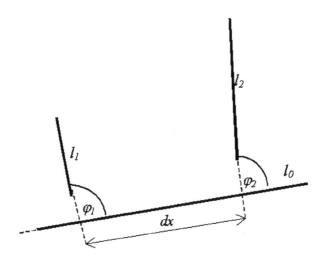

Fig. 4. An *orthogpair* ω. The distance between two orthogonal landmarks, and the relative angles between the base landmark and the orthogonal landmarks is invariant w.r.t. rotation and translation and the landmark lengths.

It consists of the disparities of the individual signature elements, given by

$$\nu_\Omega : (\Omega \times \Omega) \to R \tag{5}$$
$$\nu_\Pi : (\Pi \times \Pi) \to R \tag{6}$$
$$\nu_\Gamma : (\Gamma \times \Gamma) \to R \tag{7}$$

For example, the disparity between paral π_1 and π_2 can be determined by

$$\nu_\pi(\pi_1, \pi_2) = |dx_1 - dx_2| \tag{8}$$

The disparity for the orthog pair ω can be determined accordingly.

Since the individual signature elements can only be compared amongst equal types, the signature disparity between two landmarks is given by

$$\nu(\Sigma_i, \Sigma_j) = \sum_{k \in \{\Omega, \Pi, \Gamma\}} \alpha_k \nu_k(k_i, k_j) \tag{9}$$

where $\alpha_k \in [0, 1]$.

The disparity measure $\nu_k, k \in \{\Omega, \Pi, \Gamma\}$ is based on the best global assignment of the individual elements of k. This assignment is usually not unique. This is the case when there are several adjacent corridors of the same widths in the environment (see figure 5). The problem of finding the best global assignment can be solved by dynamic programming [3].

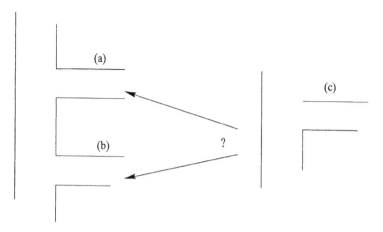

Fig. 5. The assignment dilemma, Is (a) = (c) or (b) = (c)?

4.1 Dynamic Programming

The classic dynamic programming algorithm [1] is divided into a forward step and a backtracking step. During the forward step, the cost of assigning or skipping individual signature elements is computed. At node (m, n), the decision is made whether the element m of signature Σ_i should be assigned to the element n of signature Σ_j (see figure 6). The cost at node (m, n) is given by

$$c(m, n) = min(c(m - 1, n - 1) + c^a, c(m - 1, n) + c^{na}, c(m, n - 1) + c^{na}) \quad (10)$$

where c^{na} is the cost for not assigning the signature elements and c^a is the cost for assigning the elements. Therefore the cost at any node contains the cost for all the assignments made up to that time, plus the cost of assigning the individual signature elements represented by the node.

In the backtracking step, the path in the node matrix with the minimal cost is searched for. The initial cost is given by cost at the upper right corner of the matrix:

$$v_{k,0}^{i,j} = c(M, N) \quad (11)$$

Thereafter, for each stage s, the path is selected which yields the minimum cost to the next node.

At stage s the cost accumulated thus far is given by

$$v_{k,s}^{i,j} = v_{k,s-1}^{i,j} + min(c(m - 1, n - 1), c(m - 1, n), c(m, n - 1)) \quad (12)$$

At the end of the search, the final cost is given by $v_{k,S}^{i,j}$. The match disparity $v^{i,j}$ between the two landmarks l_i and l_j, is then computed according to equation (9).

Whenever a signature element assignment is made during the search for the minimal cost path, the landmarks contained in the signature elements (see figures 3 and 4) are also assigned. Therefore computing the match disparity $v^{i,j}$ between

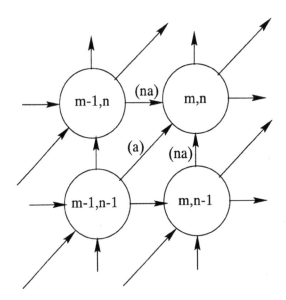

Fig. 6. The cost of assigning two signature elements.

two individual landmarks l_i and l_j, automatically generates a partial map match $\mu^{i,j}$.

Neglecting geometric aspects of the map building process, the best match is then given by

$$|\mu_0| = min(\nu^{i,j}|l_i \in L_a, l_j \in L_b) \tag{13}$$

However each map match $\mu^{i,j}$ yields a geometrical transformation $g^{i,j}$ with which the landmarks of the one set are projected onto the corresponding landmarks of the other set. Some geometrical transformations are more plausible than others (a 10m translation is more unlikely after a path length of 30m, than a 50cm translation). Therefore, based on a model of the geometrical map building error U, the best match is selected whose geometrical transformation $g^{i,j}$ is small based on the Mahalanobis Distance $d^{i,j}$ of the match.

Taking the geometric aspects into account, the optimal match between two partial maps is given by

$$|\mu_0| = min(\nu^{i,j}|l_i \in L_a, l_j \in L_b, d^{i,j} = \frac{(g^{i,j})^2}{U} < \delta) \tag{14}$$

If a match μ_0 exists, the robot has recognised a previously mapped area. The geometrical transformation $g^{i,j}$ gives an indication of the map building error that has accumulated between leaving the area and re-entering it again.

5   Experimental Results

The algorithms were tested extensively in simulation and on cleaning robot prototypes (see figure 7). On a PC with a Pentium 200 MHz processor, the matching algorithms can take several seconds, depending on the complexity of the environment. Therefore the matching can only be performed at certain instances in time.

The results presented in this paper are based on a simulation of the working environment of one of the cleaning robot prototypes (see figure 8), since the process of visualising the results on the real robot is too time consuming.

The robot started the map building at the point named *start* in figure 8. The robot then continued along the path and found a match between the old partial map (see figure 9) and the new partial map (see figure 10) at point (a). The best subset of matching features is shown in figures 11 and 12.

The robot then continued along its path and found a match between the old partial map (see figure 13) and the new partial map (see figure 14) at point (b). The best subset of matching features is shown in figures 15 and 16.

It should be noted that the partial maps differ considerably in size. Furthermore not all features have counterparts in the other partial map (in figure 15 there are no landmarks of the top "island", while in figure 16 there are no landmarks of the bottom "island". This clearly illustrates that our partial map matching algorithm is quite robust w.r.t. to missing landmarks and varying landmark lengths. It should be kept in mind that the algorithm is not designed to match all possible landmarks.

6   Conclusions

An algorithm, based on partial map matching, has been presented which enables a robot to recognise previously mapped areas during concurrent localisation and map building. The partial map match is robust w.r.t. missing landmarks, rotational and translational errors of the map building process and the considerable uncertainty in individual landmark lengths.

In future the geometrical transformation between the two partial maps will be used to correct the map building errors. This will eventually enable the robot to build up large scale consistent maps of its working environment by itself, considerably reducing the effort needed to install robots in new environments.

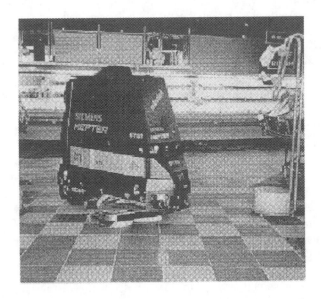

Fig. 7. Autonomous cleaning robot

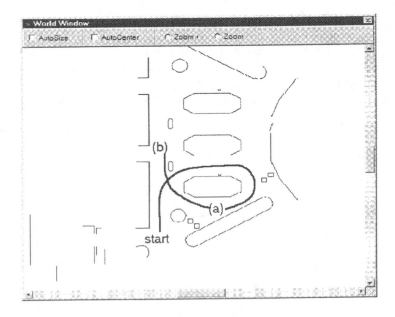

Fig. 8. Test environment

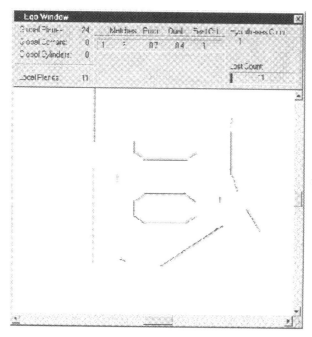

Fig. 9. Old partial map (1)

Fig. 10. New partial map (1)

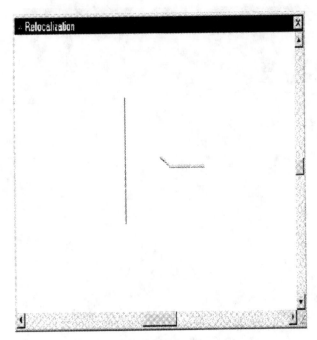

Fig. 11. Old partial map match (1)

Fig. 12. New partial map match (1)

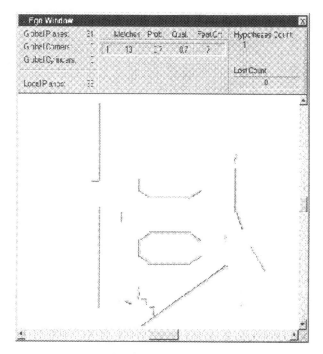

Fig. 13. Old partial map (2)

Fig. 14. New partial map (2)

Fig. 15. Old partial map match (2)

Fig. 16. New partial map match (2)

References

1. Bellman, R: Dynamic Programming. Princeton University Press, Oxford University Press, London (1957).
2. Borenstein, J, Everett, H.R., Feng, L.: Navigating Mobile Robots, Systems and Techniques. AK Peters, Wellesley Massachusetts, (1996).
3. Crowley, J.L., Mely, S., Kurek.M.: Mobile Robot Perception Using Vertical Line Stereo. Kanade, T. et al (ed). Intelligent Autonomous Systems 2 (1989), 597–607.
4. Chatila, R., Laumond, J.-P.: Position referencing and consistent world modeling for mobile robots. ICRA Proceedings, (1985).
5. Elfes, A.: Robot Navigation: Integrating Perception, Environmental Constraints and Task Execution Within a Probabilistic Framework. Dorst, L. et al (ed). Reasoning with Uncertainty in Robotics Lecture Notes in Artificial Intelligence 1093 Springer-Verlag, Berlin Heidelberg New York, (1995), 93–130.
6. Iyengar, S.S., Elfes, A. (ed): Automous Mobile Robots, Perception, Mapping, and Navigation. IEEE Computer Society Press, Los Alamitos California, (1991).
7. Karch, O., Noltemeier, H.: Robot Localisation - Theory and Practice. IROS Proceedings, (1997), 850–856.
8. Leonard, J.J., Durrant-Whyte, H.F., Cox, I.J.: Dynamic Map building for an autonomous mobile robot. Internal Journal of Robotics Research 11(4), (1992), 89–96.
9. Lu, F., Milios, E.: Globally consistent range scan alignment for environment mapping. Autonomous Robots, 4, (1997), 333–349.
10. Rencken, W.D.: Concurrent localisation and map building for mobile robots using ultrasonic sensors. IROS Proceedings, (1993), 2129–2197.
11. Thrun, S., Burgard, W., Fox, D.: A Probabilistic Approach for Concurrent Map Aquisition and Localization for Mobile Robots. Technical Report, School of Computer Science, Carnegie Mellon University, CMU-CS-97-183.
12. Weiss, G., von Puttkamer, E.: A Map based on Laserscans without Geometric Interpretation. U. Rembold et al (ed). Intelligent Autonomous Systems, IOS Press, (1995), 403–407.

Localization and On-Line Map Building for an Autonomous Mobile Robot

Ewald von Puttkamer[1], Gerhard Weiss[2], and Thomas Edlinger[3]

[1] Computer Science Department, University of Kaiserslautern
D-67663 Kaiserslautern, Germany
puttkam@informatik.uni-kl.de
[2] Fa. Hermstedt, Carl Reuther Str.3, D-68305 Mannheim, Germany
g.weiss@hermstedt.de
[3] Debis Systemhaus, Dessauer Str.6, D-80992 Munich, Germany

Abstract. Two basic tasks for an autonomous mobile robot are tackled in this paper: localization and on-line map building. For localization the robot uses a laser radar with 720 data points on a full circle with a resolution of a few mm. Localization is done by correlating radar pictures from a point P_0 to a point P_1 nearby. The radar pictures are transformed into angle histograms and correlated to give the relative rotation $\Delta\phi$ of both. The maximum angle γ in the first histogram shows the main direction in the environment The histograms are turned into this direction. Point histograms correlated give translations Δx and Δy between the poses P_0 and P_1. Rotating backwards by γ the pose P_1 is given in the coordinate system at P_0 . They are reference points. Map building uses an overlay of radar pictures from an obstacle avoiding radar to find points of interest (POI) in the near vicinity of the momentary position. They will form reference points too. A backtracking algorithm forms a topological graph of reference points exhausting the environment. The robot starts into an unknown environment and knows when it has explored it. The topological graph and the radar pictures at the nodes represent the environment.

1 Introduction

There are four basic requirements for autonomous driving: collision free movement, localization, autonomous exploration of the environment to build up an internal representation, and navigation through the explored environment. Localization and on-line exploration are dealt with in this paper. The task of localization is to keep position and orientation of a system with respect to its initial position and orientation. Outdoors this problem is solved by the global positioning system, GPS, and a compass. Indoors position and orientation from odometry is corrupted by inevitable drifts. External references are needed. They may be artificial landmarks placed into the environment as described in the book by Borenstein [3] or natural landmarks [2]. Localization may be done using Markov localization as used by Burgard [9] or by polygonal model fitting introduced by Kämpke and Strobel [7]. Once the problem of localization is solved

Christensen et al. (Eds.): Sensor Based Intelligent Robots, LNAI 1724, pp. 36–48, 1999.
© Springer-Verlag Berlin Heidelberg 1999

then map building may be done using readings from distance sensors. A map may be built fiollowing Moravec [8] as an occupancy grid of the environment. This may later be condensed into highways with maximal distance to obstacles like in the work of Elfes [6].

The rest of the paper is organized as follows: Section 2 deals with the localization, section 3 with map building and section 4 gives some results. Then a short summary is given in section 5.

2 Localization

The environment itself is taken as landmark here. The robot used for experiments is an industrial driverless vehicle of 0.63 x 1.1 m driving through an indoor environment with flat ground and enough open space to maneauvre. Its main sensors on board are a 360° laser radar (Accu Range 3000 LV) and a 180° obstacle detecting laser radar (SICK PLS) [1]. Figure 1 shows the robot with the 360° radar on top of the device and the obstacle detector at a height of ca. 30 cm above ground. Let (x, y) be the position and α the orientation of

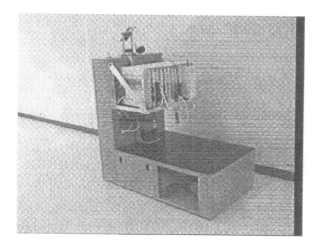

Fig. 1. Robot MOBOT IV

the robot with respect to a world coordinate system. Position and orientation are combined in the pose $P = (x, y, \alpha)$ Two 360° radar pictures taken from two poses $P_0 = (x_0, y_0, \alpha_0)$ and $P_1 = (x_1, y_1, \alpha_1)$ in near vicinity to each other will be correlated, as most of the environment is seen in both radar pictures. If the differences $\Delta x = (x_1 - x_0)$, $\Delta y = (y_1 - y_0)$, and $\Delta \alpha = (\alpha_1 - \alpha_0)$ may be calculated from the correlated radar scans then this solves the problem of localization: the position and orientation P_1 is known with respect to the initial position and orientation P_0.

2.1 Radar Pictures

The distance measuring device , an Accu-Range 300 LV laser radar sensor, takes
a 360° radar picture with a distance measured every 0.5°. The error in distances
is $\pm 5\,mm$, the movement of the turning mirror stabilized by a gyro with respect
to the movements of the robot: the system turns through 360° in 0.5 s with
respect to the environment irrespective of turning movements of the robot in
between [1]. Starting from a pose $P_0 = (x_0, y_0, \alpha_0)$ the radar points are transformed back to point P_0 interpolating the movement of the robot. The radar
picture is a set of 720 points $\{\varphi_{0,i}, r_{0,i}\}_{i=0,\ldots,719}$. Figure 2 shows an example. From a pose $P_1 = (x_1, y_1, \alpha_1)$ nearby the corresponding radar picture is
$\{\varphi_{1,i}, r_{1,i}\}_{i=0,\ldots,719}$, shown in figure 3. Seen from the positions of the robot the

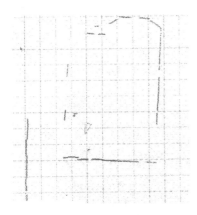

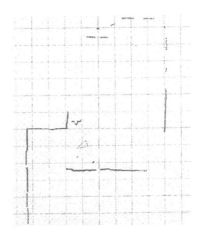

Fig. 2. Radar picture from point P_0 **Fig. 3.** Radar picture from point P_1

pictures are rotated and shifted to each other and the radar sees different items
of the environment. But the pictures are still similar: they are correlated to each
other.

2.2 Angles in Radar Pictures

Any two points in a radar picture taken at φ_i and φ_{i+j} have a fixed angular
distance $\varphi_i - \varphi_{i+j} = j \cdot 0.5°$. The angles φ_i are measured with respect to the
orientation of the robot at the start of a radar sweep. Let r_i and $(r_{i+j}$ be the
measured distances at angles φ_i and φ_{i+j} with $(i+j)$ mod 720 then the angle γ_i
between the line through the measured points and the orientation of the robot
may be calculated. In a robot fixed coordinate system the coordinates of radar
points (φ_i, r_i) and (φ_{i+j}, r_{i+j}) are

$$x_i = r_i \sin \varphi_i \qquad y_i = r_i \cos \varphi_i$$

$$x_{i+j} = r_{i+j} \sin \varphi_{i+j} \qquad y_{i+j} = r_{i+j} \cos \varphi_{i+j}$$

as shown in figure 4. The angle γ_i is given by

$$\tan \gamma_i = \frac{y_{i+j} - y_i}{x_{i+j} - x_i}$$

Two radar pictures translate into a set of angles $\{\gamma_i\}_{i=0,\dots,719}$. To eliminate noise $j = 5$ should be taken, as shown in figure 5. Taking successive points with $j = 1$

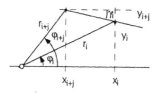

Fig. 4. Angle between radar points   **Fig. 5.** Noise in angle measurement

any measurement errors are exaggerated: on a flat wall three points in sucession will show rather different angles γ_i and γ_{i+1}. A difference $j > 5$ smoothes too much over fine details. The difference could be variable using some segmentation procedure. As primarily the orientation is of interest and not the absolute value of the angles $\gamma_i = -\gamma_i$ is taken, so $0° \leq \gamma_i \leq 180°$.

2.3   Angle Histograms

Distribute a set $\{\gamma_i\}_{i=0,\dots,719}$ into K buckets of width $\delta\gamma$. The filling of the buckets gives a histogram of the angles. With $\delta\gamma = 5°$ and an angular range of $180°$ there are $K = 36$ buckets into which 720 data points have to be ditributed. The environment as seen from a position P is condensed into an angle histogram as shown in figure 6. It is a set $\{N_k\}_{k=0,\dots,35}$ showing the distribution of angles in the environment. Peaks in a histogram reflect straight structures in the environment.

2.4   Correlation in Orientation

Let $\{N_{k,0}\}$ and $\{N_{k,1}\}$ be two angle histograms from radar pictures taken at $P_0 = (x_0, y_0, \alpha_0)$ and $P_1 = (x_1, y_1, \alpha_1)$ with

$$\tilde{N}_0 = \frac{1}{K} \sum_{k=0}^{K-1} N_{k,0} \qquad \text{and} \qquad \tilde{N}_1 = \frac{1}{K} \sum_{k=0}^{K-1} N_{k,1}$$

as median values. Then the correlation of the histograms describes the similarity of the histograms under rotation against each other

$$R_{0,1}(m) = \frac{1}{2K} \sum_{k=0}^{K-1} (N_{k,0} - \tilde{N}_0) \cdot (N_{k-m,1} - \tilde{N}_1)$$

with $(k - m) \bmod K$ as the histograms are periodic functions with period $K \cdot \delta\gamma$. Figure 7 shows a correlation. It peaks to its highest maximum between m_e and m_r. Then the peak m_a is found by a center-of-gravity approach:

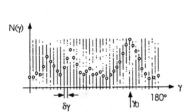

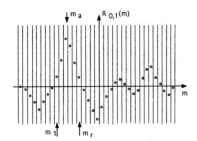

Fig. 6. Angle histogram

Fig. 7. Correlation of two angle histograms

$$m_a = \frac{\sum_{n=e}^{r} R_{0,1}(m_n) \cdot m_n}{\sum_{n=e}^{r} R_{0,1}(m_n)}$$

Both radar pictures are best correlated if rotated through an angle $\Delta\alpha = m_a \cdot \delta\gamma$ against each other. This is the difference in orientation between both positions P_0 and P_1 [4].

2.5   Correlation in Translation

Let there be a maximum in the angle histogram from P_0 at an angle γ_0. This is a main direction in the environment as seen from P_0. The radar picture at P_0 is rotated by γ_0 as shown in figure 8 into a coordinate system (x^*, y^*) adjusted to that main direction.

$$(\varphi_i, r_i)_0 \longrightarrow (\varphi_i - \gamma_0, r_i)_0$$

$$x_{0,i}^* = r_{0,i} \sin(\varphi_{0,i} - \gamma_0) \qquad y_{0,i}^* = r_{0,i} \cos(\varphi_{0,i} - \gamma_0)$$

The same is done for the radar picture at P_1; but the rotation is through an angle $(\gamma_0 + \Delta\alpha)$:

$$(\varphi_i, r_i)_1 \longrightarrow (\varphi_i - (\gamma_0 + \Delta\alpha), r_i)_1$$

$$x_{1,i}^* = r_{1,i} \sin(\varphi_{1,i} - (\gamma_0 + \Delta\alpha)) \qquad y_{1,i}^* = r_{1,i} \cos(\varphi_{1,i} - (\gamma_0 + \Delta\alpha))$$

Figure 9 shows the transformation. This leads to four sets

$$(\{x_{0,i}^*\}, \{y_{0,i}^*\}, \{x_{1,i}^*\}, \{y_{1,i}^*\})_{i=0,\dots,719}$$

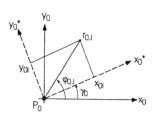

Fig. 8. Picture at P_0 rotated by γ_0

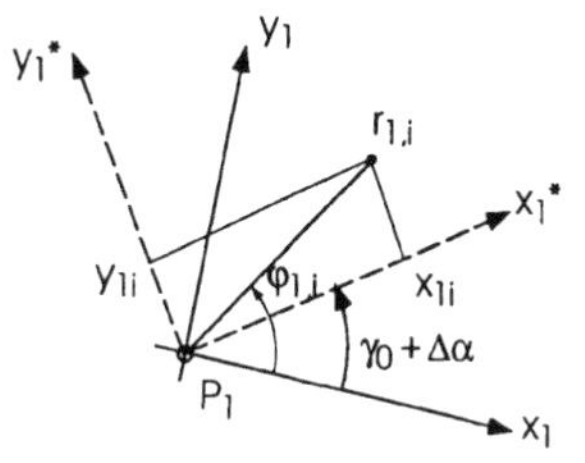

Fig. 9. Picture at P_1 rotated by $\gamma_0 + \Delta\alpha$

From each set a point histogram is formed, projecting the x- respective the y-values into buckets of width δx or δy . For $\{x*_{0,i}\}$ this is

$$H_{0x} = \{h0x_n\}_{n=-N,\ldots+N}$$

with

$$h0x_n = \natural\{x^*_{0,i}|n \cdot \delta x \le x^*_{0,i} \le (n+1) \cdot \delta x\}$$

as sketched in figure 10. Let $h0x_n = 0$ for $n > P$ or $n < -P$ then the histogram stretches from $-P \le n \le +P$. Typical values for δx and δy are 6 cm at a maximum distances of 4 m. From the four sets four point histograms are formed and correlated with respect to x and y :

$$Qx_{0,1}(m) = \frac{1}{2P} \sum_{n=-P}^{+P} h0x_n \cdot h1x_{n-m}$$

$$Qy_{0,1}(m) = \frac{1}{2P} \sum_{n=-P}^{+P} h0y_n \cdot h1y_{n-m}$$

In these correlations there are peaks at m_x and m_y respectively. This means that in the coordinate system (x^*, y^*) at P_0 the point P_1 is at $(\Delta x^*, \Delta y^*)_0$ with $\Delta x^* = m_x \cdot \delta x$ and $\Delta y^* = m_y \cdot \delta y$.

2.6   Transformation into World Coordinate System

The last step is the transformation of P_1 into the $(x, y)_0$ coordinate system around P_0 shown in figure 11. Let the coordinates of P_1 be (x_1, y_1) in the coordinate system at P_0.

$$x_1 = \Delta x^* \cos\gamma_0 - \Delta y^* \sin\gamma_0 = \Delta x \qquad y_1 = \Delta x^* \sin\gamma_0 + \Delta y^* \cos\gamma_0 = \Delta y$$

The localization of P_1 with respect to P_0 is complete now. Let

$$P_0 = (x_0, y_0, \alpha_0)$$

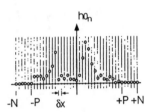

Fig. 10. Point histogram

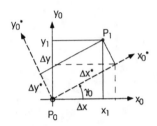

Fig. 11. Transformation into world coordinates

in world coordinates then

$$P_1 = (x_1, y_1, \alpha_1)$$
$$P_1 = (x_0 + \Delta x, y_0 + \Delta y, \alpha_0 + \Delta \alpha)$$

in the same coordinate system. Starting from a point P_0 any point within a circle around P_0 with a diameter $D \approx 2\,m$ can be correlated to P_0. Starting from this point going into one direction after $\approx 2\,m$ a new point P_1 is taken and correlated to P_0. P_1 is a reference point to P_0. Going further on from a reference point P_k a next point P_{k+1} will be reached. A set of reference points $\{P_k\}$ defines a referentiable region according to figure 12. Any point within a circle around P_k is correlated to P_0. A measure of the quality of the correlation is a closed loop shown in figure 13. The end point P_k is correlated backwards to P_0 and

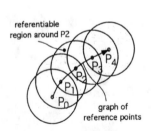

Fig. 12. Referentiable points

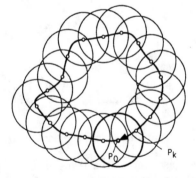

Fig. 13. Closed loop of reference points

at the same time lays within the correlation circle of P_0, thus may be directly correlated to the starting point. Any differences in the poses of P_k with respect to P_0 can be distributed back to the other points $P_1, \ldots, P_{k-1}$ and enhance the quality of correlation.

3   Map Building

Let there be no a priori knowledge of the environment. The robot starts some-where at an arbitrary position P_0. From here the environment has to be explored, using a coarse current sensor map (CSM) of the environment, formed from a fusion of the radar map and an obstacle detecting radar map with 30 sectors of $6°$ with $\pm 90°$ from the forward direction. The CSM consists of 60 sectors of $6°$. An environment map is composed as an overlay of different current sensor maps, each taken at a different reference point P_k. During exploration the robot has no chance to differentiate between temporary (moving) and permanent obstacles. Any obstacle found during exploration is taken as permanent into the environment map. Thus the environment should be kept free of persons during exploration. Both, the $360°$ radar and the obstacle avoidance sensor, detect free space. So this is known in the CSM.

3.1   Information at a Reference Point

At a reference point P_k its position and orientation is known with respect to P_0:
$P_k = (x_k, y_k, \varphi_k)_0$ Further at this point a radar map $\{r_i\}_{i=0,\dots,719}$; $\varphi_i = i \cdot \delta\varphi$; with $\delta\varphi = 0.5°$ is stored together with a current sensor map CSM according to figure 14 as a set of line endpoints $\{(x_{1n}, y_{1n})(x_{2n}, y_{2n})\}_{n=0,\dots,59}$ with

$$x_{1n} = x_k + r_n \sin(\varphi_k + n \cdot \Delta\varphi) \qquad y_{1n} = y_k + r_n \cos(\varphi_k + n \cdot \Delta\varphi)$$

$$x_{2n} = x_k + r_n \sin(\varphi_k + (n+1) \cdot \Delta\varphi) \qquad y_{2n} = y_k + r_n \cos(\varphi_k + (n+1) \cdot \Delta\varphi)$$

and $\Delta\varphi = 6°$.

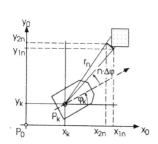

Fig. 14. Current sensor map

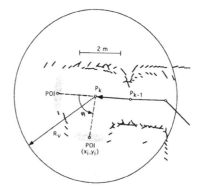

Fig. 15. Points of interest from P_k

3.2   Points of Interest, POI

From a current reference point P_k fuse the CSM with the environment map and look for free passages in the vicinity of P_k. Figure 15 shows a typical situation. The sensed area is limited to a diameter of 8 m around the robot: over a distance of $R_v = 4\,m$ the obstacle sensor "sees" any obstacle. Within this outer diameter a correlation can be done between P_k and another point. The robot then looks for throughways in the map: open space to the horizon at R_v, with a width of at least the robot breadth plus a safety margin. Then within this open space the robot can drive in principle to a point at a distance $e \approx 2\,m$ in the middle of the open space. There the robot marks a goal, a point of interest, POI, to be driven to. From a POI the robot can correlate itself to the current reference point P_k and thus POI becomes a new reference point. Points of interest not yet explored are put on a stack: their coordinates (x_i, y_i) together with the position of the reference point (x_k, y_k) and the orientation φ_i towards the POI [5].

3.3   Exploration Strategy

The exploration strategy follows a backpropagation algorithm:

```
begin
  * take a coarse current sensor map;
    fuse it to the navigation map;
      look for throughways as specified in the preceding section;
      if there are throughways not yet explored
                  then calculate POI's and pack them on a stack;
      if the stack is empty
            then end exploration
            else take a POI from the stack and drive to it;
      goto *;
end.
```

From the current position there is a path through open space to any POI taken from the stack: a chain of reference points up to the point P_k, and the direction φ_i to drive from there to (x_i, y_i).

3.4   Topological Map

Whenever the robot drives to a POI from its current position P_k then the correlation of POI with respect to P_k is calculated. Thus POI becomes a new reference point P_{k+1} and its position forms the node (k+1) in a graph connecting the reference points. If in the near vicinity of a reference point there is another one and there is free drivable space in between and no vertex between both points in the graph it is completed with a vertex. At the end of the exploration phase the environment is represented for the robot through this graph with the radar pictures and positions attached to its nodes and orientations attached to the vertices. It is a topolgical map to plan paths on it. The vertices denote paths through open space and the nodes places to correlate the momentary pose to the underlying graph. As long as the robot can correlate itself to the next nodes in the map it can drive through the environment at will.

4   Results

The robot was placed into a test environment made from tables with their small sides laid on the ground, and instructed to explore. The robot takes a current sensor map, overlays it with the map already built up and looks for POI's. Then it goes from one POI to the next and draws vertices between the nodes, representing reference points. Figure 16 shows the test environment as seen from the robot with the CSM sector maps and the topological graph. The CSM's are used only to find POI's. Their overlay gives a rough picture of the environment outlines while the graph represents the environment for the robot. As can be

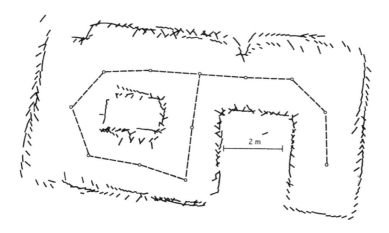

Fig. 16. Test environment

seen, there is still a slight drift in the robot: the overlay of the CSM's is not perfect. For actual driving after exploration the robot uses the topological graph and a simple extension routine from start and goal point to the respective next reference points in the graph. During driving the obstacle avoidance behaviour allows the robot to avoid static or dynamic obstacles and brings it back to the vertices of the graph again. Figure 17 shows a much larger scene: the basement of a main university building with the topological graph inserted into an overlay of the CSM's. Figure 17 shows a slight deviation from straight lines in some of the corridors in the basement: here the walls are of grey concrete and the rotating laser scanner gets a very bad image only with bad histogram peaks as consequence. The orientation detection and the correlation consequently show deviations. Figure 18 shows a floor in another building. The path explored here is about 100 m. The corridors are straigt and at right angles. The walls are made from grey painted metal sheets and the laser scanner gets better images. The map shown is built up when the robot enters unknown territory the first time. Over the total lenth of the building the deviations from straigt lines as seen by the robot are about 20 cm. The robot started its exploration at the outer end of the upper left corridor and ended at the outer end of the right corridor still keeping position and orientation. From this point the robot was commanded to drive back to the start. So it went twice through each corridor, the second time using the map.

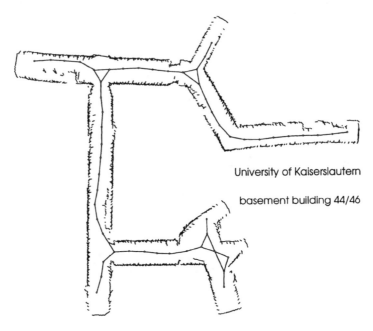

University of Kaiserslautern

basement building 44/46

Fig. 17. Basement of university building

5   Conclusions

The paper shows an algorithm to localize a robot in an indoor environment using straight structures of the environment itself as landmarks. The sensor for localization is a 360° laser radar. From radar pictures taken at positions not too far away from each other the relative rotation of the robot is deduced by angle histogram correlation. The translation is found correlating point histograms rotated into a main direction. Two points thus correlated are reference points.

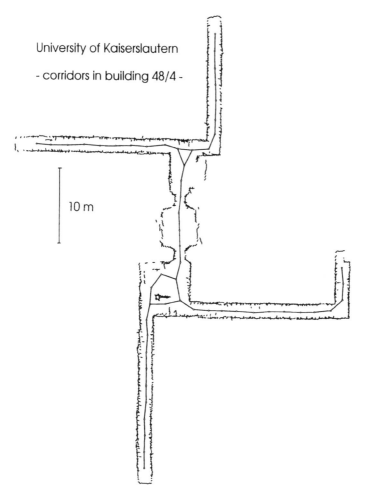

University of Kaiserslautern

- corridors in building 48/4 -

10 m

Fig. 18. Corridors in a building

A path driven is built up as a graph of pairwise correlated reference points. An exploration routine looks for possible reference points (points of interest, POI) in unexplored terrain at the same time. They are reachable from the given position and put on a stack. The robot drives to the position found at the top of stack. A backtracking algorithm spans a graph of reference points tessellating the environment. An environment map built up from an overlay of current sensor maps is not used for actual driving. It is used here for outlining the environment only. Once the graph for an environment is built up the environment is explored in the sense, that the robot can drive from any start point to a reachable point in the environment, using the graph as net of highways, planning its path on this net.

References

1. von Puttkamer, E.
   Line Based Modelling from Laser Radar Data
   in Modelling and Planning for Sensor Based Intelligent Robot Systems
   ed.Horst Bunke, Takeo Kanade, and Bernd Noltemeier
   World Scientific,1995, ISBN 98 10 22 23 86, pp 419–430
2. Shirai, Y.
   Planning with Uncertainity of Vision
   in Modelling and Planning for Sensor Based Intelligent Robot Systems
   ed.Horst Bunke, Takeo Kanade, and Bernd Noltemeier
   World Scientific,1995, ISBN 98 10 22 23 86, pp 120–139
3. Borenstein, J.
   Where am I?
   Sensors and Methods for Autonomous Mobile Robot Positioning
   University of Michigan, USA, 1994
   Technical Report UM-MEAM-94-21 Chapter 6
4. Weiss, G.; Wetzler, Ch.; von Puttkamer, E.
   Keeping Track of Position and Orientation of Moving Systems
   by Correlation of Range-Finder Scans
   Proc. 1994 IEEE Int. Conf.on Intelligent Robots and Systems
   (IROS'94) Munich, Germany, 1994, pp 595–601
5. Edlinger,Th.;Weiss,G.
   Exploration,Navigation and Self Localization
   in an Autonomous Mobile Robot
   Fachgespraech Autonome Mobile Systeme, AMS'95
   Karlsruhe,Germany, Nov.30 – Dez.1, 1995, pp 142–151
6. Elfes,A.
   Incorporating Spatial Relations at Multiple Levels
   of Abstraction in a Replicated Multilayered
   Architecture for Robot Control
   in Intelligent Robots, Sensing, Modeling and Planning
   ed. R.C. Bolles, H. Bunke, H. Noltemeier
   World Scientific,1997, ISBN 98 10 23 18 57, pp 249–266
7. Kämpke,T.; Strobel,M.
   Navigation nicht-kreisförmiger mobiler Roboter
   in hindernisreichen Umgebungen
   in Autonome Mobile Systeme,14. Fachgespräch, Karlsruhe 1998
   ed.H.Wörn, R.Dillmann,D. Henrich
   Springer, 1998, ISBN 3 540 65192 6, pp   156–163
8. Moravec,H.P.
   Sensor Fusion in Certainity Grids for Mobile Robots
   AI Magazine, pp 61–74, summer 1988
9. Burgard,D.;Fox,D.;Hennig,D.; Schmidt,T
   Estimating the Absolute Position of a Mobile Robot
   Using Position Probability Grids
   Proc.of the 14th Nat. Conf. on Artificial Intelligence,
   pp 896–901,1996

Interactive Learning of World Model Information for a Service Robot

Steen Kristensen, Volker Hansen, Sven Horstmann, Jesko Klandt,
Konstantin Kondak, Frieder Lohnert, and Andreas Stopp

Daimler-Benz Research and Technology, Intelligent Systems Group
Alt-Moabit 96A, D-10559 Berlin, Germany
{kristens,hansen,horstman,klandt}@dbag.bln.daimlerbenz.com
{kondak,lohnert,stopp}@dbag.bln.daimlerbenz.com

Abstract. In this paper, the problem of generating a suitable environment model for a service robot is addressed. For a service robot to be commercially attractive, it is important that it has a high degree of flexibility and that it can be installed without expert assistance. This means that the representations for doing planning and execution of tasks must be taught on–line and on–site by the user. Here a solution is proposed where the user interactively teaches the robot its representations, using the robot's existing navigation and perception modules. Based on a context adaptive architecture and purposive sensing modules it is shown how compact, symbolic representations sufficient for planning and robust execution of tasks can be generated.

1 Introduction

The general idea with robots is to relieve human beings of unwanted work tasks, this may be because they are tiring, dangerous, difficult, or simply too expensive to perform with human labour. Our work in the Advanced Servicing Robot (ASR) project is concerned with the development of a versatile service robot for a broad range of applications and environments. By a service robot we understand the following (adapted from [1]):

A service robot is a programmable movement mechanism partly or fully autonomously performing services. Services are activities not directly contributing to the production of industrial goods but to the performance of humans and installations.

The goal of the ASR project is to develop a set of perception and control modules (skills) plus the necessary system framework (planning, skill coordination, communication) for an advanced service robot. It has been chosen to use an architecture with a standardised but highly specialised set of perception and control modules since we believe that in order to make the development and employment of service robots commercially attractive, there is a need to develop a standard hardware platform that can easily be configured to a wide set of tasks

Christensen et al. (Eds.): Sensor Based Intelligent Robots, LNAI 1724, pp. 49–67, 1999.

only by changing a set of likewise standard software modules. We furthermore think it is important that the model representations used by the perception and control modules are such that they can be taught by the user and do not have to be programmed by a system expert.

The fact that autonomous service robots are going to be deployed in environments potentially much more diverse, unstructured, and dynamic than those of, e.g., industrial robot manipulators, means that advanced sensing and reasoning capabilities are a must for these robots. This however also further stresses the need for easily adaptable robot architectures since a service robot system capable of handling a prototypical scenario (cleaning, transportation, supervision etc.) in contrast to a manipulator for, e.g., welding, does not apply to a wide enough range of task/environments to make it commercially attractive to develop this functionality independently. In other words, the complex and diverse scenarios for service robot deployment not only imply that the functionality becomes expensive to develop but also that it will apply to a small set of tasks/environments and thereby few customers.

The challenge of producing commercially attractive service robots is therefore to be able to create systems that are highly adaptable to their specific working conditions and yet so uniform that it is economically and technically feasible to build and maintain them. We believe that the best way to make the systems adapt themselves to the working conditions is to make them modular and capable of learning the representations necessary for planning and executing their tasks in the given environments.

In this paper, we show how a modular design can not alone support the robust execution of tasks but also how it can be directly exploited for teaching of the representations needed for the planning and execution of these tasks. In Section 2 some more background in terms of related work and project specific constraints are outlined. In Section 3 the architecture we have developed to support the modular and teachable robot system is explained. In Section 4 it is explained how the representations necessary for the planning and execution of tasks are taught. In Section 5 an example of system operation is presented. Finally in Section 6, the results are discussed and summarised.

2   Background and Related Work

2.1   Architectures

Flexibility and modularity can be seen as two different ways to achieve a versatile robot system. In principle, a system can have only a single, extremely flexible control/perception module to cope with all situations or, as the other extreme, it can have a large number of specific modules that in combination can solve any task (if one knows how to combine them). The first strategy is along the lines of the traditional AI approach and tends to produce systems that are hard to (knowledge) engineer, maintain, and understand and that do not really interact well with the environment due to limited knowledge representations and long computation times. The pros of the AI approach is that the systems are theoretically well understood and predictable. The main advantages of the second, "behaviour based", approach is that it in principle needs no representations ("the world is its own best model") and that it interacts well with a dynamic environment through fast responses. The problem with the approach is, however, that it has proven extremely difficult to dynamically combine more than a few modules in a systematic way in order for the robot to adapt to the given task and environment. Also, without any representations is it not possible to plan and thus make the system fulfil a variety of just moderately complex tasks. Most research in autonomous robots has therefore implicitly or explicitly led to the conclusion that a practical system design is somewhere between the two extremes outlined above [2]. Fig. 1 shows a commonly used "generic" robot navigation architecture. The *mission planner* is a high–level symbolic planner dividing the user–specified task up into smaller parts that are then transferred to the *navigator*. The navigator combines the symbolic task description with sensory information in order to produce goal–points for the *pilot* which is a sub–symbolic, reactive module actually commanding the platform.

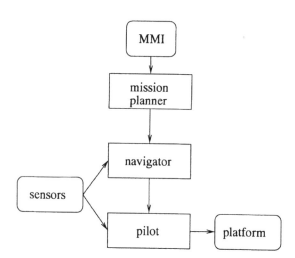

Fig. 1. Generic hierarchical autonomous robot navigation architecture (after [3]).

The schematic in Fig. 1 is well suited for understanding current robot architectures on a very high level. Experience and the bulk of literature in the robotics field shows, though, that below this abstraction level, systems tend to be quite different. One reason for this is that many systems are still research systems where there is a direct demand for trying out various approaches and hardware configurations. However, we find that another important reason is that the robot systems are highly tailored to their specific task and environment. This is by no means surprising nor un–desirable since examples from both robotics and biology clearly show that the more adapted to the conditions an agent/organism is, the better it performs.

Since every mobile robot system has some kind of architecture, there has been a substantial amount of work done in the area. However, for many systems, the architecture seems to have been developed in an ad hoc manner with no explicit considerations about modularity, flexibility, re–usability etc. One of the early noteworthy exceptions is Firby's reactive action packages (RAPs) [4] that provides a systematic way of describing tasks and their decomposition. Similar to RAP is Simmon's Task Control Architecture (TCA) [5] which, however, has more operating system–like features thereby also providing the underlying infrastructure for executing the given tasks. TCA has been successfully used on several robot prototypes. A direct descendant of RAP is Gat's Conditional Sequencing [6] which is one of the few architectures where an effort has been made to directly quantify the architecture's merit in terms of reliability improvement and system development time.

The OpenR Architecture [7] deals explicitly with system extension and reconfiguration capabilities for mechanical, electrical and software systems. It provides a distributed, object oriented Operating System allowing to configure modules in a plug and play fashion together with a graphical programming environment for the programmer of an entertainment robot.

One thing common to these architectures is that they are all strictly task–driven in the sense that the specifications, i.e., the RAPs and TCA's task trees and OpenR's MoNets (Motion Networks), are all based on a specific task and have to be especially tailored by an expert designer. We would like to improve this in the sense that a layman should be able to do the "teach–in" of a new task without having to know anything about the task specific problems. What a non–expert is likely to be able to teach the robot is not *how* to perform various tasks (with all exceptions) but rather *where* and with *what* objects to perform them, for instance the environment where the robot must move around and by what objects to dock/do manipulation. This means that we want an architecture where the specific task *context* determines the operational modus f.ex. by dynamically generating relevant task trees.

Although this seems to be a trivial observation we believe it has far–reaching consequences and poses a significant challenge for the underlying system architecture and its representations.

2.2 Learning of Environment Representations

All but the most simple, purely reactive robots internalise some kind of environment representation and can thus be said to "learn" their environment. Typical is the generation of an evidence grid representation of the obstacle/freespace conditions in the environment [8]. The kind of representation we want the robot to learn, however, needs to fulfil the following conditions:

- It should be usable and thus valid over extended periods of time so that the robot only has to be taught once or whenever the environment has changed "significantly". However, it should also be possible to adapt and modify existing models (either automatically or manually) to improve system performance.

- It should facilitate the user specification and automatic planning of mission commands such as "bring the cup to the kitchen", i.e., it needs to be a (partly) symbolic representation.
- It should contain the necessary information for the execution of planned tasks. This means that information about landmarks and objects must be present in the model plus information about what robot control strategies to invoke where.

A type of representation which satisfies the conditions is the annotated topological graph. In this graph representation, nodes represent distinctive places and edges the existence of traversable paths between the places. This purely topological representation is often annotated with metric information about the location of the places plus other information such as what local control strategies (LCSs)[1] to employ when traversing a given path. As mentioned above, we furthermore would like to include information about landmarks and objects.

Topological maps are known from cognitive science but has also been widely employed in mobile robotics [10] [11] [12] [13]. All of the here referenced research also shows how it is possible to automatically learn the topological graphs with exploration strategies. For most terrestrial applications, however, we think it is more sensible to let the robot learn the environment in interaction with the user since autonomous exploration has a number of problems:

- It cannot be guaranteed that the robot will see all the relevant parts of the surroundings due to obstacles, closed doors, people moving about etc.
- There may be parts where the robot is not allowed/supposed to go. This will often be the case for, e.g., office and factory applications.
- The resulting topological graphs are "un–grounded" in the sense that nodes do not have (for the user) meaningful symbolic names such as "kitchen", "printer room" etc. This can of course be added later, but then again there is no guarantee that the nodes correspond to what the user intuitively would define as "places" or "rooms".

Although our architecture by no means hinders autonomous exploration it has therefore been chosen to let the robot learn the topological graph in interaction with the user. This, however, does only mean that the user should be able to direct the robot through the environment and point out/name specific places and objects, i.e., it does not mean that the user should also point out appropriate landmarks nor provide metric information or information about relevant LCSs. How this information is taught is explained in Section 4.

---

[1] After [9].

3   An Architecture for the Execution of Missions Using Learned World Models

As previously mentioned does the fact that we want the user to configure the system by (only) teaching a topological graph representation of the environment pose some challenges to the underlying system architecture. The major challenge is that the system must be able to configure itself to solve tasks in environments that are not a priori known and where the only context information the robot has is what has been taught and stored in the world model. We call this kind of automation configuration for *context adaptive operation* (CAO).

The number of types of environments the robot can face—and thus the number of different sensing and planning modules it will need—is of course open ended and we therefore want to be able to easily adapt the robot by adding new planning and perception modules. Also this poses some demands for the architecture.

3.1   System Overview

From an engineering point of view, the problem with a context adaptive system is to keep it structured and understandable although the exact configuration at some point in time cannot be a priori known. It has therefore been chosen to functionally divide the system up into *competences*, of which we currently have four; the navigation competence, NAV, the manipulation competence, MAN, the relocalisation competence, REL, and the world modelling/teaching competence MOD. Coordinating the activation of these competences is a mission planner and a mission executor. A functional[2] block diagram of the system is shown in Fig. 2.

As illustrated in the figure, the following four levels exist:

Mission planning level This is the symbolic level where missions are planned using a standard A^* planner. A mission, entered by the user in the MMI, specifies the goal state and possibly a number of intermediate states to be reached, whereas the current state of the system defines the initial state. A state transition is realised by executing a *task* (see below). The set of possible tasks, i.e., state transitions, for bringing the system from the initial state to the goal state are partly stored in the database (e.g., manipulation tasks) and partly dynamically extracted from the current topological graph in the world model (navigation tasks). This implies that when new information has been learned it is immediately available to the planner. The output from the mission planner is a *mission timeline* which specifies the sequence of tasks to be performed by the competences in order to carry out the given mission. The *mission executor* groups consecutive tasks for the same competence into

---

[2] We distinguish between *functional* and a *structural* descriptions. Functional descriptions illustrate abstract functionalities whereas structural descriptions illustrate the actual processes and their connections.

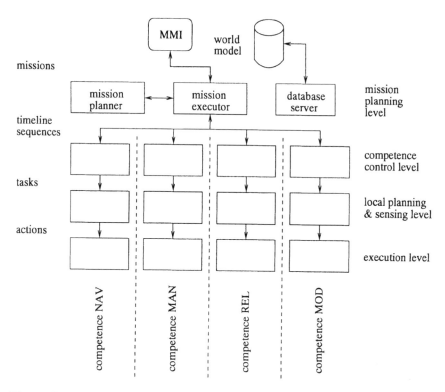

Fig. 2. Functional overview of the ASR system. Note that a box corresponds to a single, abstract functionality, not to a single process. See text for further explanations.

packages (timeline sequences). These task packages are sequentially transferred to—and executed by—the competences. If an error occurs, the mission planner is requested to do a replanning without including the failing task in the timeline. The world model is, due to the close connection with the mission planner, illustrated on the mission planning level, although it is also accessible from the other levels.

Competence control level This level constitutes the interface to the symbolic planning level and handles the overall control of the competences. In general it extends the purely symbolic timelines with relevant metric information from the world model and thereby generates tasks which can be executed by the lower levels. A task, T, is defined as the functionality which can be performed by a competence, thus the union of all tasks constitutes the system's instruction set at the timeline level. Note that one competence may be capable of performing several tasks. A typical task for NAV would be "do room–navigation to point $< X, Y >$" or "dock at object $< Z >$". A list of the currently implemented tasks is given in Table 1.

Local planning and sensing (LOPAS) level At this level, higher level sensing and high level non–symbolic planning is performed. For a docking task, for instance, the sensing part would be to recognise and localise the in the task specified docking object and the planning part would be to generate a series of via points for the robot to traverse in order to end up adequately docked by the object. The output from the LOPAS level is *actions*, which are sub–symbolic commands.

Execution level At this level the actions are executed, typically by real–time, reactive modules directly commanding the hardware.

Table 1. The list of currently defined tasks.

competence	task
NAV	room_navigation
NAV	corridor_navigation
NAV	door_traversal
NAV	docking
MAN	pickup
MAN	place
MAN	move
REL	start_reloc
REL	end_reloc
MOD	start_model
MOD	end_model

What is important to note about the overall system structure is its uniformity. Such uniformity may at a first glance seem artificial and awkward as for example with the REL competence which has no actuators to control. However, it can still be said to have an execution level in the form of the Kalman filter fusing the results from the sensing modules and continously correcting the platform's position estimate. We have found that using such a nomenclature and forcing the system into this functional structure has greatly improved the clarity of the design and not least the amount of software modules shared across the competences.

From a structural point of view, it is evident that the connection between competences is not—as shown in the functional view in Fig. 2—only at the timeline level, but also at lower levels. For example must the competences NAV and REL share information about the robot's position. The most significant difference between the functional and the structural organisation of the system is however at the LOPAS level where there exists a pool of sensory modules, called recognisers, which are dynamically activated and shared across competences. The generic processing structure for the LOPAS level is shown in Fig. 3. For each context type, C, there exists an instantiation of this structure. We define a context type as a tuple, $C = (\mathcal{E}, \mathcal{M}, \mathcal{T})$ where $\mathcal{E}$ is an environment type, $\mathcal{M}$ is the available model information, and T is a task type. In our system, there is one local planner for each T whilst the set of recognisers depends on $\mathcal{M}$ as well. How the set of recognisers is instantiated is described below.

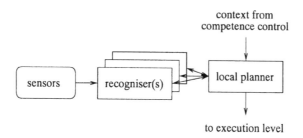

Fig. 3. The generic processing structure used at the LOPAS level. Several recognisers can be working in a parallel or a serial configuration. For the NAV and MAN competences, the local planner computes via points based on sensory information. In the REL competence, it selects relevant landmarks to relocate with.

By using the generic processing structure, we have achieved that also the LOPAS level, although specific for the context, is uniform, extendable and easy to understand and maintain.

3.2   Context Adaptive Operation

By context adaptive operation (CAO) we as previously mentioned understand the fact that the system changes its workings depending on the context, $\mathcal{C}$.

A prerequisite for context dependently using different modules is, of course, that the single modules have standard interfaces ensuring that is does not matter which module is used, as long as it is the right type. To that end we have developed standard interfaces for the recognisers in the system.

These interfaces are inspired by Henderson's *logical sensor systems* [14] where each sensor data processing unit is considered a sensor on its own from which other processes can get their input. In that sense, all recognisers are logical sensors and produce a standardised *output vector* consisting of a head and a body. The structure of the head is identical for all recognisers and says something about the type, origin, and the status of the data while the body is sensor type specific. The idea is that any client, which can be another recogniser, should be able to quickly evaluate the information in the head to see if the data is of any relevance and act accordingly.

Each recogniser can in principle work data *and* model driven. If the recogniser is *activated* with a set of models, it will try to match these models against the extracted objects, and if no models are given, it will simply start extracting and delivering hypothesises for whatever kind of objects it is designed to recognise. Furthermore, every recogniser can be a server as well as a client in the sense that a recogniser can register by any other recogniser in order to get sensor/object input. When a client registers by a recogniser, it normally also activates it. The registration is automatically performed by the system by means of a *channel structure* and a simple configuration file where it says what recognisers are capable of recognising what type of objects. This means that a client in general registers for a *type* of object (by all recognisers of this kind), and not by some named recogniser(s). This has the advantage of making the addition and deletion of recognisers in the system fully transparent to the client processes and furthermore of conveniently implementing redundancy and thereby fault tolerance.

The overall algorithm of a recogniser is as follows:

> wait for activation
> register by/activate relevant "sub–recognisers"
> while activated
>   wait for input data
>   extract and send objects to clients
> de–register from/de–activate "sub–recognisers"

From this simple algorithm it can be seen that the activation of recognisers spreads out in a tree–like manner[3] with the sensor servers as leaf nodes and the end users, for example the relocalisation process, as root nodes. In this way, what recognisers are activated and when, i.e., the configuration of the system, is implicitly determined by the current task, $\mathcal{T}$, and what kind of model objects, $\mathcal{M}$, are present and thus need to be extracted/recognised. The next section more concretely describes examples of where and how we employ this kind of context adaptive operation.

3.3   Examples of CAO

In this section concrete examples of how CAO is used in the system are presented.

Model information is used in the system by sensory modules primarily for landmark recognition. For example, there exists a number of recognisers that can recognise doors in laser (lidar) scans and match these against stored model doors in a world model. These recognisers are activated via the REL local planner when the robot is in a part of the world where doors are stored in the world model and stopped when there are none. The same applies to all other landmark types which we can recognise and store in the world model (currently walls and corners). In this way it is generally achieved that relocalisation is automatically performed with whatever recognisable landmarks are at hand without any supervisor process or system designer needs to explicitly tell what landmarks to use and when. The drawback of the method is that some recognisers may start processing without there being any real need for their results. In the future we will handle this by introducing a utility based resource management [15] from

---

[3] Formally, the activation net can be modelled as a directed graph with no cycles.

which each recogniser will have to get permission to use system resources (sensors, CPU time, etc.) before starting processing. This will for example mean that a recogniser doing landmark recognition will only be allowed to run when the platform position is uncertain since the utility of doing pose estimation when the platform position is quite accurately known is small.

As previously mentioned, recognisers are shared across competences. An example of this is that the door recognisers used by the REL competence are also used by the NAV competence when the robot is driving through doorways. This is due to the fact that the door passage is cast as a servoing problem and thus a recogniser that will deliver the position of the door relative to the robot is called for. The NAV and MAN competences also share 3D–object recognisers for the purposes of docking and grasping, respectively.

Another place where CAO is explicitly used is in the pilot, i.e., the execution level of the NAV competence. In contrast to the LOPAS level which is quite modular (see Fig. 3) the pilot is realised as a monolithic block. This is due mainly to safety reasons: we will rather have one well-tested, "proven" module that controls the robot all the time than a number of specific modules switching control between them. Furthermore, the task of the pilot, namely to drive the robot to subgoals while avoiding obstacles, is sufficiently narrow and well–defined for it to be implemented as a single module. The requirement is, however, that this module can adapt itself to the given requirements of the environment and the task. For example, if the robot drives in a narrow corridor, it is normally desired to servo to the middle in order to keep as much freespace between the robot and the walls as possible. In larger rooms, this behaviour can lead to very inefficient trajectories while during docking tasks it is directly wrong since the purpose here is exactly the opposite: to come close to and maybe even touch some given object. Here corridor vs. room navigation is an example of the environment, $\mathcal{E}$, determining the desired operation while it for docking is the task, $\mathcal{T}$.

In order to make the pilot adaptable to the environment and task requirements, we have chosen to use the *dynamic window approach* [16] which is an action space method where various objectives such as velocity, freespace, and heading towards goal are explicitly evaluated and weighted to determine the "optimal" action at a given time, i.e., given the current goal point and sensory readings. By creating a set of weights, W_T, for each navigation context type, $\mathcal{C}_{NAV}$, it is then possible to control the operation of the pilot. In other words, each W_T implements a local control strategy for the robot. In order to limit the number of W_T's we have in this case collapsed $\mathcal{C}$ into $\mathcal{T}$ by plainly defining the task associated with, e.g., a room environment as room navigation (see Table 1). In the cases where this association does not hold, as by e.g., docking in a room environment, the task (docking) anyway dominates the environment (room) with respect to how the pilot should select its actions.

After describing the architecture that allows the system to work with taught environment information, we will in the next section describe how the teaching is done.

4   Interactive Teaching

The basic concept behind the teaching of world model information is that it should be interactive (for reasons explained in Sect. 2) and that it should be done on–line and on–site with the real system. The latter has the advantage that the user gets direct feedback from the system regarding whether the robot can do what it is intended to do. Furthermore, it ensured that the representations are perfectly matched to the system in the sense that for example features detected and entered into the world model by a recogniser is guaranteed to be detectable by at least one recogniser, namely the one originally entering it.

4.1   The World Model

In Sect. 2.2 it was argued that the annotated topological graph is a suitable choice of model representation because it satisfies the demands for longevity and because it supports the planning and execution of missions. In the following, the components of the implementation of the topological graph, we have chosen, will be elaborated.

Nodes The nodes represent "places" in the environment. Since the nodes are used for the mission planning as well as for the navigation, we will define places to be rooms, special locations of important objects, plus for the navigation significant points such as doorways and intersections. The nodes have as annotation a 2D position and a covariance matrix representing the uncertainty of the position.

Edges The edges represent traversable paths between the places. The edges have as annotations the for the traversal appropriate navigation task, $\mathcal{T}_{NAV}$.

Features We define features as characteristics of the environment which can be used for navigational purposes and that can be sensed "on the fly" by the robot's recognisers. Typical examples for indoor environments are doors and walls. Features have similar to nodes a position and the corresponding uncertainty associated.

3D–Objects These objects are special objects used for tasks where the 3D shape is of importance. In our system, these are docking objects and objects to be grasped and transported. 3D–Objects can—depending on the robot's capabilities—also be features, but due to the normally rather limited on–board processing power of mobile robots, it is not feasible to continously recognise these objects.

Links The links aré used to relate the features and 3D–objects to the nodes. Features are linked to the nodes that have associated edges (paths) from which they can possibly be seen. This has two purposes; one is to be able to efficiently index the relevant features for landmark recognition, depending on where the robot currently is. The other purpose is to be able to rectify the world model by modifying the node positions (for example with the method described in [17] and [18]) and still keep the local consistency of the model. In this connection are the links used to express the relative position of features and nodes and thereby also what features should move with the nodes when they are displaced.

An example of a simple world model is shown in Fig. 4. The nodes are shown as small circles, the edges as lines between the nodes. Shown on the edges are mnemonics for the associated tasks. As features as shown doors and walls, which are symbolised as black bars and lines, respectively. 3D–objects are shown as small squares. The links are not shown in this plot.

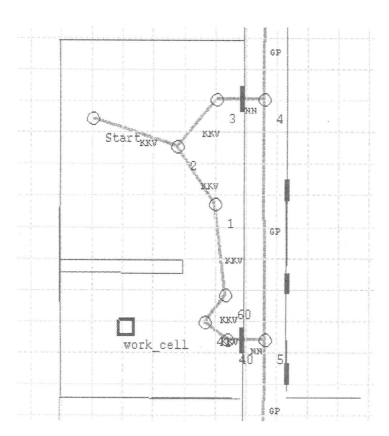

Fig. 4. A simple example of a world model. This world model was taught interactively and on–line but in a simulator with the CAD data of the modelled building.

4.2   Teaching with the MOD Competence

The purpose of the MOD competence is to allow the user to easily teach the robot the world model representation described above. The basic representation, the graph, is taught by simply leading the robot through the environment using a virtual joystick. As this joystick, we use a display with an evidence grid showing the local surroundings of the robot. The evidence grid, shown in Fig. 5, is generated using the robot's laser scanner and thus delivers a very reliable and easily understandable "radar view". The user can command the robot by simply clicking with a pointing device in the grid where he/she wants the robot to go. Currently a small pop–up window prompts the user for the relevant LCS, but in a future version this will be suggested automatically by a neural net analysing the topology of the surroundings. The advantage of using the virtual joystick versus a conventional joystick is that the commands from the user in the former case are converted into commands identical to those that will later be generated by the NAV competence controller when traversing the taught graph. Thus it is immediately confirmed if the robot can autonomously navigate to the commanded position using the suggested LCS. When the robot has reached the next "place" the user tells this by clicking a button in a simple GUI thus causing the MOD controller to generate a node in the world. The edges are automatically generated by the controller by simply tracking between what nodes the user has driven the robot with what LCSs.

Fig. 5. Evidence grid used as virtual joystick. White denotes occupied space, black freespace, and grey unknown. In the plot, the robot is standing in a room with a doorway approximately 2.5 meters in front of it.

The features are automatically learned when the user is driving the robot around, teaching the graph. The MOD competence achieves this by simply starting the respective recognisers and registering for the features they deliver. When a feature has been seen N times, it is entered into the world model by the controller. Currently we let the user interactively confirm the door hypothesises in the GUI, since also phantoms can occur due to door–similar structures like windows. The links to the features are automatically generated by the MOD controller which keeps track of between what nodes the features were seen by the recognisers.

The 3D–objects are also taught interactively by the user. So far, we have only polyhedral 3D–objects in our world model. These are taught using a teachbox with which the user can request a 3D depth image of some specified part of the environment. The 3D image is segmented into planar surfaces [19] and the result is presented to the user in the teachbox interface. A typical example is shown in Fig. 6. With a pointing device the user can now select the regions belonging to an object, name this, and define possible docking and grasp positions relative to the object. When done, the object is stored away in the database and can subsequently be referenced in the world model at the positions where an object of this type exists.

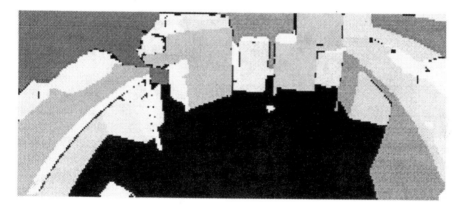

Fig. 6. Range image segmented into planar surfaces.

The here described interactive teaching method has been proven to be adequate for the fast and reliable generation of world models. An example of such a model is presented in the next section.

5   Experiments

In this section, we will present an example of a world model which has been taught by leading the real robot, shown in Fig. 5, around in a previously unknown but standard office environment. The purpose of the experiment is to show what such a model looks like and what the capabilities and the limits of the currently generated models are.

In this experiment, the robot was driven from a starting point in a room out into a hallway which forms a rectangular loop. Halfway around this loop the robot was driven into another room where it was taught a 3D–object. Then

Fig. 7. The experimental platform used in the ASR project. The robot consists of a re–fitted LabMate base equipped with a 7–DOF Amtec manipulator and a Sick lidar plus a set of colour CCD cameras mounted on a pan/tilt unit. The robot has two on–board P200 single board computers and radio Ethernet providing a facility for sending status information to a stationary workstation.

the robot was driven further around the loop and back to the room where the teaching was started. The resulting model is shown in Fig. 8. In order to keep the model simple enough for also showing the links, it was chosen to exclude the line features which can, however, like any other model components, seamlessly be added later by letting the robot drive through the environment once again. For the shown model, teach–in time was 23 minutes and total path length is approximately 72 meters. The teacher in this case was a person with thorough knowledge of the system, which is due to the fact that the user interfaces for the teaching are still under development and the system does not yet provide the full degree of support for the user.

What can be seen from the plot in Fig. 8 is that odometry error is a problem for the model generation which is due to the bootstrapping–type problem it is to generate a globally consistent model without having global landmarks to relocate with. This means that in the model shown in Fig. 8 the nodes labelled 3 and 18 physically correspond to the same place and thus the model is here *locally* inconsistent. Otherwise the model is at all places consistent and it has been shown with thorough experiments that the robot is capable of planning and executing all missions that do not imply the traversal of the "virtual" path

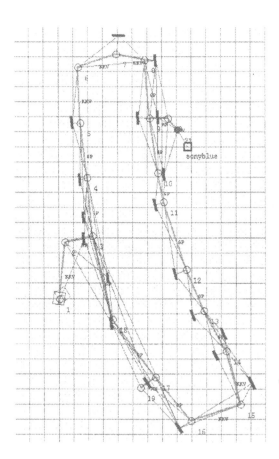

Fig. 8. World model taught by leading the robot through a previously unvisited office environment. The hallway constitutes a rectangularly shaped loop, which due to odometry error has been somewhat distorted. Grid size is 1 meter. See text for further details.

between nodes 3 and 18. Current work along the lines of [18] is concerned with the automatic detection and correction of inconsistencies in the world model.

6   Discussion and Conclusion

In this paper, we have described the interactive teaching of world model information developed in the Advanced Servicing Robot project at Daimler-Benz AG. An important criteria when designing service robots is that they need to be very flexible to be commercially attractive and yet very specialised in order to be safe and robust enough. It was argued that an important aspect of achieving this is to enable the robot to learn the representations needed in order to perform its tasks. The fact that only the representation type but not the concrete environment and its model is a priori known sets some demands for the architecture with respect to flexibility regarding model information and the given environment. The first problem was addressed by introducing the concept of context adaptive operation, the latter by using a highly standardised, modular, client–server based design.

After explaining the system architecture, we presented the method for teaching the model information. The basic idea behind the teaching is to let the user guide the robot through the environment as if it was driving autonomously and automatically record information about places, paths, and landmarks. Furthermore, the user must interactively teach special 3D–objects. This produces compact models that are directly verified because the robot has just been at the places, driven the paths, and seen the landmarks and objects that are stored in the model.

With an experiment we showed that the information taught by the user together with the CAO is sufficient to plan missions and to make the robot navigate robustly in a previously unknown but standard type of in–door office environment. On–going work in the area of detecting and correcting inconsistencies in the generated models is expected to solve current problems with inconsistencies appearing when loops exist in the environments and thus in the models. With other experiments, not reported here, we have shown that the framework is conveniently transferred to other types of in–door scenarios including a factory and a restaurant setting.

Further work includes the improvement of the user interface and more support from the system in the teaching process, for example by letting the system generate the navigationally important nodes and relevant LCSs.

Acknowledgements

This research was partly sponsored by the BMBF[4] project NEUROS.

References

1. In: Hägele, M. (ed.): Serviceroboter – ein Beitrag zur Innovation im Dienstleistungswesen. Fraunhofer–Institut für Produktionstechnik und Automatisierung (IPA) (1994) Foreword. In German
2. Gat, E.; Three–Layer Architectures. In: Kortenkamp, D., Bonasso, R.P., Murphy, R.: Artificial Intelligence and Mobile Robots. AAAI Press/The MIT Press (1998) 195–210
3. Meystel, A.: Autonomous Mobile Robots. World Scientific Press, Singapore (1991)
4. Firby, R.J.: Adaptive Execution in Dynamic Domains. Yale University, Department of Computer Science (1989)
5. Simmons, R.G.: Structured Control for Autonomous Robots. IEEE Trans. on Robotics and Automation, Vol. 10(1) (1994) 34–43

---

[4] BMBF is the German Ministry for Education, Science, Research, and Technology.

6. Gat, E.; Robot Navigation by Conditional Sequencing. In: Proc. of the 1994 IEEE Int. Conf. on Robotics and Automation, San Diego, California. IEEE Computer Society Press (1994) 1293–1299

7. Fujita, M., Kageyama, K.: An Open Architecture for Robot Entertainment. In: Proceedings of the First Int. Conf. on Autonomous Agents. ACM (1997) 435–442

8. Elfes, A.: Sonar–Based Real–World Mapping and Navigation. IEEE Journal of Robotics and Automation, Vol. 3(3) (1987) 249–265

9. Kuipers, B.J., Buyn, Y-T.: A Robust, Qualitative Method for Robot Spatial Learning. In: Proceedings of the Seventh National Conference on Artificial Intelligence. AAAI Press, Menlo Park, Calif. (1988) 774–779

10. Kuipers, B.J., Buyn, Y-T.: A Robot Exploration and Mapping Strategy Based on a Semantic Hierarchy of Spatial Representations. Robotics and Autonomous Systems, Vol. 8 (1991), 47–63

11. Koenig, S., Simmons, R.G.: Xavier: A Robot Navigation Architecture Based on Partially Observable Markov Decision Process Models. In: Kortenkamp, D., Bonasso, R.P., Murphy, R.: Artificial Intelligence and Mobile Robots. AAAI Press/The MIT Press (1998) 91–122

12. Thrun, S., Bücken, A., Burgard, W., Fox, D., Fröhlinghaus, T., Hennig, D., Hofmann, T., Krell, M., Schmidt, T.: Map Learning and High–Speed Navigation in RHINO. In: Kortenkamp, D., Bonasso, R.P., Murphy, R.: Artificial Intelligence and Mobile Robots. AAAI Press/The MIT Press (1998) 21–52

13. Gutmann, J-S., Nebel, B.: Navigation mobiler Roboter mit Laserscans. In: Autonome Mobile Systeme. Springer (1997), in German.

14. Henderson, T., Shilcrat, E.: Logical Sensor Systems. Journal of Robotic Systems Vol. 1(2) (1984) 169–193

15. Kristensen, S.: Sensor Planning with Bayesian Decision Theory. Robotics and Autonomous Systems, Vol. 19 (1997) 273–286

16. Fox, D., Burgard, W., Thrun, S.: The Dynamic Window Approach to Collision Avoidance. IEEE Robotics and Automation Magazine, Vol. 4(1) (1997) 23–33

17. Lu, F., Milios, E.: Robot Pose Estimation in Unknown Environments by Matching 2D Range Scans. Journal of Intelligent Robotic Systems, Vol. 18 (1997) 249–275

18. Lu, F., Milios, E.: Globally Consistent Range Scan Alignment for Environment Mapping. Autonomous Robots, Vol. 4(4) (1997) 333–349

19. Jiang, X. Bunke, H.: Fast Segmentation of Range Images into Planar Regions by Scan Line Grouping. Machine Visions and Applications, Vol. 7(2) (1994) 115–122

MAid: A Robotic Wheelchair Operating in Public Environments

Erwin Prassler[1], Jens Scholz[1], Matthias Strobel[1], and Paolo Fiorini[2]

[1] Research Institute for Applied Knowledge Processing (FAW), P.O. Box 2 060
D-89010 Ulm, Germany
Phone: +49 731 501-621 Fax: +49 731 501-999
{prassler,scholz}@faw.uni-ulm.de

[2] Jet Propulsion Laboratory, California Institute of Technology
Pasadena, CA 91109, USA
Phone: +1 818 354-9061 Fax: +1 818 393-5007
fiorini@jpl.nasa.gov

Abstract. In this paper we describe the hardware design, the control and navigation system, and our preliminary experiments with the robotic wheelchair MAid (Mobility Aid for Elderly and Disabled People). MAid's general task is to transport people with severely impaired motion skills such as, for example, paraplegia, multiple sclerosis, poliomyelitis, or muscular dystrophy. Following the advice of disabled people and physicians we did not set out to re-invent and re-develop the set of standard skills of so-called intelligent wheelchairs, such as *FollowWall, FollowCorridor, PassDoorway* which are commonly described in the literature. These maneuvers do not always require fine motion control and disabled people, in spite of their disability, are often well capable of navigating their wheelchair along a corridor and actually eager to do it. In our work we focused instead on maneuvers which are very burdensome because they take a long time and require extreme attention. One of these functions is deliberative locomotion in rapidly changing, large-scale environments, such as shopping malls, entry halls of theaters, and concourses of airports or railway stations, where tens or hundreds of people and objects are moving around. This function was not only acknowledged as being very useful but also very entertaining, because MAid often had to work very hard to find its way through a crowd of people. MAid's performance was tested in the central station of Ulm during rush-hour, and in the exhibition halls of the *Hannover Messe '98*, the biggest industrial fair worldwide. Altogether, MAid has survived more than 36 hours of testing in public, crowded environments with heavy passenger traffic. To our knowledge this is the first system among robotic wheelchairs and mobile robots to have achieved a similar performance.

1 Introduction

Nowadays, the freedom and capability to move around unrestrictedly and head for almost any arbitrary location seems to be an extremely valuable commodity.

Christensen et al. (Eds.): Sensor Based Intelligent Robots, LNAI 1724, pp. 68–95, 1999.

Mobility has become an essential component of our quality of life. A natural consequence of this appreciation of the good mobility is the negative rating of the loss of mobility caused, for example, by a disease or by age. The loss of mobility represents not only the loss of a physiological function, but often a considerable social descent. People with severely impaired motion skills have great difficulties to participate in a regular social life. Not seldom a loss of mobility leads to a loss of contacts to other non-disabled people or makes it at least difficult to establish such contacts. A loss of mobility, may it be due to an injury or to advanced age, is always accompanied by a loss of autonomy and self-determination, it creates dependence and in extreme cases it may even affect the individual intimacy and dignity.

The loss of one's mobility may be seen as a difficult individual fate. However, there are two aspects which may make it not only into an individual but also into a general problem of our society. First, the average age in western societies is increasing dramatically. As a natural consequence the number of people suffering from severe motion impairment will increase too. At the same time, we can observe an equally dramatic increase of the expenditures for health care and nursing and furthermore a reduction of nursing staff in order to limit the cost explosion. The results of these developments are foreseeable: the quality of health care will decay, individual care will become still more expensive and less affordable for people with medium and lower income, elderly will then be sent to nursing homes much earlier than nowadays in order to get sufficient care.

A way out of this unpleasant development may be through the development of robotic technologies. Many activities which a person with severe motion impairment is unable to execute may become feasible by using robot manipulators and vehicles as arms and legs, respectively. Lifting things, carrying and manipulating things and moving around in ones own home this way becomes feasible without the assistance and help of a nurse. People with motion impairment get back a certain amount of autonomy and independence and can stay in their familiar environment. Expenditures for nursing personnel or an accommodation in a nursing home can be avoided or at least limited.

In this paper we describe a robotic wheelchair MAid (Mobility Aid for Elderly and Disabled People) whose task is to transport people with severely impaired motion skills and to provide them with a certain amount of autonomy and independence. The system is based on a commercial wheelchair, which has been equipped with an intelligent control and navigation system.

Robotic wheelchairs have been developed in a number of research labs (see our review in Section II.) The common set of functions provided by most of those systems consists of *AvoidObstacle FollowWall*, and *PassDoorway*. In conversations with disabled and elderly people, and with physicians we learned that not all of these functions are of equal interest for people with motion impairment. Particularly *FollowWall* and *PassDoorway* are maneuvers which most disabled people still want to execute themselves provided they have the necessary fine motor control.

Following this advice, in our work we focused on different types of motion skills. Our system has two modes of operation, a semi-autonomous and a fully autonomous mode. In the semi-autonomous mode the user can command MAid to execute local maneuvers in narrow, cluttered space. For example, the user can command MAid to maneuver into the cabin of a restroom for handicapped people. Maneuvers in narrow, cluttered space require extreme attention and often lead to collisions, particularly if the patient lacks sufficient fine motor control. We denoted this type of maneuver in small, narrow areas as NAN (narrow area navigation), and the implementation of this capability is described in [12,14].

In the second mode, MAid navigates fully autonomously through wide, rapidly changing, crowded areas, such as concourses, shopping malls, or convention centers. The algorithms and the control system which enable MAid to do so are described in the following. We denoted this latter type of motion skill as WAN (wide area navigation). The only action the user has to take is to enter a goal position. Planning and executing a trajectory to the goal is completely taken care by MAid. MAid's capability of navigating in rapidly changing environments was not only acknowledged as being very useful but also very entertaining. MAid often had to work very hard to find its way through a crowd of people and our test pilots were often very curious to see what MAid would do next, bump into a passenger - very rarely it did - or move around.

MAid's performance was tested in the central station of Ulm during rush-hour and in the exhibition halls of the *Hannover Messe '98*, the biggest industrial fair worldwide. Altogether, MAid has so far survived more than 36 hours of testing in public, crowded environments with heavy passenger traffic. To our knowledge there is no other robotic wheelchair and no other mobile robot system which can claim a comparable performance.

Note that at first sight the two types of motion skills, NAN and WAN, which MAid is capable of, have little in common with the navigation skills of other intelligent wheelchairs. Quite the opposite is the case. In terms of its performance WAN can be seen as a superset of functions such as *AvoideObstacle* or *FollowWall*. When WAN is activated and a destination at the opposite end of a hallway is specified then MAid will automatically show a wall following behavior and at the same time avoid obstacles, although there is no explicit implementation of such a behavior in the WAN module. Likewise, passing a door or docking at a table are typical instances of NAN maneuvers.

The rest of this paper is organized as follows. In the next section we give an overview of the state of the art in the development of robotics wheelchairs. MAid's hardware design is described in Section 3. In Section 4, we then describe the software architecture and the algorithms which enable MAid to navigate in a wide, rapidly changing, crowded environment. Note that although MAid's capability to navigate in narrow, partially unknown, cluttered environment is mentioned several times below, the focus in this paper is on MAid's WAN skill. We will not go into the details of MAid's NAN skill, presented in [12,14].

2   Related Work

In recent years there has been the development of several intelligent wheelchairs. A first design concept for a self-navigating wheelchair for disabled people was proposed by Madarasz in [8]. The vehicle described there used a portable PC (320 KB of memory) as on-board computer. The sensor equipment of the wheelchair included wheel encoders, a scanning ultrasonic range-finder and a digital camera. The system was supposed to navigate fully autonomously in an office building. To find a path to its destination it used a symbolic description of significant features of the environment, such as hallway intersections or locations of offices. The path computed by the path planner consisted of a sequence of primitive operations such as *MoveUntil* or *Rotate*.

In [3], the system NavChair is described. NavChair's on-board computer is also a portable IBM compatible PC. An array of 12 Polaroid ultrasonic sensors at the front of the wheelchair is used for obstacle detection and avoidance. NavChair's most important function is automatic obstacle avoidance. Other functions include wall following and passing doorways.

Hoyer and Hölper [7] present a modular control architecture for an omni-directional wheelchair. The drive of this system is based on Meccanum-wheels. The wheelchair is equipped with ultrasonic and infrared sensors and a manipulator. A low-level control unit is in charge of the operation of the sensor apparatus, the actual motion of the vehicle and the operation of the manipulator. This control unit is realized on an VME-Bus-system using pSOS+. A high-level PC/UNIX based planning module consists of a path and a task planner to execute task oriented commands.

A hybrid vehicle RHOMBUS for bedridden persons is described in [9]. RHOMBUS is a powered wheelchair with an omni-directional drive which can be automatically reconfigured such that it becomes part of a flat stationary bed. The bedridden person does not have to change seating when transferring between the chair and bed.

Mazo et al. [10] describe an electrical wheelchair which can be guided by voice commands. The wheelchair recognizes commands such as *Stop, Forward, Back, Left, Right, Plus, Minus* and turns them into elementary motion commands. The system also has a control mode *Autonomous*. In this mode the wheelchair follows a wall at a certain distance.

Miller and Slack [11] designed the system Tin Man I and its successor Tin Man II. Both systems were built on top of a commercial pediatric wheelchair from Vector Wheelchair Corporation. Tin Man I used five types of sensors, drive motor encoders, eight contact sensors used as whiskers, four IR proximity sensors distributed along the front side of the wheelchair, six sonar range sensors, and a flux-gate compass to determine the vehicle's orientation. Tin Man I had three operation modes: *human guided with obstacle override, move forward along a heading,* and *move to (x, y)*. These functions were substantially extended in Tin Man II. Tin Man II capabilities include *Backup, Backtracking, Wall Following, Passing Doorways, Docking* and others.

Wellman [15] proposes a hybrid wheelchair which is equipped with two legs in addition to the four regular wheels. These legs should enable the wheelchair to climb over steps and move through rough terrain. A computer system consisting of a PC 486 and a i860 co-processor for the actuator coordination is used to control the wheelchair.

3   Hardware Design

Our system MAid (see Fig. 1) is based on a commercial electrical wheelchair type SPRINT manufactured by MEYRA GmbH in Germany. The wheelchair has two differentially driven rear wheels and two passive castor front wheels. It is powered by two 12 V batteries (60 Ah) and reaches a maximum speed of 6 km/h. The standard vehicle can be manually steered by a joystick.

Fig. 1. The robotic wheelchair MAid

The goal of the work presented here was to develop a complete navigation system for a commercial wheelchair, such as SPRINT, which would enable it to automatically maneuver in narrow, cluttered space as well as in crowded large-scale environments. The hardware core of the navigation system developed for the task is an industrial PC (Pentium 166MHz) which serves as on-board computer. The computer is controlled by the real-time operating system QNX.

MAid is equipped with a variety of sensors for environment perception, such as collision avoidance, and position estimation. In particular, MAid's sensor apparatus includes the following devices:

- a dead-reckoning system consisting of a set of wheel encoders and a optical fiber gyroscope (Andrew RD2030),
- a modular sonar system consisting of 3 segments each equipped with 8 ultrasound transducers and a micro-controller, mounted on an aluminum frame which can be opened to enable the user to sit in the wheelchair,
- two infrared scanners (Sharp GP2D02 mounted on servos) for short range sensing,
- a SICK 2D laser range-finder PLS 200 mounted on a removable rack.

The dead-reckoning system, which integrates over the distance traveled by the vehicle and over its orientation changes, provides elementary position and orientation information. This information is rather inaccurate with errors accumulating rapidly over the traveled distance but it is available at low cost and at all times.

The sonar system and the laser range finder are the sensors which MAid uses to actually perceive the surrounding environment. This perception has the form of two dimensional range profiles and gives a rather coarse picture of the environment. From the range profiles MAid extracts the basic spatial structure of the environment, which is then used for two purposes: the *avoidance of stationary and moving objects*, and the *estimation of MAid's position* in the environment.

For the latter, we apply an Extended Kalman filter, provided we have a description of the environment. Based on such a description, on a model of MAid's locomotion, and its last position this filter produces a set of expectations regarding MAid's sensor readings. The deviation between these expectations and the true sensor readings is then used to compute correction values for MAid's estimated position.

It should be mentioned that in a wide, crowded, rapidly changing, mostly unknown environment there is little advantage in using a Kalman filter since one of its essential ingredients, namely the *a priori* description of the environment is not available. For the navigation in such an environment, we have to rely exclusively on the position information provided by the dead-reckoning system.

While all other components of MAid's navigation system are low cost or can at least be substituted by cheaper components without reducing MAid's performance, the laser range-finder is undoubtedly an expensive sensor (approx. US$ 4000). However, as we have argued in [13] there is no other sensor which is equally suited for detecting and tracking a large number of moving objects in real-time. This function in turn is essential for the navigating in wide, crowded, rapidly changing environments.

Except for the laser range finder, the other sensors are connected to, and communicate with, the on-board computer using a *field bus* as shown in Fig. 2. The interface between these devices and the field bus is implemented by a number of micro-controllers (68HC11). Due to the high data rates of the laser range-finder, this device is directly connected to the on-board computer by a serial port. The motion commands computed by the navigation system are also sent over the field bus to the motion controller, which powers the wheel motors.

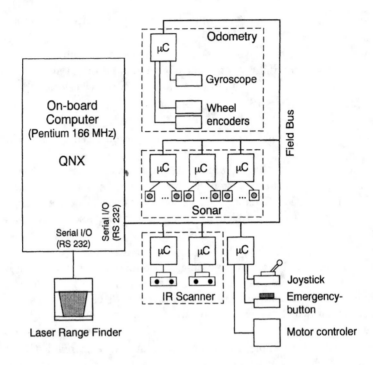

Fig. 2. Hardware architecture of MAid's control system

The user interface of MAid's navigation system consists of the original wheelchair joystick and of a notebook computer. With the joystick the user points to the desired motion direction. The notebook is used to select MAid's operation mode and to enter the goal position. We are planning to enhance this interface with a commercial speech recognition system, similar to those currently used in the automobile industry. It is obvious however, that in order to be useful for a severely disabled person, MAid's user interface has to be adapted to that person's specific disability.

4   Control Architecture

MAid has a hierarchical control architecture consisting of three levels: a basic control level, a tactical level, and a strategic level. The components of this control system which contribute to MAid's capability of navigating in a wide, crowded, rapidly changing area (WAN) are shown in Fig. 3. Note that in the following we simply denote these components as *WAN module*. For the sake of clearity, in Fig. 3 we omit the parts of the control system, which implement MAid's navigation skills for narrow, cluttered, partially unknown areas (NAN). These are described in [12].

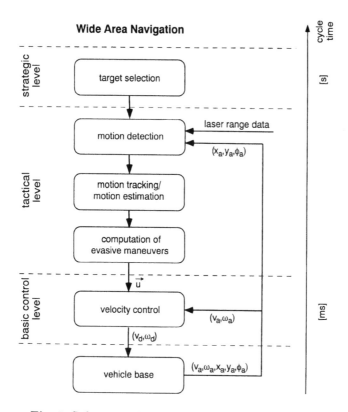

Fig. 3. Software architecture of MAid's WAN module

On the basic control level we compute the values of the control variables which put the vehicle into motion. The *velocity control* module on this basic control level receives as input a velocity vector u describing the target velocity and heading and the actual values of the translational and rotational (v_a, ω_a) of the vehicle. The velocity vector is converted into target values for the translational and rotational velocity. Out of these target values and the actual values provided by the vehicle's dead-reckoning system the velocity controller computes appropriate correction values which are then fed to the motor controllers.

On the tactical level, which essentially forms the core of the WAN module we have three submodules, a *motion detection* module, a module for *motion tracking and estimating object velocities and headings*, and a module for *computing evasive maneuvers*. In the following paragraphs we give a brief description of the interaction of these submodules. The methods which they actually implement are described in detail in Section 5.

In order to be able to react to a rapidly changing environment with potentially many moving objects, MAid continuously observes the surrounding world with a 2D laser range-finder. The range data provided by this range-finder essentially represent MAid's view of the world. In the continuous stream of range data MAid tries to detect the objects in its environment and to identify which of these objects are stationary and which are in motion (see [13] for more details). From the stream of range data MAid further derives estimates for the motion direction and velocity of the objects by extrapolating their past motion trajectory and velocity.

Based on these predictions and on its own motion direction and velocity MAid then determines if it is moving on a collision course with one or several of the moving objects. After an analysis of so-called *Velocity Obstacles* [5] MAid computes an avoidance maneuver, which is as close to its original heading as possible but does not lead to a collision with the objects moving in the vicinity of MAid.

Motion detection, motion prediction, the computation of collision courses and the computaiton of the avoidance maneuver take approximately 70 ms. If we include the time for a sensor observation (recording of a range image) the cycle time increases to 0.3 sec, thus MAid is able to compute a new maneuver every 0.3 s. This is primarily due to the low transmission rate of the range-finder.

MAid's main task while it navigates in a wide, crowded, rapidly changing area is to reach a specific goal at some distance from its present position. In the current design it does not pursue any more complex, further reaching plans such as visiting a sequence of intermediate goals. Accordingly, the strategic level consists of the selection of the next goal, which is left to the user. At a later point, the strategic level will be expanded by a path planner, for example, which will provide the WAN module with a sequence of intermediate goals.

5   Navigation in Rapidly Changing, Crowded Environments

In this section, we describe the methods and components which contribute to MAid's capability of navigating in a wide, crowded, rapidly changing area (WAN). Amongst existing robotic wheelchairs this capability is rather unique. The part of MAid's control system which implements this capability essentially consists of three components: an algorithm for motion detection, an algorithm for motion tracking, and an algorithm for computing evasive courses, which is based on the Velocity Obstacle (VO) approach [6].

5.1   Motion Detection and Motion Tracking

A rather obvious approach to identify changes in the surrounding environment is to consider a sequence of single observations and to investigate where these observations differ from each other. A discrepancy between two subsequent observations is a strong indication of a potential change in the environment. Either an unknown object has been discovered due to the self-motion of the observer or an already discovered object has moved by some distance. In the following sections we discuss how this simple idea can be used in a fast motion detection and tracking algorithm.

Sensor selection and Estimation of Self-Motion The sensor which we use for motion detection and tracking in crowded, rapidly changing environments is a laser range-finder (SICK PLS 200). In [13] we argue that this is the most appropriate sensor for the given task. The device works on a time-of-flight principle. It has a maximum distance range of $d = 50$ m with an accuracy of $\sigma_d \approx 50$ mm and an angular range of ± 90 deg with a resolution of 0.5 deg. The device is currently operated at a frequency of 3 Hz providing three range images per second.

The range data provided by the laser range-finder are naturally related to the local frame of reference attached to the sensor. In order to compare two subsequent local range images and to compute a differential image, it is necessary to know precisely which motion the sensor has undergone between two observations, how far it has moved from one viewpoint to the next and how far it has turned. This information is provided by the *dead-reckoning system*, which enables the wheelchair to keep track of its position and orientation over some limited travel distance with reasonable accuracy. With the information about the current position and orientation of the vehicle it is straightforward to transform the local range images from earlier measurements into the actual frame of reference.

Representations of Time-varying Environments A very efficient and straightforward scheme for mapping range data is the *occupancy grid representation* [4]. This representation involves a projection of the range data on a two-dimensional rectangular grid, where each grid element describes a small region in the real-world.

While investigating the performance of existing grid based mapping procedures, we noticed that most of the time was spent for mapping free space. Particularly, the further away the observed objects were, the more time it costs to map the free space between the sensor and the object. Also, before a range image could be assimilated into a grid, the grid had to be completely initialized, that is, each cell had to be set to some default value. For grids with a typical size of several tens of thousands of cells these operations became quite expensive.

Now, mapping large areas of free space is rather useless for detecting and tracking moving objects. To avoid this, we devised an alternative representation in which we map only the cells observed as occupied at time t, whereas all other cells in this grid remain untouched. We call this representation a *time stamp map*.

Compared to the assimilation of a range image into an occupancy grid the generation of a time stamp map is rather simplified. Mapping a range measurement involves only one single step, i.e. the cell coinciding with the range measurement is assigned a time stamp t. This stamp means that the cell was *occupied at time t*. No other cell is involved in this operation. Particularly, we do not mark as *free* any cell which lies between the origin of the map and the cell corresponding to the range measurement.

The time variation of the environment is captured by the sequence TSM_t, $TSM_{t-1}, \ldots, TSM_{t-n}$ of those time stamp maps. An example of such a sequence is shown in Fig. 4 a) - c). These pictures show three snapshots of a simple, time-varying environment with a moving and a stationary object in a time stamp map representation. The age of the observation is indicated through different gray levels where darker regions indicate more recent observations. Note that the maps are already aligned so that they have the same orientation. A translation by a corresponding position offset finally transforms the maps into the same frame of reference. The aligned maps are shown in Fig. 4 d). The assimilation of a range image into a 200×200 time stamp map takes 1.5 ms on a Pentium 166Mhz.

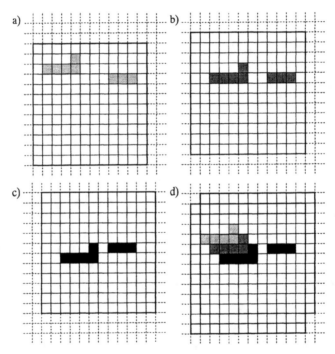

Fig. 4. A sequence of time stamp maps describing a toy-environment. The age of the observation is indicated through different gray levels. The more recent the observation the darker is the gray level which marks the object contours.

An Approach to Fast Motion Detection Motion detection in a sequence of time stamp maps is based on a simple heuristic. We consider the set of cells in TSM_t which carry a time stamp t (*occupied at time t*) and test whether the corresponding cells in TSM_{t-1} were occupied too, i.e., carry a time stamp $t-1$. If corresponding cells in TSM_t, TSM_{t-1} carry time stamps t and $t-1$, respectively, then we interpret this as an indication that the region in the real world, which is described by these cells has been occupied by a stationary object. If, however, the cells in TSM_{t-1} carry a time stamp different from $t-1$ or no time stamp at all, then the occupation of the cells in TSM_t must be due to a moving object. The algorithm which implements this idea is described in pseudo-code notation in Table 1.

procedure *detectMotion;*
    for each *cell ensemble* $cs_{x,t}$ *describing an object* x *in* TSM_t
        for each *cell* $c_{i,t}$ *in* $cs_{x,t}$
            for each *corresponding cell* $c_{i,t-1}, \ldots, c_{i,t-k}, \ldots, c_{i,t-n}$ in
                $TSM_{t-1}, \ldots, TSM_{t-k}, \ldots, TSM_{t-n}$
                if $c_{i,t-k}$ *carries a time stamp* $t-k$
                    then c_i *is occupied by a stationary object*
                    else c_i *is occupied by a moving object*
        if *majority of cells* $c_{i,t}$ *in* $cs_{x,t}$ is moving
            then *cell ensemble* $cs_{x,t}$ *is moving*
            else *cell ensemble* $cs_{x,t}$ *is stationary*

Table 1. A motion detection algorithm based on a sequence of time stamp maps.

As we pointed out earlier, the time stamp representation of a time-varying environment is more efficient for motion detection than commonly used grid representations. Particularly, the time stamp representation allows us to use a sequence of maps in a round robin mode without a need to clear and initialize the map which is used to assimilate the new sensor image. Outdated time stamps which originate from the mapping of previous sensor images do not have to be deleted but are simply overwritten. This procedure leaves the map receiving a new sensor image polluted by outdated information. However, this is not only efficient - as we save an expensive initialization operation - but is also correct. Cells which are marked by an outdated time stamp are simply considered as free space, which has the same effect as assigning some default value.

Motion Tracking and Estimation of Object Trajectories Although kinematic and dynamic models of human walking mechanisms and gaits have been developed, there is no analytical model of human purposive locomotion, which would allow us to make inferences about the motion of a person over longer distances. Therefore, the best we can do to track a moving object in environment such as a crowded concourse in a railway station is to collect information about its past motion and to extrapolate this past motion into the near future, if necessary. For this purpose we consider the sequence of recent sensor images and extract the information about motion direction, velocity, or acceleration describing the motion history of the moving objects from the spatial changes which we find in the mappings of these sensor images.

Note that while it is sufficient for motion detection to investigate only the mapping of two subsequent sensor images, provided the objects move at a sufficient speed, establishing a motion history may require to consider a more extended sequence of sensor images. We assume that the cells describing distinct objects are grouped into ensembles, and we also assume that these ensembles and their corresponding objects are classified either as *moving* or as *stationary* by the motion detection algorithm described above.

The first step in establishing the motion history of an object is to identify the object in a sequence of mappings. Once we have found this correspondence it is easy to derive the heading and the velocity of a moving object from its previous positions. To find a correspondence between the objects in the mappings of two subsequent sensor images we use a nearest-neighbor criterion. This criterion is defined over the Euclidean distance between the centers of gravity of cell ensembles representing distinct objects. For each cell ensemble representing an object at time t we determine the nearest ensemble in terms of the Euclidean distance in the map describing the environment at the preceding time step $t-1$. Obviously, this operation requires the objects to be represented in the same frame of reference.

If the distance to the nearest neighbor is smaller than a certain threshold then we assume that both cell ensembles describe the same object. The threshold depends on whether the considered objects and cell ensembles are stationary or moving. For establishing a correspondence between the two cell ensembles describing a stationary object we choose a rather small threshold since we expect the cell ensembles to have very similar shapes and to occupy the same space. Currently, we use a threshold of 30 cm for stationary objects. For a correspondence between the cell ensembles describing a moving object this value is accordingly larger. Here we use a threshold of 1 m which is approximately the distance which a person moving at fast walking speed covers between two sensor images.

A description of the above algorithm in pseudo-code notation is given in Table 2. On a Pentium 166MHz a complete cycle involving both detecting and tracking any moving objects takes approximately 6 ms. For a more detailed description and discussion of our motion detection and tracking method we refer to [13].

The algorithms in Table 2 allow us to establish a correspondence between the objects in two or even more subsequent time stamp maps. Having found this correspondence it is straightforward to compute estimates of the heading and the velocity of the objects. Let $cog(o_{i,t})$ and $cog(o_{i,t-1})$ be the centers of gravity of the visible contour of the object o_i at times t and $t-1$. Estimates for the velocity v and the heading ϕ of o_i are given by

$$v(o_{i,t}) = \frac{\| \boldsymbol{u}_{i,t} \|}{\Delta t} \quad \text{and} \quad \phi(o_{i,t}) = \text{atan}(\Im(\boldsymbol{u}_{i,t}), \Re(\boldsymbol{u}_{i,t})),$$

where

$$\boldsymbol{u}_{i,t} = cog(o_{i,t}) - cog(o_{i,t-1}).$$

```
procedure findCorrespondence;
    for each object o_{i,t} in TSM_t
        for each object o_{j,t-1} in TSM_{t-1}
            CorrespondenceTable[i,j] = corresponding(o_{i,t}, o_{j,t-1});

function corresponding(o_{i,t}, o_{j,t-1});
    if o_{i,t} is stationary and o_{j,t-1} is stationary
        then ϑ = ϑ_s;  (threshold for stationary objects)
        else ϑ = ϑ_m;  (threshold for moving objects)
    if d(o_{i,t}, o_{j,t-1}) < ϑ
        and not_exists o_{k,t} :   d(o_{k,t}, o_{j,t-1}) < d(o_{i,t}, o_{j,t-1})
        and not_exists o_{l,t-1} :  d(o_{i,t}, o_{l,t-1}) < d(o_{i,t}, o_{j,t-1})
        then return true;
        else  return false;
```

Table 2. An algorithm for tracking moving objects in a crowded environment.

As the above equations reveal, we use a very simple model for predicting the velocity and heading of an object. In particular, we assume that the object o_i moves linearly in the time interval from $t-1$ to t. Apparently, this may be a very coarse approximation of the true motion. The approximation, however, has proven to work sufficiently well at a cycle time of less than 100 ms for motion detection, motion estimation, and computation of an evasive course. At a slower cycle time a more sophisticated, nonlinear model for estimating the velocity and heading of an object may be appropriate.

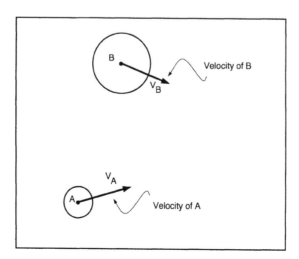

Fig. 5. The mobile robot A and the moving obstacle B.

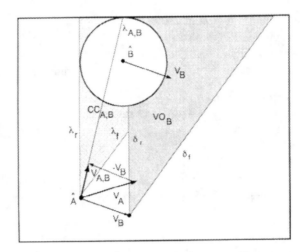

Fig. 6. The Relative Velocity $\mathbf{v}_{A,B}$, the Collision Cone $CC_{A,B}$, and the Velocity Obstacle VO_B.

5.2   Motion Planning Using Velocity Obstacles

In this section, we briefly summarize the concept of *Velocity Obstacle* (VO) for a single and multiple obstacles. For simplicity, we model the robotic wheelchair and the obstacles as circles, thus considering a planar problem with no rotations. This is not a severe limitation since general polygons can be represented by a number of circles. Obstacles move along arbitrary trajectories, and their instantaneous state (position and velocity) is estimated by MAid's sensors, as discussed earlier.

To introduce the Velocity Obstacle (VO) concept, we consider the two circular objects, A and B, shown in Fig. 5 at time t_0, with velocities $\mathbf{v}_A$ and $\mathbf{v}_B$. Let circle A represent the mobile robot, and circle B represent an obstacle. To compute the VO, we first map B into the *Configuration Space* of A, by reducing A to the point $\widehat{A}$ and enlarging B by the radius of A to $\widehat{B}$, and represent the state of the moving object by its position and a velocity vector attached to its center. Then, the set of colliding relative velocities between $\widehat{A}$ and $\widehat{B}$, called the *Collision Cone*, $CC_{A,B}$, is defined as $CC_{A,B} = \{\mathbf{v}_{A,B} \mid \lambda_{A,B} \cap \widehat{B} \neq \emptyset\}$, where $\mathbf{v}_{A,B}$ is the relative velocity of $\widehat{A}$ with respect to $\widehat{B}$, $\mathbf{v}_{A,B} = \mathbf{v}_A - \mathbf{v}_B$, and $\lambda_{A,B}$ is the line of $\mathbf{v}_{A,B}$. This cone is the light grey sector with apex in $\widehat{A}$, bounded by the two tangents λ_f and λ_r from $\widehat{A}$ to $\widehat{B}$, shown in Fig. 6. Any relative velocity that lies between the two tangents to $\widehat{B}$, λ_f and λ_r, will cause a collision between A and B. Clearly, any relative velocity outside $CC_{A,B}$ is guaranteed to be collision-free, provided that the obstacle $\widehat{B}$ maintains its current shape and speed.

The collision cone is specific to a particular pair of robot/obstacle. To consider multiple obstacles, it is useful to establish an equivalent condition on the *absolute*

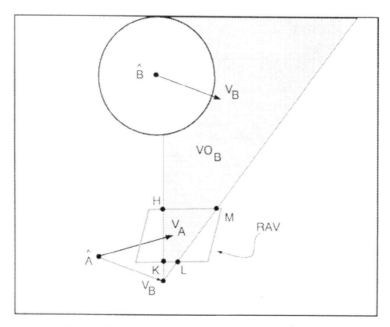

Fig. 7. The reachable avoidance velocities RAV.

velocities of A. This is done simply by adding the velocity of B, $\mathbf{v}_B$, to each velocity in $CC_{A,B}$ and forming the *Velocity Obstacle VO*, $VO = CC_{A,B} \oplus \mathbf{v}_B$ where $\oplus$ is the Minkowski vector sum operator, as shown in Fig. 6 by the dark grey sector. The VO partitions the absolute velocities of A into *avoiding* and *colliding* velocities. Selecting $\mathbf{v}_A$ outside of VO would avoid collision with B. Velocities on the boundaries of VO would result in A grazing B.

To avoid multiple obstacles, we consider the union of the individual velocity obstacles, $VO = \cup_{i=1}^{m} VO_{B_i}$, where m is the number of obstacles. The avoidance velocities, then, consist of those velocities $\mathbf{v}_A$, that are outside all the VO's.

In the case of many obstacles, obstacles avoidance is prioritized, so that those with imminent collision will take precedence over those with long time to collision. Furthermore, since the VO is based on a linear approximation of the obstacle's trajectory, using it to predict remote collisions may be inaccurate, if the obstacle does not move along a straight line. By introducing a suitable *Time Horizon* T_h, we limit the collision avoidance to those occurring at some time $t < T_h$.

The Avoidance Maneuver An *avoidance maneuver*, consists of a one-step change in velocity to avoid a future collision within a given time horizon. The new velocity must be achievable by the moving robot, thus the set of avoidance velocities is limited to those velocities that are physically reachable by robot A at a given state over a given interval. This set of *reachable velocities* is represented schematically by the polygon $KLMH$ shown in Fig. 7. The set of *reachable avoidance velocities*, RAV, is defined as the difference between the reachable velocities and the velocity obstacle. A maneuver avoiding obstacle B can then be computed by selecting any velocity in RAV. Fig. 7 shows schematically the set RAV consisting of two disjoint closed subsets. For multiple obstacles, the RAV may consist of multiple disjoint subsets.

It is possible then to choose the type of an avoidance maneuver, by selecting on which side of the obstacle the mobile robot will pass. As discussed earlier, the boundary of the velocity obstacle VO, $\{\delta_f, \delta_r\}$, represents all absolute velocities generating trajectories tangent to $\widehat{B}$, since their corresponding relative velocities lay on λ_f and λ_r. For example, the only tangent velocities in Fig. 7 are represented by the segments KH and LM of the reachable avoidance velocity set RAV. By choosing velocities in the set bound by segment HK or ML, we ensure that the corresponding avoidance maneuver will avoid the obstacle from the rear, or the front, respectively.

The possibility of subdividing the avoidance velocities RAV into subsets, each corresponding to a specific avoidance maneuver of an obstacle, is used by the robotic wheelchair to avoid obstacles in different ways, depending on the perceived danger of the obstacle.

Computing the avoidance trajectories A complete trajectory for the mobile robot consists of a sequence of single avoidance maneuvers that avoid static and moving obstacles, move towards the goal, and satisfy the robot's dynamic constraints. A global search have been proposed for off-line applications, and a heuristic search is most suitable for on-line navigation of the robotic wheelchair. The trajectory is generated incrementally by selecting a single avoidance velocity at each discrete time interval, using some heuristics to choose among all possible velocities in the reachable avoidance velocity set RAV.

The heuristics can be designed to satisfy a prioritized series of goals, such as survival of the robot as the first goal, and reaching the desired target, minimizing some performance index, and selecting a desired trajectory structure, as the secondary goals. Choosing velocities in RAV (if they exist) automatically guarantees survival. Among those velocities, selecting the ones along the straight line to the goal would ensure reaching the goal, the TG strategy shown in Fig. 8. Selecting the highest feasible velocity in the general direction of the goal may reduce motion time, the MV heuristics shown in Fig. 8. Selecting the velocity from the appropriate subset of RAV can ensure a desired trajectory structure (front or rear maneuvers), the ST heuristics shown in Fig. 8. It is important to note that there is no guarantee that any objective is achievable at any time. The purpose of the heuristic search is to find a "good" local solution if one exist.

In the experiments described in the following section, we used a combination of the TG and the ST heuristics, to ensure that the robotic wheelchair moves towards the goal specified by the user. When the RAV sets include velocity vectors aiming directly to the goal, the largest among them is chosen for next control cycle. Otherwise, the algorithm computes the centers of the RAV sets,

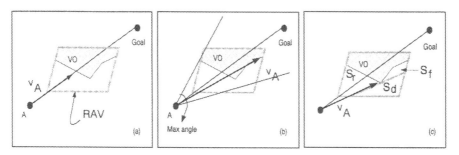

Fig. 8. a: TG strategy. b: MV strategy. c: ST strategy.

and chooses the velocity corresponding to the center closest to the direction to the goal. This heuristics adds also an additional safety margin to the mobile robot trajectory, since the velocity chosen is removed from the boundary of its *RAV* set, thus accounting for unmodelled uncertainties on the obstacle shapes and trajectories.

6   Experimental Results

MAid's performance was evaluated in two steps. Before taking the system to a realistic environment such as the concourse of a railway station, we conducted extensive tests under the simplified and controlled conditions of our laboratory. The laboratory embodiment of a "rapidly changing environment" consisted of an empty, delimited, and locked area where a second mobile robot moved on prescribed paths with known velocity profile, or groups of three and four people were asked to walk at moderate speed in front of MAid.

6.1   Experiments under Laboratory Conditions

To examine MAid's motion detection and tracking capability a commercial mobile robot Nomad XR4000 was programmed to move along a rectangular trajectory in front of the robotic wheelchair, equipped with the laser range-finder, at a distance of 2 m. The Nomad robot followed a velocity profile that made its center follow the polygonal trajectory represented by the solid line shown in Fig. 9. The wheelchair is identified by the cross mark at coordinates $(0,0)$ in Fig. 9, and its position or orientation were not changed during the experiment.

The dotted line in the figure represents the motion of the Nomad robot as it was sensed and tracked by MAid. The estimated trajectory shows a maximum tracking error of less than 15 cm, which can possibly reduce the performance of the navigation algorithm. However, this is not the case, since the error is always reducing the estimate of the obstacle distance and therefore increases the navigation safety margins. The nature of the error can be easily understood by noticing that the trajectory estimation is carried out by tracking the center of the visible

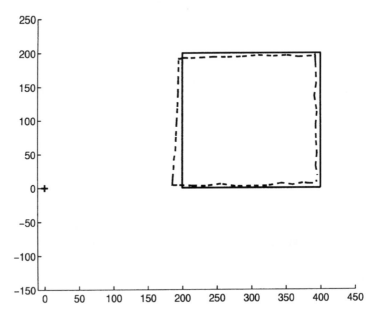

Fig. 9. Tracking a single moving object with ground truth.

contour of mobile obstacle. Since the the visible obstacle is smaller than the true obstacle, its center will always result closer than its real position. Furthermore, the estimation error only affects the magnitude of the avoidance velocity and not its direction, which is computed using the left and right boundaries of the visible obstacle.

In a second set of experiments we asked a number of people to move at a comfortable walking speed along prescribed trajectories in an experimental area of approximately $4{\times}7\,\mathrm{m}^2$. The wheelchair with the range-finder was again kept stationary. The results of this set of experiments are shown in Fig. 10a) - d).

During the first experiment, a single person was asked to walk along a given rectangular trajectory in the area facing the range-finder sensor. After several laps, the person headed back to his initial position. The tracking algorithm tracked his motion in real time without any problem. The trajectory estimated by the tracking algorithm is shown in Fig. 10a).

In the second experiment, three people moved across the field of view the wheelchair along straight lines, more or less parallel to each other, and the left most person made a sudden left turn and headed back to the wheelchair, as shown in Fig. 10b). The subjects moved at slightly different speeds, so that their complete walk was visible by the range-finder. As shown in the figure, the tracking algorithm could easily track the motion of the three people. In the experiment shown in Fig. 10c), we tracked the motion of three subjects moving again on parallel straight lines directly away from the wheelchair. This time the subjects moved at a similar speed.

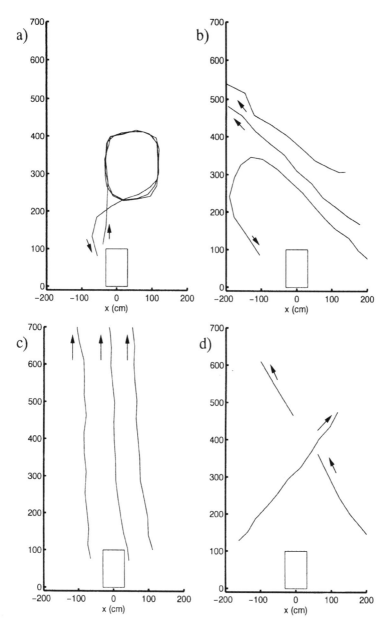

Fig. 10. Tracking a group of people in a lab environment.

The last experiment deserves a more detailed discussion. We let two subjects move along straight lines which crossed each other in front of the wheelchair. Accordingly, for a short period of time one person was occluding the other from the range-finder view. Apparently then, the algorithm was unable to track the occluded person during this period of time. This loss of tracking manifests itself as the interruption of one of the trajectories, as shown in Fig. 10d). Our algorithm lost the occluded person for two time steps. It detected and continued to track the motion after the subject became visible again.

Tracking moving objects whose trajectories cross each other is a very general problem and is not a specific problem of our tracking algorithm. Problems of this type cannot be eliminated even by very sophisticated methods such as those described in [2], which assume the knowledge of a model of the motion of the objects to be tracked. As mentioned above, we cannot make such an assumption since valid models of human motion are not available for our application domain. Experimental results further showing the performance of MAid's motion detection and tracking algorithm in a real environment are described in [13].

To complete the laboratory experiments, we evaluated the performance of the complete system, including motion detection, tracking and computation of avoidance maneuvers, under controlled conditions. This is a difficult task, since no metric is available to quantify the behavior of on-line algorithms reacting to unpredictable external events. Our experiment consisted of asking two subjects to approach the wheelchair at walking speed (approx. 1 m/s). The wheelchair's initial velocity was set to 0.5 m/s. The reaction of the system after it noticed the approaching objects is shown in a sequence of snapshots in Fig. 11. In the figure the wheelchair is represented by a rectangle, whereas the two subjects are represented by circles. The arrows attached to the rectangle and the circles represent the motion direction and velocity of MAid and the two people, respectively. The length of the each arrow represents to the distance traveled in one second. The entire experiment lasted less than five seconds as can be seen from the time stamps attached to the snapshots.

Before time 1.54 sec the two subjects moved in a safe direction without the danger of a collision. At time 1.54 sec one person changed direction and directly headed for the wheelchair. As we can observe, MAid reacted to this new situation by reducing its velocity and turning right. At time 3.1 sec the danger of a collision had disappeared again and MAid turned back to its initial direction and accelerated to its previous velocity. At time 4.14 sec one person had already left MAid's perceptual field when the other person suddenly made a turn and directly headed for MAid. Since this would have lead to an immediate collision MAid reduced its velocity to zero and stopped. Half a second later - the person had slowed down too and turned right a little - MAid accelerated again in a direction which allowed it to finally pass the person.

6.2   Experiments in the Concourse of a Railway Station

After MAid had successfully passed a number of laboratory experiments similar those described above, the time had come to confront the real world. The real world was the concourse of the central station in Ulm, a hall of approximately $25 \times 40 \, m^2$. First test runs were conducted during the morning rush hours. We thought that this would represent the worst scenario MAid would ever have to face. In fact, after the arrival of a commuter train typically up to several hundred

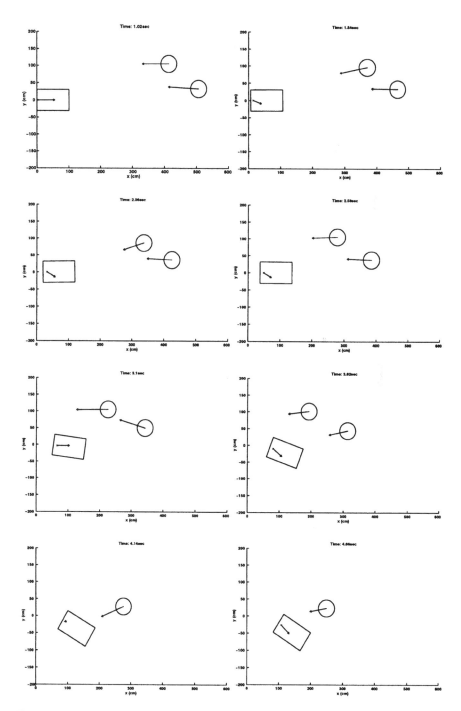

Fig. 11. Lab experiment: MAid on a collision course with two approaching people.

Fig. 12. MAid traveling in the concourse of a railway station.

people moved through the concourse within two or three minutes. We counted up to 150 people crossing the concourse within about a minute. After two or three minutes however, the concourse was practically empty again, thus leaving not enough time for conducting experiments. The railway station manager, who was observing our experiments with great interest, finally told us that the ideal time for our tests would have been Friday noon, which does not exhibit the densest but the most continuous passenger traffic in the concourse. During the period between 11:00am and 1:30pm in fact, typically several tens of people stay and move around in the concourse, thus making it very suitable for the navigation experiments.

To test MAid's navigation performance we let it cross the concourse in arbitrary directions. MAid is put in motion by pushing the joystick shortly into the direction of the target location and by entering a travel distance. The wheelchair then starts moving in the desired direction as long as there is no collision imminent. If a collision with an object or a person is impending, MAid, while continuing its motion, determines a proper avoidance maneuver and follows this new direction until it can turn back and head again to the goal location. Snapshots of MAid's test runs in the crowded concourse are shown in Fig. 12. The passenger traffic in these images is moderately dense, which actually facilitated recording the pictures. When the passenger traffic became too dense, MAid occasionally simply stopped, and did what a human operator would have also probably done in that situation: it waited until the group of people blocking its way had passed and then it continued its journey.

MAid's performance is demonstrated in the diagrams of Fig. 13, by showing the relations between some of the navigation variables such as, for example, wheelchair velocity, relative velocity between wheelchair and nearest objects, and clearance between the wheelchair and the nearest object. Fig. 13a) shows the wheelchair velocity plotted over the distance between the wheelchair and the nearest object which is on a collision course with the wheelchair. The data in this diagram indicate that with decreasing clearance the wheelchair velocity drops to zero. Similarly, the velocity increases as the distance to the nearest approaching obstacle becomes larger. It is important to note that there is no unique causal relation between the wheelchair velocity and the clearance with the environment. Rather, the wheelchair's velocity depends on a number of quantities, such as the object motion direction, the object velocity, and the number of objects in MAid's proximity. This explains the variations in the data set. Note also that the distance to the nearest object is measured with respect to MAid's vertical axis and not with respect to the boundary of the wheelchair.

An equivalent dependency is shown in Fig. 13b). There, the relative velocity between the wheelchair and the nearest object is plotted against the distance between the two objects. A negative value of the relative velocity means that the object is approaching the wheelchair, whereas a positive value means that the object is moving away from it. According to the data, the velocity of the nearest object relative to the wheelchair velocity decreases as the distance between the two objects decreases. This dependency describes the combined effect of motion

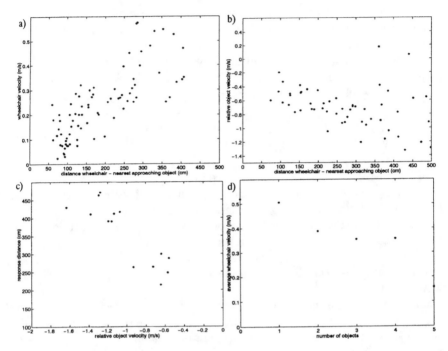

Fig. 13. Illustration of MAid's performance in terms of wheelchair velocity and distance between the vehicle and the nearest object.

planning and control algorithms, which reduces MAid's velocity whenever an object approaches the wheelchair. Note that this is not a unique causal correlation either.

In Fig. 13c) we show the relation between the relative velocity of the nearest object and the distance at which MAid starts an evasive maneuver. The larger the relative velocity of an approaching object, the sooner MAid initiates an avoidance maneuver, and the slower an object is approaching the wheelchair, the shorter is distance at which MAid starts to get out of its way. Fig. 13d) finally shows MAid's average velocity plotted over the number of approaching objects. We can see that MAid decreases its speed the more objects are approaching on a collision course.

During our experiments in the concourse MAid collided several times with objects. Usually these objects were bags or a suitcases lying on the floor invisible to MAid's laser range-finder and its sonar sensors. To discover small obstacles in front of the wheelchair we mounted two extra sonar sensors to the foot rests of the wheelchair.

So far, MAid has survived about 18 hours of testing in the concourse of the central station in Ulm and we plan to continue conducting experiments in this environment.

MAid was presented to a wider audience during the Hannover Fair '98. The Hannover Fair is the largest industrial fair worldwide. In Hannover, MAid drove through the exhibition halls for seven days between two and three hours per day at regular visiting hours. Altogether MAid has successfully navigated in crowded, rapidly changing environments for more than 36 hours.

7   Conclusion

In this paper, we presented the hardware and software design of the navigation system of our robotic wheelchair MAid. This navigation system enables MAid to move through crowded, rapidly changing environments, such as shopping malls and concourses of railway stations or airports, and also through narrow, cluttered, partially unknown environments. In this paper we only described the first of these two capabilities, which we denoted as WAN (wide area navigation). Three components essentially contribute to the capability to navigate in a wide, crowded, rapidly changing area: an algorithm for motion detection, an algorithm for motion tracking, prediction and the computation of potential collisions, and finally an algorithm for computing the avoidance maneuvers.

The algorithms for motion detection and tracking use the range data provided by a 2D laser range-finder. This sensor was chosen to facilitate the real-time capability of the tracking system. By using a laser range-finger our approach differs from the majority of known methods for motion detection and tracking which are based on visual information.

The time variation of the environment is captured by a sequence of temporal maps, which we call time stamp maps. A time stamp map is the projection of a range image onto a two-dimensional grid, whose cells coinciding with a specific range value are assigned a time stamp. Based on this representation we have discussed simple algorithms for motion detection and motion tracking, respectively. One complete cycle involving both motion detection and tracking takes approximately 6 ms. Our algorithms for motion detection and tracking do not presuppose the existence of kinematic and dynamic models of purposive human locomotion. Those models are not available in an environment such as a concourse of a railway station. With a cycle time of 6 ms for motion detection and tracking however, our approach is definitely "quick" and assures the required real-time capability.

The avoidance maneuvers are computed using the Velocity Obstacle approach, which allows the fast computation of the wheelchair velocity avoiding all static and moving obstacles. To take into account the environment uncertainty, an avoidance maneuver is computed at each sampling time, thus modifying in real time the nominal trajectory of the wheelchair. The complete trajectory to the goal is then computed incrementally, by selecting the avoidance velocities according to appropriate heuristics. The most commonly used heuristics has been to select an avoidance velocity in the general direction of the goal, to ensure that the wheelchair does not stray to far from its nominal trajectory, and can re-acquire its original goal after the obstacle avoidance.

MAid has undergone rather extensive testing. Its performance was tested in the central station of Ulm during rush-hour and in the exhibitions halls of the *Hannover Messe '98*, the biggest industrial fair worldwide, during regular visiting hours. Altogether, MAid has survived more than 36 hours of testing in public, crowded environments with heavy passenger traffic. To our knowledge there is no other robotic wheelchair, and no other mobile robot, that can claim a similar performance.

Acknowledgment

This work was supported by the German ministry for education, science, research, and technology (BMB+F) under grant no. 01 IN 601 E 3 as part of the project INSERVUM. The development of the Velocity Obstacle approach has been carried out in part at the Jet Propulsion Laboratory, California Institute of Technology, under a contract with the National Aeronautics and Space Administration.

References

1. J. Barraquand, J.-C. Latombe. Nonholonomic Multibody Mobile Robots: Controllability and Motion Planning in the Presence of Obstacles. *Algorithmica*, 10, pp. 121-155, Springer-Verlag, 1993.
2. Y. Bar-Shalom, T.E. Fortmann. *Tracking and Data Association*. Academic Press, 1987.
3. D.A. Bell, J. Borenstein, S.P. Levine, Y. Koren, and L. Jaros. An Assistive Navigation System for Wheelchairs Based upon Mobile Robot Obstacle Avoidance. In *Proc. of the 1994 IEEE Int. Conf. on Robotics and Automation*, San Diego, 1994.
4. A. Elfes. *Occupancy Grids: A Probabilistic Framework for Robot Perception and Navigation*. PhD thesis, Electrical and Computer Engineering Department/Robotics Institute, Carnegie-Mellon University, 1989.
5. P. Fiorini, Z. Shiller. Motion Planning in Dynamic Environments Using the Relative Velocity Paradigm. In *Proc. of the 1993 IEEE Int. Conf. on Robotics and Automation*, Atlanta, 1993.
6. P. Fiorini, Z. Shiller. Motion Planning in Dynamic Environments Using Velocity Obstacles. *International Journal of Robotics Research*, July 1998, Vol. 17, No. 7, pp. 760-772.
7. H. Hoyer, R. Hölper. Open Control Architecture for an Intelligent Omnidirectional Wheelchair. In *Proc. of the 1st TIDE Congress*, Brussels, pp. 93-97, IOS Press, 1993.
8. R.L. Madarasz, L.C. Heiny, R.F. Cromp, N.M. Mazur. The Design of an Autonomous Vehicle for the Disabled. IEEE Journal of Robotics and Automation, Vol. RA-2, No.3, 1986.
9. S. Mascaro, J. Spano, H. Asada. A Reconfigurable Holonomic Omnidirectional Mobile Bed with Unified Seating (RHOMBUS) for Bedridden Patients. In *In Proc of the 1997 IEEE Int. Conf. on Robotics and Automation*, Albuquerque, pp. 1277-1282, 1997.

10. M. Mazo, F.J. Rodriguez, J.L. Lazaro, J. Urena, J.C. Garcia, E. Santiso, P.A. Revenga, and J.J. Garcia. Wheelchair for Physically Disabled People with Voice, Ultrasonic and Infrared Sensor Control. *Autonomous Robots*, 2, pp. 203-224, 1995.
11. D. Miller, M. Slack. Design and Testing of a Low-Cost Robotic Wheelchair Prototype. *Autonomous Robots*, 2, pp. 77-88, 1995.
12. E. Prassler, J. Scholz, M. Strobel. MAid: Mobility Assistance for Elderly and Disabled People. In *Proc. of the 24th Int. Conf. of the IEEE Industrial Electronics Soc. IECON'98*, Aachen, Germany, 1998.
13. E. Prassler, J. Scholz, M. Schuster, D. Schwammkrug. Tracking a Large Number of Moving Objects in a Crowded Environment. In *IEEE Workshop on Perception for Mobile Agents, Santa Barbara*, June 1998.
14. M. Strobel, E. Prassler, D. Bank. Navigation of non-circular mobile robots in narrow, cluttered environments (in preparation).
15. P. Wellman, V. Krovi, V. Kumar. An Adaptive Mobility System for the Disabled. In *In Proc. of the 1994 IEEE Int. Conf. on Robotics and Automation*, San Diego, pp. 2006 - 2011, 1994.

Interactive Robot Programming Based on Human Demonstration and Advice

Holger Friedrich, Rüdiger Dillmann, and Oliver Rogalla

University of Karlsruhe, Institute for Process Control & Robotics, D-76128
Karlsruhe, Germany
{friedric,dillmann,rogalla}@ira.uka.de,
http://wwwipr.ira.uka.de

Abstract. Service robots require interactive programming interfaces
that allow users without programming experience to easily instruct the
robots. Systems following the *Programming by Demonstration (PbD)*
paradigm that were developed within the last years are getting closer
to this goal. However, most of these systems lack the possibility for the
user to supervise and influence the process of program generation after
the initial demonstration was performed. In this paper an approach is
presented, that enables the user to supervise the entire program generation process and to annotate, and edit system hypotheses. Moreover,
the knowledge representation and algorithms presented enable the user
to generalise the generated program by annotating conditions and object
selection criteria via a 3D simulation and graphical user interface. The
resulting *PbD*-system widens the *PbD* approach in robotics to programming based on human demonstrations and user annotations.

1   Introduction

The development of service robots is one of the main topics in robotics research.
One of the major problems to be solved in order to successfully apply manipulators to service tasks is the problem of providing a proper programming interface
for unexperienced users. Interactive programming interfaces are required that allow users that know how to perform a task to be programmed to easily instruct
a robot without having to have a programming education.

In recent years several robot programming systems were developed that follow the *Programming by Demonstration (PbD)* paradigm [11,13,16,3]. Most of
these systems are focused on the task of reconstructing the trajectories and manipulations a user performs. Their goal is to reconstruct demonstrations or at
least a set of environmental states with the highest accuracy possible. In order
to achieve this the focus is set on powerful sensor data processing techniques and
the use of highly accurate sensor systems, such as multiple vision systems, laser
sensors, structured light, and in some cases especially designed input devices. In
general these systems have three characteristics, that do limit their applicability
to end-user programming end service tasks.

Christensen et al. (Eds.): Sensor Based Intelligent Robots, LNAI 1724, pp. 96–119, 1999.
© Springer-Verlag Berlin Heidelberg 1999

1. Most of the state-of-the-art systems don't employ further user interaction in the programming process beyond the physical demonstration itself. This limits the information the programming system's data processing steps are based on to the trajectories of the users extremities and the objects and their properties. The user's intention that stands behind the observable actions has to be guessed or is not taken into account at all. The systems work based on the assumption that all effects occurring in the course of a demonstration due to the user's actions are intended and should be reproduced by the generated program.

2. The systems do not provide any supervision and monitoring interfaces for the user besides the actual execution of the generated program itself. Thus, wrong system hypotheses derived in the course of the program generation process can't be identified and corrected by the user. This can lead to incorrect programs that don't match the user's intention.

3. The generated programs are either inflexible to changes in the execution environment against the demonstration environment since the program accurately replicates the user's demonstration, or they are generalised based on heuristics or inductive learning techniques without really knowing the user intention.

Summarising, part of the state of the art systems are either very sophisticated accurate visual teaching systems which are easy and comfortable to use but do lack the power to generate flexible programs. The others do allow the generation of flexible robot programs but do not guarantee that these match the users' intentions. Both results are not fulfilling the requirements posed by the task of programming service robots. The major requirements are:

- intuitive interaction modes and interfaces,
- processing processes of the programming system should be transparent for the user,
- system hypotheses have to be matched against user's intention,
- programs shall be generalised w.r.t. user's intention,
- partial solutions should be reusable, and
- programs should be represented in a robot independent way.

Inflexible programs will fail in unstructured, changing service environments and flexible programs derived from demonstrations whose accordance with the user intentions can not be guaranteed are not acceptable if not even potentially dangerous.

In the following sections an approach is presented that allows the representation of actions with different levels of accuracy, as well as the flexible representation of object selection conditions for program execution. Thereafter, a method is given that allows the generation of programs following several processing and user interaction phases starting with the initial user demonstration.

The approach that will be presented in this paper is divided into several processing steps. Firstly, the user's behavior is recorded via a sensor system. The sensor system consist of a data glove with a 6DOF magnetic field position sensor and a vision based trinocular camera system. For experiments described in this paper only data from the position sensor and the data glove is used [1]. These recordings provide data from the user's hand trajectory and finger angles. Secondly, the trajectory is analysed and segmented. Important segments as grasping, un-grasping and basic movements are represented in symbolic identifiers, called *Elementary Operations (EOs)*. This symbolic representation allows abstraction from the raw sensory data to a reusable operator structure. Since the structure of these Elementary Operations do play an important role for the system's behavior a detailed formal definition is presented in section 2. The generated Elementary Operations can be compiled into a new macro operator which represents a symbol for a whole action sequence. In case the Elementary operations can be mapped directly into robot commands the derived task can be executed on a robot system afterwards.

Figure 1 shows the overall process with it's basic components. Each step is explained in the following sections. For better understanding the structure of the symbolic representatives of actions, the Elementary Operators, section 2 starts with a formal description. Sections 3 4 5 explain how the sensory data can be compiled into symbolic identifiers.

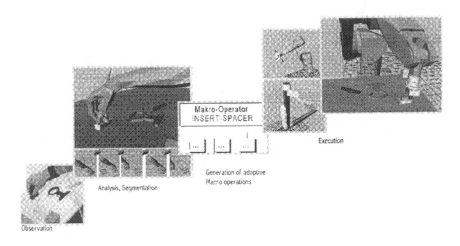

Fig. 1. Programming by demonstration process

2   Program Representation

The basic requirement for service task programs is the *adaptability of the program to the user's needs and intentions*. This means, that a generated program has

---

[1] This means that positions are estimated only within the virtual reality environment that will be presented later.

to be *as accurate as required and as flexible as possible*. Therefore, important features of the program representation are the following.

1. It has to be interpretable by the user.
2. Actions have to be represented accurately, but also flexible to changes in object locations at execution time.
3. It has to contain object selection criteria that allow the instantiation of variables representing the objects to be manipulated in different environments.

In order to cover all of these aspects the following operator based representation was developed.

Definition 1. *An **Operator** $\mathcal{OP}$ is a 5-tuple $(\mathcal{N}, \mathcal{O}, \mathcal{C}, \mathcal{T}, \mathcal{P}, \mathcal{B})$.*
The five attributes contain:

$\mathcal{N}$ = *operator's name being an unambiguous identifier.*
$\mathcal{O}$ = *list of variables representing the objects the operator is applied to (called object list).*
$\mathcal{C}$ = *condition that enables the instantiation of the object list's variables.*
$\mathcal{T}$ = *list of object relative frames representing a trajectory to be driven.*
$\mathcal{P}$ = *list of parameters e.g. grasping force, acceleration profile.*
$\mathcal{B}$ = *the operator's body (see definition 2).*

The different attributes of operators serve different purposes. The following features and advantages are implemented by using values stored in the above presented operator structure:

- *Object selection conditions (stored in $\mathcal{C}$)* are the key to program generalisation, instantiation and execution in different environments. The object selection conditions for relevant objects of a manipulation are represented by logic terms given as disjunctive normal form of spatial relations. This representation is efficient since it allows the adjustment of a programs applicability and therefore flexibility to the user's intention by adding or deleting relations. Different conjunctions in a term representing a selection condition do specify alternative environmental states in which the program or one of its sub-steps shall be executed.
  In the system well-known spatial relations as *in, on, aligned*, and object related relations like *type, colour, lot* are used. Based on these relations object configurations and features that represent object selection criteria and thus carry the user intention are semantically modelled and described. Due to the modular structure of the implemented system the set of relations and features can easily be extended and exchanged for different task domains, e.g. one could use contact relations as proposed by Ikeuchi [6]. To add other object related relations regarding specific product oriented information like *stiffness* or *roughness* is straight forward. Thus the means available for describing object selection criteria and thereby representing the user intention can easily be tailored to the specific task requirements.

- *Object relative trajectory representation (stored in T)* allows us to store trajectories in an object dependent way. Each frame which is to be visited in the course of following the trajectory is given relative to an object using a common frame notation. If for example a spacer is to be picked up, the trajectory to the spacer's location is given relative to it. Each sample is given in the form *(spacer T)*, where *spacer* denotes the object and T the frame given relative to the object. During execution the trajectory towards the object is driven as desired, regardless whether the object is situated at exactly the same location as during initial demonstration or not.

  A problem occurs when the reference object changes between two trajectory samples and the objects' positions at execution time differ from the ones they were at during demonstration. In this case methods for local path planning and interpolation have to be applied in order to transfer the "gap" in the trajectory that results from the different object configuration.

- *Scalable accuracy and flexibility* of the actions performed in the course of program execution is achieved, by mapping the user demonstration on a set of *Elementary Operations (EOs)*. Using this concept [4] provides several advantages. Firstly, standard robot motion strategies can be used, e.g. *linear-move*, *spline-move*, which eases the mapping of a demonstration on standard manipulators. Secondly, the accuracy of the trajectory's representation can be adjusted to the user's intention. An *exact-move*, that exactly replays the original trajectory provides the highest accuracy. Besides motion operations also different grasp operations can be represented by different operators. In principle, different grasp classification hierarchies [1,8] can be used by implementing operators for these.

- *Reuse of programmed task solutions* is enabled by compiling a sequence of elementary operations that was derived from an initial demonstration into a new macro operator. The generated macro contains the respective EO sequence as its execution code [3]. EOs that have been used for implementing the macro operator are stored in the operator's body (see definition 2).

- *Robot independent representation of programs* is also achieved by the operator based representation. This feature is realised through the concept of using different bodies for elementary and macro operators.

  Definition 2. *Operator's body: An operator's body B contains the code to be executed. This code can be*
  - *an executable, robot dependent program in case of an elementary operator or*
  - *a list of operators in case of a macro operator.*

  Moreover, a macro operator's body contains the information of how the operators contained are to be parametrised

  Following this definition EOs provide robot independent shells for the programming process, while serving as an interface to robot dependent implementations of skills [4].

The programming process which is depicted in figure 2 exploits the benefits of the program representation. It consists of several processing phases.

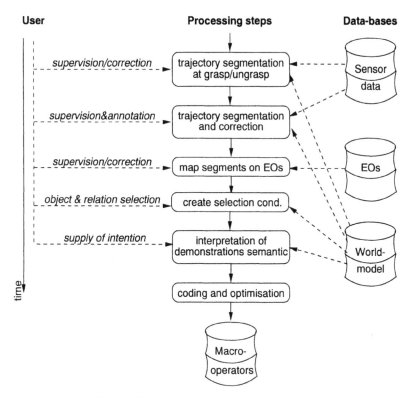

Fig. 2. Interactive programming process

User demonstrations are recorded, segmented, analysed, and mapped on a sequence of elementary operations. Finally, the EO sequence is compiled into a new macro operator representing the final program. Figure 2 shows the user is integrated in all processing steps. In the following sections the different programming phases are described in detail.

3   Sensors and Demonstration Recording

The developed system uses a 22 sensor data glove and magnetic field based $6D$ position sensor to record physical user demonstrations (figure 3). Therefore, the trajectory of the user's motions and his fingers' poses are given directly but due to the position sensor's measuring principle with limited accuracy.

However, in principle the methods presented work on arbitrary spatial trajectories, regardless whether they were recorded with a position sensor or whether they were derived from highly accurate vision or laser based sensor recordings. Figure 4 shows the trajectory recorded from the demonstration of a bag packaging task, where 3 infusion bags were stacked in a cardboard box.

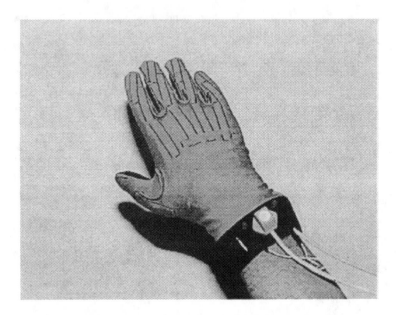

Fig. 3. Data-glove with the tracking system's receiver

4   Trajectory Segmentation and Grasp Analysis

The first programming step consists of the segmentation of the demonstration into different phases. Segmentation is done in order to divide the demonstration in meaningful phases that are associated with the different manipulations that the user performed. In order to achieve this segmentation the grasp and release actions that took place during the demonstration are identified. Like in other programming systems that are based on the analysis of demonstrations [7] the method which is realised here is based on the analysis of the distance of the user's hands to objects, the velocity of the user's hand and the changes in its finger's posture.

4.1   Detection of Grasp Actions

Grasp actions are detected following a three step approach. In the first step the trajectory of the recorded demonstration is followed until the distance between the user's hand and an object falls below a threshold $thresh_{grasp}$ which was determined experimentally. When the hand approached an object like this a grasp action is assumed. By analysing the distance of the user's hand to the closest object the search interval $I_{grasp} = i, i+1, \cdots, j-1, j$ of demonstration samples is determined such that

$$\forall k \in I_{grasp} : min(distance(hand, all - objects)_k) < thresh_{grasp} \qquad (1)$$

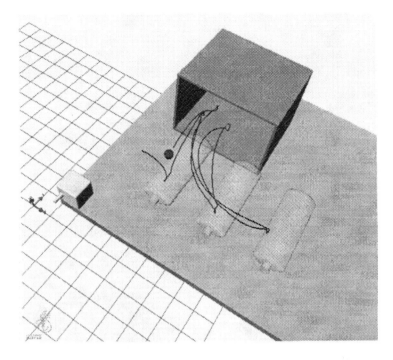

Fig. 4. Trajectory sampled with $20Hz$

holds. Within this interval the index i_{grasp} is determined at which the hand's velocity combined with the fingers' flexion is minimal. The flexion of all fingers at time t is represented through the sum of the measured flexion angles of all finger joints $\Sigma joints_t$. After determining i_{grasp} the following equation holds.

$$\forall j \in I_{grasp} : j \neq i_{grasp} : v_j - \frac{\Sigma joints_j}{10} \geq v_{i_{grasp}} - \frac{\Sigma joints_{i_{grasp}}}{10} \qquad (2)$$

Figure 5 illustrates the motivation for this approach. It shows that grasping actions do occur when the hand barely moves and the fingers close until a steady grasp is established. Three grasp/release actions are shown w.r.t. the hand's velocity (*light grey*) and the fingers' flexion (*dark grey*). Hypotheses about the points in time when grasp actions took place are given as the ascending edges of the *grey* rectangles.

Once a hypothesis about a grasp action is derived a candidate for the grasped object is determined. The object which is closest to the local coordinate system of the hand model at the point of grasp is chosen.

4.2   Detection of Release Actions

Hypotheses for release actions in the course of a demonstration are derived based on the assumption that they do occur when an object is already grasped, the

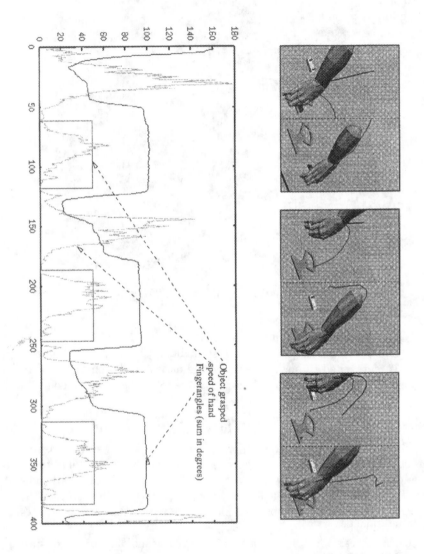

Fig. 5. Detection of grasp and release actions

hand opens, and the hand's velocity is low. The recorded data samples are analysed starting from the last detected grasp action searching for the point in time when the sum of the fingers' joint angles increases w.r.t. to the value calculated at the grasping point for the given grasp. In order to be able to cope with sensor noise and small spontaneous finger motions the fingers have to open above an experimentally determined threshold in order to trigger the release detection. From the sample representing the point in time were the hand opens above the threshold the sequence of recorded data samples is processed backwards until the next local velocity minimum is encountered. This sample serves as hypothesis for the point in time when a release action took place. In figure 5 the descending edges of the grey rectangles depict the hypotheses derived w.r.t. release actions.

Due to the use of the sum of finger joint angles as segmentation criterion for release actions the approach presented here is limited to the application of demonstrations containing static grasps.

Once a hypothesis about a release action is derived a hypothesis about what the reference object for the released object might be is generated. The trajectory of the motion towards the release position will be stored relative to this reference object. As candidate the object which is closest to the released one is chosen.

4.3   Classifying Grasps

In order to represent different ways of grasping objects grasp hierarchies are often used. The root of this approach lies in the work of Schlesinger [15]. Following Schlesinger's and Napier's [12] analysis, Cutkosky developed a grasp hierarchy which offers a classification scheme for typical grasps occurring in manufacturing and assembly tasks [1] (s. fig. 6).

State-of-the-art grasp recognition schemes mostly do rely on the analysis of contact points between the human hand and the grasped object and the analysis of the hand posture itself (e.g. [8,9]). Although this approach offers a very powerful and precise way to classify and recognise different grasps it has several limitations. In order to compute the contact points between the user's hand and grasped objects exact geometric models of these are required. Since human hands vary strongly in size and shape it is very difficult to provide an exact hand model. Moreover, models for non-rigid objects are difficult to provide and to maintain. Furthermore, sensor noise limits the applicability of the method of analysing the hand's posture w.r.t. the object since it requires exact data about the hand's and the object's position as well as the fingers' posture. Finally, a high computational effort has to be spent in order to compute the hand/object contact points.

In the developed system a different approach is followed in order to classify grasps w.r.t. to the grasp hierarchy which was introduced by Cutkosky. The basic idea is to perform the classification step on data about an hand's posture only. No complicated geometric computations is required whatsoever. The 20 finger flexion and abduction values are used as input of the grasp classification process since these describe the posture of the user's hand. Based on this data neural network classifiers for grasp classification are trained.

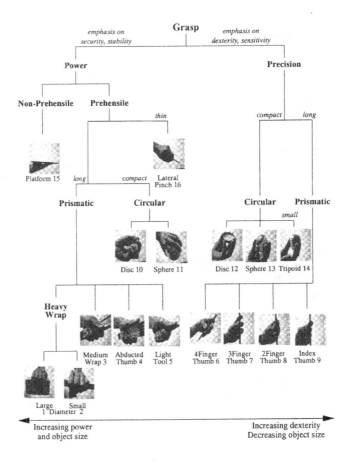

Fig. 6. Cutkosky's grasp hierarchy

The data pre-processing, training and classification scheme as well as the results obtained and limitations discovered are discussed in [5].

4.4   User Interaction w.r.t. Segmentation Hypotheses

Once hypotheses about the occurrence of grasp and release actions as well as types of grasps and relevant objects were derived automatically based on the demonstration and the world model it's the user's turn to confirm or correct these. In order to provide a user friendly way to edit the system hypothesis the interface displayed in figure 7 was designed.

By either choosing from menus in the graphic interface or picking objects directly in the 3D simulation window the user can edit hypotheses about the grasped object or reference objects respectively. Moving the sliders in the graphic interface window the user can edit the grasp and release action hypotheses. The 3D simulation is constantly updated w.r.t. to the users' actions.

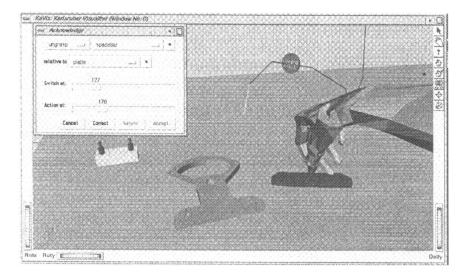

Fig. 7. Interactive confirmation and correction of segmentation hypotheses

4.5   Correction of Positions of Released Objects

Due to sensor noise, variations of users' hands and user mistakes, the final position of manipulated and released objects in the updated world model is in general not optimal. Since the desired goal position depends on the task and the user's intention a purely automatic method for correcting positions of objects that were manipulated does not make sense. Therefore, a variation of three different correction methods was integrated into the proposed programming system.

Interactive correction: The proposed method allows the interactive correction of objects' positions by manipulating their geometric models directly in the 3D-simulation window. Graphic interaction components are added to the visualised objects (e.g. the wire-cube in figure 8b). By picking, moving, and rotating these

a) uncorrected position of bags   b) correction using an interaction object   c) corrected bag positions

Fig. 8. Interactive correction of object positions

interaction components the position of the associated object is altered. Due to the chosen graphical interaction mode this method suits the strong visual orientation of human users. Figure 8 shows the result of the demonstration of a bag packaging task before, during and after position corrections were made.

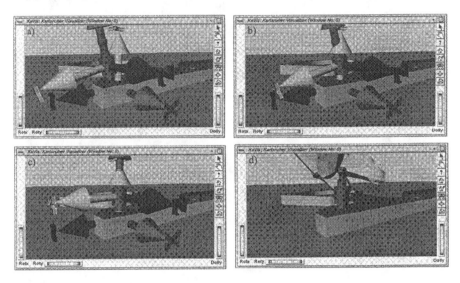

Fig. 9. Automatic correction of object positions for insertion actions; a) initial situation, b) translational correction, c) rotational correction, d) final situation

Automatic correction: Whenever background knowledge about objects and possible configurations of these is available, this knowledge can be used for automatic correction of the position of released objects. An example is the automatic correction of objects after insertion operations. Insertion axes of real objects that can be inserted as well as of holes are modelled as attribute/value pairs. On the basis of this representation the correction is done in three processing steps. After the targeted hole is identified the inserted object's local coordinate system is projected on the holes insertion axis (s. fig. 9 a & b). Thereafter the object's insertion axis is matched with the hole's insertion axis by rotating the inserted object appropriately (s. fig. 9 c & d).

Iconic annotation of object positions: As third technique for the correction of object positions an approach was developed which relies on user annotations and implements a semi-automatic position correction. The user interface is based on a set of graphic icons representing spatial relations between objects (s. fig. 10).

Firstly, the user specifies the direction in which an object's position shall be corrected. Directions can be translations and rotations along the axis of the world coordinate system or the object's local coordinate system respectively.

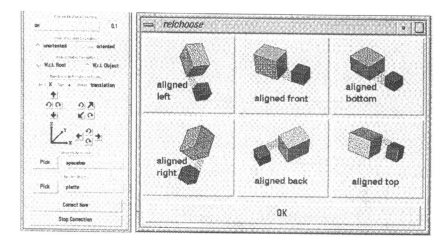

Fig. 10. Iconic user interface for interactive position correction

Furthermore, the user specifies a reference object to which the intended spatial relation shall be established. Finally, the user chooses the spatial relation to be established from icon menus. After this annotation phase the system moves the object whose position is to be corrected until the chosen relation to the selected reference object gets valid. With this annotation based correction technique the user is able to position objects such that the relations he intended to establish with his demonstration are valid (s. fig. 11).

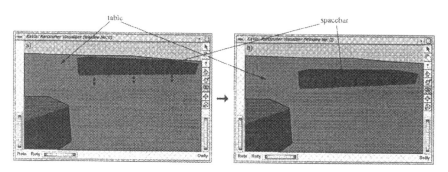

Fig. 11. Correction of *spacebar*'s position such that *on(spacebar, table)* holds

5   Mapping of Trajectory Segments on EOs

After the demonstration was segmented the decision has to be made by what elementary motion operations the trajectory the user's hand has travelled in the course of the demonstration is to be represented.

Segment-wise analysis of the hand's trajectory follows the segmentation and correction phases. The identified segments between grasp/release operations are analysed w.r.t. linearity. The analysis is performed with the *iterative end-point fit* algorithm [10] which originally was developed as an edge reconstruction algorithm for sensor data processing. For this task the algorithm was extended from handling 2*D* to 3*D* data. As result the processed trajectory segment is given in a piece-wise linear representation. Depending on the threshold parameters used for the algorithm for line segment of about 50cm length between 5-20 line segments are produced. The resulting set of lines is shown in figure 12.

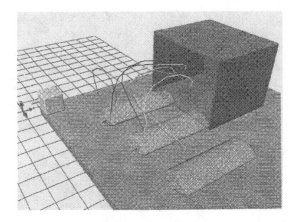

Fig. 12. Results of the trajectory linearisation

Mapping trajectory segments on EOs is a crucial step, since the degree of accuracy the trajectory is represented with in the final program is determined. The system generates hypotheses regarding the desired accuracy of reconstruction based on

1. the speed and acceleration trajectory w.r.t. the corresponding space trajectory, and
2. the length and angular deviations of the successive linear segments that were identified in the previous steps.

Currently, hypotheses regarding $linear - moves$ and $free - moves$ are generated, where $linear - moves$ are characterised by low speed and accelerations on the according linear segments.

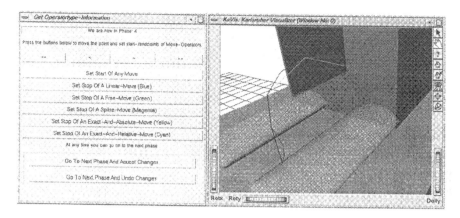

Fig. 13. Menu for motion EO selection (left); visualised mapping hypotheses (right)

User interaction w.r.t. EO mapping hypotheses Thereafter, the hypotheses are presented to the user giving him the possibility to acknowledge or edit the system's choices using the menu and $3D$ simulation based interface shown in figure 13. The data stored varies for different EOs. For *free − moves* and

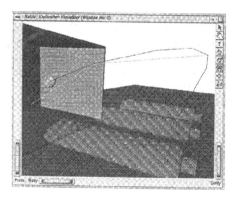

Fig. 14. Identified motion operators; left: system's hypotheses, right: user's choice

linear − moves only start and end points of the corresponding line are stored, whereas for a *spline − move* the end points of all included lines are stored as supporting knots of the spline. The most accurate representation of the demonstration trajectory is achieved using *exact − moves*. For these all samples of the demonstration trajectory that are included in the chosen segment are stored.

The modular design of the system allows the easy extension with more EOs that might be needed for different task domains, e.g. *circular – move* or *insertion*.

6   Acquisition of the User's Intention

During execution it has to be ensured, that the objects chosen for manipulation and configurations chosen as destinations match with the user intentions. This is essential, since at execution time multiple objects in different locations and configurations could serve as candidates for manipulation. Therefore, following, the representation of the trajectory with the user intended accuracy, the object selection conditions are determined. They enable correct instantiation w.r.t. the user's intention in case of execution. Since the required information can't be guessed by the programming system it has to be acquired from the user.

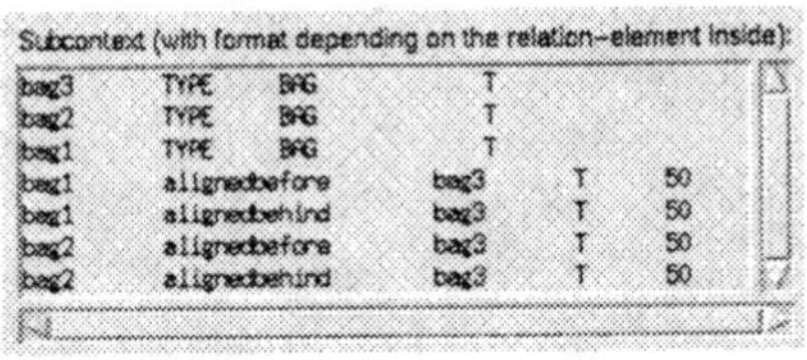

Fig. 15. Condition for the final release action

Object selection conditions are acquired for each grasp/release phase. For grasp phases they serve as representation of the spatial conditions an object has to be in and the features it has to have in order to be a candidate for manipulation during program execution. For ungrasp phases they do describe the spatial object configuration the currently manipulated object is to be put in and the features of the objects building this configuration.

Since the computation of all valid relations between all objects in the environment would be very time consuming, and the selection among these by the user would be very uncomfortable, the user is asked to specify significant objects and relations in advance. The objects involved in the spatial context are chosen by the user by simply clicking on them in the 3D-simulation. Thereafter, the user chooses the object features and spatial relations that are of general interest in the context at hand from menus of the user interface.

Based on the specified sets of objects and relations the system computes the values for all relations and permutations of objects. The result of this computation is presented to the user in an interface menu that allows the specification of object selection conditions, by consecutively selecting relations from the menu and adding them to the already selected ones. In order to clarify the meaning of spatial relations and to point out the objects involved, the concept of 3D-Icons [2] is used. The user can trigger this visualisation of relations (see fig. 16),

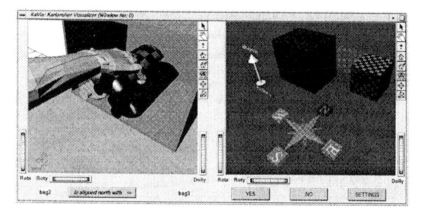

Fig. 16. Visualising the *aligned*(*bag*1, *bag*3) relation with a 3*D*-icon

whenever in doubt about their semantics or the related objects. Finally, when an object selection condition is completely specified (see fig. 15), the programming system stores it in the respective slot of the grasp and release operators.

Thereafter, the program is complete and the generated operator sequence is stored in the program data-base for further use.

7 Experiments

In the following at first the presented segmentation approach is evaluated. There-after, the experiment of programming an insertion task based on a user demon-stration is presented.

Performance of the segmentation method. In order to evaluate the proposed segmentation method the experiment of picking and inserting a peg was demon-strated 20 times. All demonstrations were recorded with sample frequence f of $6.6Hz$. Thereafter, the demonstrations were segmented in two different ways.

1. The demonstration segmentation was performed automatically.
2. The automatically generated segmentation hypotheses were edited by human experts using the developed graphical interaction interface.

	mean		$\sqrt{MSE}$		minimal deviation		maximal deviation	
	samples	in [ms]	samples	in [ms]	samples	in [ms]	samples	in [ms]
grasping	2.75	413	3.31	496	0	0	11	1650
releasing	3.40	510	3.52	527	0	0	12	1800

Table 1. Analysis of deviations regarding manual vs. automatic segmentation

The mean values and rooted mean square error for the deviation of system hypotheses vs. experts segmentation are given in table 1.

The numbers show that the occurrence of grasp and release actions that was detected with the proposed method deviates about 3 frames from the actions determined by human experts. With the chosen sample rate this is a deviation of approx. half a second. Further analysis of the recorded data traces showed that during grasping and releasing of objects users' hands usually stand still for a short period of time. Both, the system's as well as the expert's segmentation hypotheses fall into these periods of standstill. The difference is that the system detects the actions at the beginning of the period whereas the human expert prefers the period's center.

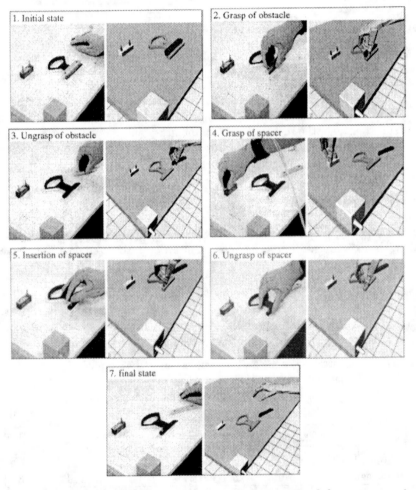

Fig. 17. Removing obstacle & inserting peg; demonstration left; reconstruction right

Programming an assembly task. In the following experiment one spacer of the well-known Cranfield Benchmark is inserted into the base-plate. Figure 17 shows demonstration and reconstruction of the removal of an obstacle and the insertion of the spacer.

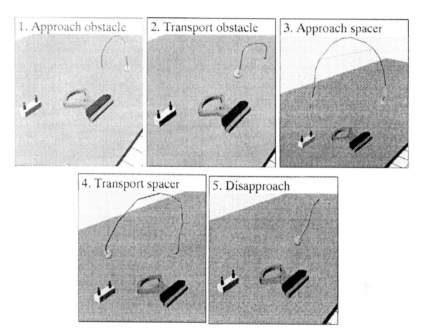

Fig. 18. Mapping of trajectory on motion EOs for each segment

Due to inaccuracies in the model of the user's hand and especially the sensor noise of the magnetic field based position sensor the reconstructed position of the inserted peg is not optimal. Therefore, the presented position correction techniques are applied in order to correct the demonstration's reconstruction. Thereafter, the trajectory segments are mapped on elementary motion operators. The resulting EO sequences are shown in figure 18.

Selection conditions that allow the applicability of the resulting program for inserting spacers in each of the four holes of the base-plate are specified interactively (Table 2). By specifying the terms $\neg on(k, y)$ and $\neg partly\_in(k, y)$ the user explicitly specifies that the hole has to be empty and unobstructed. This condition can be used to generate a branch in the program which triggers the removal of the obstacle only when necessary [3].

Figure 19 shows again the peg insertion experiment. It presents side by side the user demonstration on the left, the reconstruction in the simulation in the middle, and finally the simulated execution of the generated EO sequence by a Stäubli $RX90$ robot on the right hand side[2].

---

[2] For the simulating the robotic execution the simulation library and tool *KLIMT* (*K̲inematic L̲ibrary for S̲imulating M̲anipulators*) was used [14]

Grasping the obstacle	
Spatial context	on(spacebar, hole1) ∧ on(hole1, plate) ∧ type(hole1, hole) ∧ type(plate, base-plate)

Releasing the obstacle	
Spatial context	on(spacebar, table1) ∧ type(table1, table)

Grasping the peg	
Spatial context	partly_in(peg1, res) ∧ type(peg1, peg) ∧ type(res, reservoir)

Releasing the peg	
Spatial context	partly_in(peg1, hole1) ∧ completely_in(hole1, plate) ∧ on(plate, table1) ∧ ¬on(k, hole1))) ∧ ¬partly_in(k, hole1))) ∧ type(peg1, peg) ∧ diameter(peg1, small) ∧ type(hole1, hole) ∧ type(plate, base-plate) ∧ type(table1, table)

Table 2. Specified spatial contexts serving as selection conditions

8   Summary and Conclusion

In this paper a program representation was presented that is based on a set of elementary operations. It fulfils the requirements of service task programming. It provides freely scalable accuracy and flexibility of the stored trajectories. Flexible representation of conditions specifying required geometric and spatial object configurations and required object features allow the instantiation of generated programs in different environments.

Furthermore an interactive programming system was presented that generates programs in the developed representation in a sequential process based on a physical user demonstration and annotations. All processing and programming steps can be monitored and influenced by the user. System hypotheses regarding segmentation, grasp types, motions, conditions, etc. are to be acknowledged, edited or actively specified using a comfortable menu driven user interface and direct interaction with a 3D simulation environment.

The presented representation in combination with the interactive programming system is an extension of the currently available methods and systems towards more flexible end-user programming. It is not only concentrating on the exact replication of a physical demonstration but also taking into account and representing the user's intention.

Although the system's capabilities are considerably high, improvements can be achieved by adding additional sensor sources. This will be on the one hand a trinocular movable camera head and on the other hand force sensors on the data glove's fingertips. This will lead to a more precise environment model for better performance. Additionally the physical connection to a robot system must be further investigated and validated.

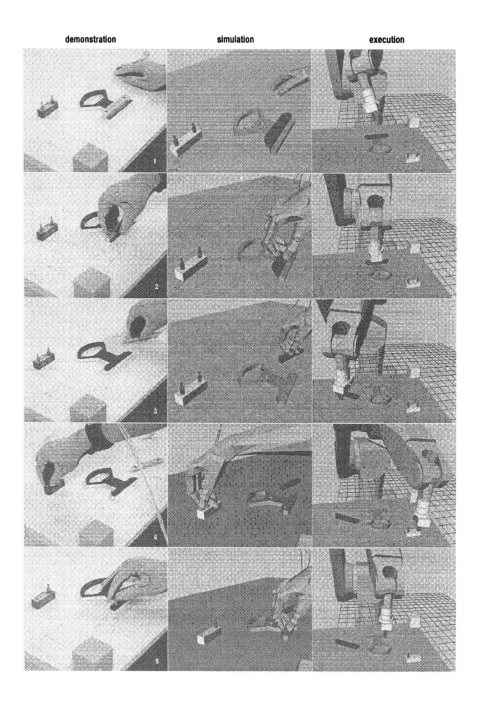

Fig. 19. The complete peg insertion experiment (demonstration, simulation, execution)

Acknowledgements

This work has partially been supported by the DFG Project "Programmieren durch Vormachen". It has been performed at the Institute for Real-Time Computer Systems and Robotics, Prof. Dr.-Ing. H. Wörn, Prof. Dr.-Ing. U. Rembold and Prof. Dr.-Ing. R. Dillmann.

References

1. M. R. Cutkosky. On grasp choice, grasp models, and the design of hands for manufacturing tasks. *IEEE Trans. Robotics and Automation*, 5(3):269–279, 1989.
2. H. Friedrich, H. Hofmann, and R. Dillmann. 3d-icon based user interaction for robot programming by demonstration. In *Proceedings of the International Symposium on Computational Intelligence in Robotics and Automatio (CIRA'97)*, Monterey, Kalifornien, USA, 10-11 Juli 1997.
3. H. Friedrich, S. Münch, R. Dillmann, S. Bocionek, and M. Sassin. Robot programming by demonstration: Supporting the induction by human interaction. *Machine Learning*, pages 163–189, Mai/Juni 1996.
4. H. Friedrich, O. Rogalla, and R. Dillmann. Integrating skills into multi-agent systems. *Journal of Intelligent Manufacturing*, 1998.
5. Volker Grossmann. Erkennung menschlicher Griffarten mittels neuronaler Netze. Master's thesis, Universität Karlsruhe, Fakultät für Informatik, Institut für Prozeßrechentechnik und Robotik, 1998.
6. K. Ikeuchi and T. Suehiro. Assembly task recognition using face-contact relations. In *Proceedings of the IEEE International Conference on Robotics and Automation (ICORA'92)*, volume 3, pages 2171–2177, Nizza, Frankreich, 1992.
7. S. B. Kang. *Robot Instruction by Human Demonstration*. PhD thesis, Carnegie Mellon University, Pittsburgh, PA, 1994.
8. S. B. Kang and K. Ikeuchi. A grasp abstraction hierarchy for recognition of grasping tasks from observation. In *Proceedings of the IEEE/RSJ International Conference on Intelligent Robots and Systems (IROS'93)*, volume 1, pages 621–628, Yokohama, Japan, Juli 26-30 1993.
9. S. B. Kang and K. Ikeuchi. Toward automatic robot instruction from perception – mapping human grasps to manipulator grasps. *IEEE Transactions on Robotics and Automation*, 13(1), Februar 1997.
10. T. Knieriemen. *Autonome Mobile Roboter - Sensordateninterpretation und Weltmodellierung zur Navigation in unbekannter Umgebung*, volume 80 of *BI - Wissenschaftverlag*. K.H. Böhling, U. Kulisch, H.Maurer, 1991.
11. Yasuo Kuniyoshi, Inaba Masayuki, and Hirochika Inoue. Learning by watching: Reusable task knowledge from visual observation of human performance. *IEEE Transactions pn Robotics and Automation*, 10(6):799–822, December 1994.
12. J. R. Napier. The prehensile movements of the human hand. *The journal of bone and joint surgery*, 38B(4):902 – 913, 1956.
13. G. V. Paul and K. Ikeuchi. Modelling planar assembly tasks: Representation and recognition. In *Proceedings of the IEEE/RSJ International Conference on Intelligent Robots and Systems (IROS'95)*, volume 1, pages 17–22, Pittsburgh, Pennsylvania, USA, August 5-9 1995.
14. Kilian M. Pohl. Modellierung von Robotersystemen. Master's thesis, Universität Karlsruhe, Fakultät für Informatik, Institut für Prozeßrechentechnik und Robotik, 1998.

15. G. Schlesinger. *Ersatzglieder und Arbeitshilfen*, chapter Der mechanische Aufbau der künstlichen Glieder, pages 21 – 600. Springer-Verlag, Berlin, Deutschland, 1919.

16. T. Takahashi. Time normalization and analysis method in robot programming from human demonstration data. In *Proceedings of the IEEE International Conference on Robotics and Automation (ICORA '96)*, volume 1, pages 37–42, Minneapolis, Minnesota, USA, April 1996.

Towards Smarter Cars

Karin Sobottka[1], Esther Meier[2], Frank Ade[2], and Horst Bunke[1]

[1] Institute of Computer Science and Applied Mathematics, University of Bern,
CH-3012 Bern, Switzerland,
{sobottka,bunke}@iam.unibe.ch
[2] Communication Technology Lab, Image Science, Swiss Federal Institute of
Technology, CH-8092 Zurich, Switzerland,
{ebmeier,ade}@vision.ee.ethz.ch

Abstract. Most approaches for vision systems use greyscale or color
images. In many applications, such as driver assistance or presence detection systems, the geometry of the scene is more relevant than the
reflected brightness information and therefore range sensors are of increasing interest.

In this paper we focus on an automotive application of such a range
camera to increase safety on motorways. This driver assistance system
is capable of automatically keeping the car at an adequate distance or
warning the driver in case of dangerous situations.

The problem is addressed in two steps: obstacle detection and tracking.
For obstacle detection two different approaches are presented based on
slope evaluation and computation of a road model. For tracking, one
approach applies a matching scheme, the other uses a Kalman filter.
Results are shown for several experiments.

1 Introduction

Modern everyday life, characterized by an ever increasing interaction between
man and machine, leads to a growing number of potentially harmful situations
through accidents, malfunctions or human oversight. Related to this is our investigation into reliable presence detection systems based on coarse range images.

In the project MINORA, which is part of the Swiss priority program OP-
TIQUE II, a universal miniaturized optical range camera is developed [18,19].
This new sensor, working in the near infrared, will be fast, cheap and can supply
3D information with high accuracy. Necessary trade-offs however, mean that the
sensor provides range images which are coarse (resulting from the need for inex-
pensive sensor and computing hardware) and incomplete (due to insufficient or
saturated reflections from targets). For measuring distances ranging from a few
meters to a few hundred meters the sensor is restricted to a narrow field of view.
As a consequence the entire road width is within the sensor's field of view at a
distance of about 20 meters. Accordingly, we focus on well ordered environments
such as motorways where this restriction is tolerable.

Two main applications of the universal range camera are targeted in the
project MINORA:

Christensen et al. (Eds.): Sensor Based Intelligent Robots, LNAI 1724, pp. 120–139, 1999.
© Springer-Verlag Berlin Heidelberg 1999

1. Automotive applications such as driver assistance systems with distance measurements of some hundred meters and a relatively narrow field of view to increase safety on motorways.
2. Safety and surveillance applications, such as the control of automatic sliding or revolving doors with distances of a few meters and a wide field of view.

While other project partners work on the electro-optical and computing components of the proposed sensor, our task is the development of algorithms for the interpretation of coarse and incomplete range image sequences. Although the sensor used in both the safety/surveillance and the automotive applications is the same, a discussion of the theory behind both applications is beyond the scope of a single paper. In the following we focus on the automotive application.

The aim of this paper is to present and discuss different methods for obstacle detection and tracking for a driver assistance system based on coarse range image sequences. Robust obstacle detection and tracking is performed to ensure collision avoidance. All potential obstacles have to be recognized within the scene. Two obstacle detection methods based on vertical slope evaluation and computation of a road model are presented. Furthermore, obstacles have to be tracked over time employing a matching scheme or Kalman filtering to predict dangerous situations. Due to the availability of 3D information and low resolution of range images a processing in real-time is ensured for all approaches at any discrete point of time.

We show results with simulated and scaled range images. Real data will be available later in the project.

2   Types of Range Images

Range imaging is a key technology for many applications where it is important to know the geometrical description of the observed scene. Although several measuring techniques exist, for example, stereo, structured light, triangulation, structure from motion, range from focusing, and time of flight, the literature is not too abundant. The techniques can be classified as either active or passive, where active methods involve a controlled energy source. Some ranging systems have already been successfully applied to niche applications, but none can claim to be a universal sensor.

Only little research was done on the interpretation of range image sequences for outdoor applications such as intelligent cruise control. Table 1 gives examples of current research on this topic and summarizes the advantages and disadvantages of different kinds of range sensors. To the knowledge of the authors, they have not been described and discussed in sufficient details in the open literature.

As can be seen in Table 1, the optical range camera that is developed within the MINORA project is, due to synchronous image sensing, well suited for automotive applications. Like other laser range scanners it is insensitive to ambient light and weather conditions and provides range data with a high accuracy in a short acquisition time. But novel algorithms have to be developed to handle the low resolution and incompleteness of the range data.

Table 1. Comparison of different kind of sensors

Sensor	Lit.	Advantages	Disadvantages
Stereo vision	[3]	- high resolution - brightness and range information available	- sensitive to illumination - calibration necessary - correspondence problem
Light stripe range scanner	[1]	- high resolution - dense depth map	- restricted to indoor appl. - calibration necessary - sensitive to illumination - long acquisition time
Depth from motion	[2]	- low cost	- low accuracy - dependent on motion segm. - sparse depth map
Millimeter- wave radar	[15], [21]	- long distance detection - high accuracy - insensitive to weather cond.	- long acquisition time (for imaging radar) - incomplete data - high cost
Ultrasonic sensor	[14]	- low cost	- low lateral accuracy - sensitive to wind - short distance measure- ment - incomplete data
Mechanically scanned optical radar	[5], [20]	- long distance measurement - high accuracy - insensitive to weather cond.	- long acquisition time - incomplete data - high cost
Optical range camera with synchronous image sensing	[11],[16], [17]	- low cost - high accuracy - insensitive to weather cond. - short acquisition time	- low resolution - incomplete data

3   An Automotive Application

Vision based driver assistance improves traffic safety by warning the driver, e.g. by an acoustic or visual signal, in case of dangerous situations. An example for a scenario, in which range imagery is used for vision based driver assistance, is depicted in Fig. 1. As can be seen, a range camera is fixed near the front bumper of a vehicle and acquires a stream of range images over time. The field of view of the range sensor is described by the angular width of image sensing. Assuming a geometry of the sensor as shown in Fig. 2, ϕ denotes half of the angular width in vertical direction and θ denotes half of the angular width in horizontal direction. Collision avoidance, then, is a sequence of operations including obstacle detection and tracking. In obstacle detection, the vision system checks if there is any object in the sensor's field of view that interferes with the path of the sensor vehicle. Such obstacles may be static or mobile. For obstacle tracking a sequence of frames is considered and feature correspondences are determined to

Fig. 1. Using range imagery for vision based driver assistance (with friendly permission of A.D.C. GmbH, Germany)

derive a robust estimation of the positions of obstacles and information about their temporal behaviour.

3.1 Obstacle Detection

Obstacle detection schemes for natural environments should be able to detect obstacles of any shape, e.g. cars, pedestrians, posts, trees, bushes and traffic signs. For that purpose a general obstacle model seems to be more appropriate than several specific 3D object models.

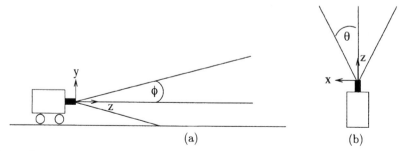

Fig. 2. Geometry of sensor: (a) vertical angular extent and (b) horizontal angular extent (top view)

In the context of automotive applications, an obstacle can be defined as any region the sensor car should not traverse. This criterion can be examined in two different ways. On the one hand, the traversability of the environment in the sensor's field of view can directly be checked by slope evaluation. In this case a region with steep slope indicates the presence of an obstacle. On the other hand obstacles lie on top of the road. From an estimate of the location of the road we can separate the image into ground and obstacles. Both approaches allow robust obstacle detection and, combined with a tracking scheme, they can be employed for collision avoidance.

Radial Slope A straightforward approach to checking the traversability in the sensor's field of view is the evaluation of the slope of a region. It can be computed very efficiently in radial direction. Thus determining a region with steep slope indicates the presence of an obstacle.

Also in [5] obstacles are defined as untraversable areas. An ideal obstacle detection method which uses a complete vehicle model is applied to continuous range data. Specific features, such as discontinuities and slope, are extracted to indicate an obstacle. In [22] partial derivatives of the range in a spherical coordinate system are used to indirectly measure the Cartesian slopes. This approach is used for obstacle detection by an autonomous car.

Based on the sensor geometry illustrated in Fig. 2, for each pair of two vertically adjacent laser beams with depth $r(\phi)$ and $r(\phi - \Delta\phi)$, respectively, slope β can be computed based on the following relationships:

$$\beta = \psi - \phi \tag{1}$$

$$\psi = 180^o - \gamma \tag{2}$$

$$tan\psi = -tan\gamma = \frac{r(\phi - \Delta\phi) \cdot sin\Delta\phi}{r(\phi - \Delta\phi) \cdot cos\Delta\phi - r(\phi)} \tag{3}$$

The geometrical relationships for radial slope computation are illustrated in Fig. 3. In our approach obstacles are detected in range image sequences by first computing the radial slope in vertical direction for each pixel. If the slope value of a pixel exceeds a predefined threshold, the pixel is labeled as belonging to an obstacle. Experiments showed that the definition of this threshold, which characterizes non-traversable object areas, is not critical. We have decided to define it as 25 degrees. A region growing step is applied next to determine regions of steep slope. Since obstacles may have flat surface parts in between steep surface parts, the detected regions are taken as start regions for a connected component analysis. The objective of our connected component analysis scheme is to determine point clouds that are connected in 3D space and that have at least one part with steep slope. For that we consider neighborhood in depth and neighborhood in 2D image space as well as height information as features. Thus flat surface parts of an obstacle are merged as well as steep surface parts.

Results for an example scene are shown in Fig. 4. In Fig. 4a the original range image, with range values encoded as greylevels, is shown. The scene contains two

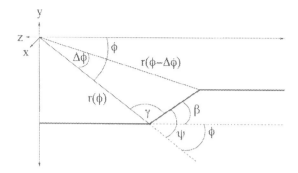

Fig. 3. Geometrical relationships for radial slope computation

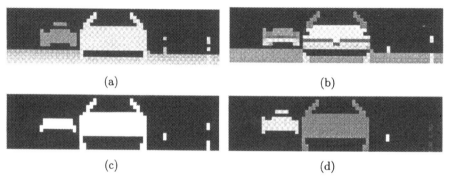

Fig. 4. (a) Range image, (b) radial slope image, (c) thresholded slope image, (d) connected components

cars with one of them driving ahead of the sensor car and the other car driving in the oncoming traffic. On the right side of the road, two posts can be seen. The computed radial slope image is illustrated in Fig. 4b. In this image, slope is encoded as greylevel: A slope of 0° is represented as medium greylevel, and positive (negative) slope as brighter (darker). It can be seen that, depending on the shape of a car, different parts of the same car may have different slope values. The road has a slope value of almost zero degrees. The thresholded slope image is shown in Fig. 4c. As can be seen, pixels belonging to the cars are labeled as obstacles. Also the posts are detected as obstacles. Road pixels are eliminated and thus correctly classified as belonging to a traversable object. The detected connected components are shown in Fig. 4d. Both cars and the posts are correctly detected as obstacles. Due to the fact that slope values can be computed in radial direction, the obstacle detection scheme using radial slope is of low computational complexity. By performing a connected component analysis it is ensured that obstacles are detected in their entire size even if obstacles have flat surface parts in between. Furthermore the connected component analysis in 3D space increases the robustness against noise since simple slope computations are

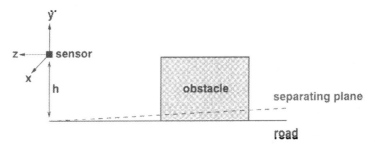

Fig. 5. Separation of image feature

sensitive to noise. However at least two valid depth measurements in neighboring rows are necessary to detect an obstacle. A more detailed description of the radial slope approach for obstacle detection can be found in [17].

Separating Plane In this alternative approach, the knowledge of the geometry of the sensor with respect to the road is used to separate image features into ground and obstacles introducing a separating plane. We make the assumption that the road can be locally modeled as a plane, and determine a depth map of a slightly steeper plane than the current road model (rotated by 0.5°). The center of rotation is placed vertically under the range sensor (see Fig. 5). Upon initialization the road model is assumed to be horizontal and the sensor has a height h. Using an inclined plane has the advantage that noisy ground pixels and small road elevations are less likely erroneously identified as obstacles. It is better to use a rotated instead of a vertically shifted plane as our expected noise depends on the depth. Consequently, near and potentially dangerous obstacles are completely detected whereas objects far away may be lost. Although hardware limitations allow the development of only a simple road model (e.g. limited field of view and range), this is sufficient for our purpose.

Similar systems, also using a road model, have been proposed in the literature. The approaches described in [4] and [10] use the disparity obtained from stereo image pairs to distinguish features painted on the road and obstacles lying on the ground. The lane markings are detected and used to update the road model and the geometric parameters of the stereo cameras. In comparison to these methods we process much smaller images and 3D information for each pixel is directly available. On this basis, all the extracted 3D ground pixels are used to update our simple road model. Given our coarse images, it is not feasible to use a more complex road model like the one described in [6].

We determine a depth map by calculating the distances the sensor would measure if only the separating plane were seen. By comparing the measured depth values (see Fig. 6a) with those from the depth map, we can separate image features into ground and obstacles. Pixels with a smaller distance belong to an obstacle, while larger distances are assigned to the ground (see Fig. 6b,c).

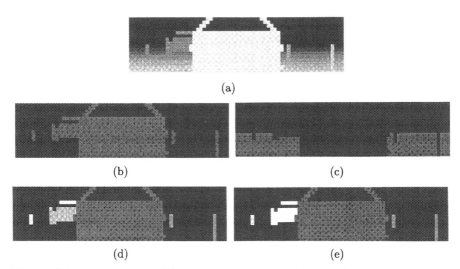

Fig. 6. (a) Range image, (b) obstacle detection, (c) road/ground detection, (d) connected components, (e) connected regions

However, the car suspension and the road inclination may alter the geometry of the road model, with small differences potentially producing large errors in the separation of the image features. Therefore, the parameters are updated by fitting a plane through all pixels marked as ground. We use the least squares method which minimizes the sum of the squares of the residuals.

As we are dealing with incomplete data, we might not always have enough pixels to get a robust result. Even if we have enough pixels to perform the fit (> 3) it can still be insufficient because of incorrectly classified pixels. Therefore, we calculate the mean distance q of the extracted ground pixels to the fitted plane

$$q = \sqrt{\frac{\sum_{i=1}^{N} d_i^2}{N}}, \tag{4}$$

where N is the number of defined ground pixels and d_i is the distance of each pixel to the plane. If this measure is higher than a specified threshold, the plane fit is insufficient and the update of the geometric parameters is not reasonable. In this approach the threshold is set to 0.5m.

An update of the current road model is not necessary for each frame. Noise and roughness of the road should not cause an adaption of the geometric parameters. The following two properties are used to decide if an update should be made:

1. the number of ground points
2. the quality of the plane fit.

All distances which are smaller than the depths of the separating plane characterize obstacles which we group into different objects. This is done by connected component analysis (see Fig. 6d). We use spatial coherence to identify regions from the depth information. First, we are looking for connected components in the 8-neighborhood in the image plane. In the depth dimension a neighboring pixel is connected if the difference in depth is less than a threshold.

As we have to deal with incomplete data and occlusion, it is possible that an obstacle will be split into different disjoint regions. If two regions belong to the same obstacle, they must be close to each other in 3D space. For each pair of regions we determine a probability measure by the *proximity* and *depth similarity*. This measure corresponds to the likelihood that the regions belong to the same obstacle. They are merged if this probability measure is larger than a specified threshold. In Fig. 6e we see the result after combining close regions where the roof and the body of the oncoming car are merged to one object.

Within the separating plane method, the choice of a rotated plane eliminates the problems of noise and uneven roads. The method can also adapt itself to changes in the environment. On the other hand valid ground pixels are required to perform an update of the current road model. A flat angle of incidence or a rough road surface may cause a insufficient reflection of the light. More detailed information about this approach can be found in [12].

3.2   Tracking

Robust tracking schemes for driver assistance and traffic surveillance have to cope with several problems, such as occlusion, new appearance, and disappearance of scene objects. Furthermore, system and measurement noise corrupts the data and thus a tracking mechanism should not rely on perfect data or perfect low level processing.

However, the use of 3D data allows to treat these problems since accurate information about the geometrical arrangement is available. Thus the search space for finding correspondences between consecutive frames can be restricted to a small depth slice. In this paper two tracking schemes are reported. The first is a matching scheme using temporally local and global processes. Secondly Kalman filtering is described. It finds the most likely estimates of the states of obstacles. Both approaches results in a list of tracked obstacles.

Due to the low resolution of our data, the tracking schemes have to deal with the problem that narrow obstacles, e.g. posts and trunks, are only visible at times. Such obstacles can be tracked by means of aging. That means each obstacle is attached an age as attribute. Age is initialized by zero and reset to zero if a correspondence is found. If an obstacle disappears, its age is increased by one and its parameters are updated with estimated values which can be computed by the Kalman filter or by a weighted regression method. The tracking process continues if the obstacle reappears within a certain time interval. Otherwise, if its age has exceeded a certain threshold, it will be removed from the list of tracked obstacles.

Results of both tracking schemes show that obstacles can be tracked even under difficult conditions.

Matching features A successful way of determining the position of an obstacle at the next frame is to find feature correspondences in consecutive frames. In this procedure the use of 3D data is an enormous advantage since the search for correspondences can be restricted to a small depth slice. To allow robust long-term tracking of obstacles, knowledge over larger time intervals has to be acquired as well. Thus our tracking scheme employs temporal global processes, such as the prediction of 3D positions, aging, and global merging of obstacles.

A region-based matching approach is also presented in [1]. At each point of time height maps are computed and segmented into unexplored, occluded, traversable or obstacle regions. Then height maps of consecutive frames are matched in order to estimate the vehicle motion and to find moving objects. Also in [13] feature correspondences are determined to track independently moving and deformable objects. The correspondence problem is solved locally by assigning individual trackers to each feature and searching for the correspondence in a small region-of-interest around the feature's next predicted position.

In our approach correspondences between obstacle hypotheses H_i^{t-1}, $i = 1, .., m$ that were stored at time $t - 1$ and connected components C_j^t, $j = 1, .., n$ that were detected as obstacles at time t, are determined. To simplify the correspondence problem, we make use of the maximal possible displacement (Fig. 7a), of similarity in object parameters (Fig. 7b), of continuity in motion (Fig. 7c) and consistency of matching (Fig. 7d). In the first step, connected components that are located in the relevant depth slice and image space are determined for each hypothesis H_i^{t-1}. The boundaries of the relevant depth slice can be determined based on the maximum speed of cars and the frame rate. Assuming a maximum speed of 200 km/h and a frame rate of 25 frames/sec, a displacement of 2.22 meters for vehicles driving ahead and a displacement of 4.44 meters for oncoming vehicles is possible. Thus the relevant depth slice can be restricted to [−4.44 meters, 2.22 meters] surrounding the obstacle hypothesis.

The definition of the relevant image space depends on the sensor geometry and the distance of the hypothesis. The search window is large for close objects and small for distant objects. After all relevant connected components are determined, a grouping step is applied next. Thus groups G_k^t of connected components are built, which probably belong to the same object. This is done to take into account that, due to segmentation errors, noise or poor reflection properties, objects could be split into several parts over time and then are separately detected as connected components. Afterwards all relevant matches $H_i^{t-1} \longrightarrow G_k^t$ are assessed. The assessment is defined based on the weighted sum of individual assessments, which describe the distance in horizontal and vertical direction, the similarity with the predicted position, the similarity in object height, the similarity in object width, and the importance of the match based on the volumes of both. The best five matches are stored for each hypothesis H_i^{t-1}.

Because each hypothesis is considered independently up to this step, it may happen that the same connected component is assigned to different hypotheses. To ensure consistent matching, a consistent subset of assignments is selected. Based on this subset, a list of hypotheses for obstacles at time t is stored. Con-

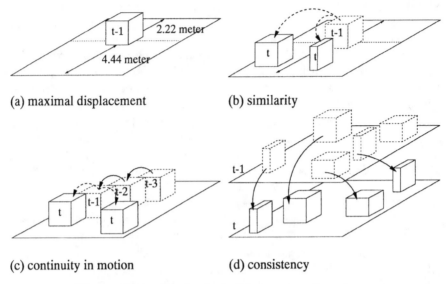

(a) maximal displacement   (b) similarity

(c) continuity in motion   (d) consistency

Fig. 7. Constraints for finding feature correspondences

nected components that were detected at time t, but that were not assigned to one of the hypotheses at time $t - 1$, are added to the list as new hypotheses. Temporally global processes increase the robustness of tracking significantly by considering larger time intervals. In our tracking scheme we employ aging, the prediction of 3D positions of obstacles, and global merging of hypotheses as global processes.

As stated above, aging means that hypotheses for which no correspondence was found by local matching are kept for a few frames. Thus at each discrete point of time, it is checked for a hypothesis, if its age has exceeded a maximum threshold. If this is the case, the hypothesis is deleted. An example is shown in Fig. 8. Four consecutive frames are illustrated showing one obstacle in the tracking process. For clarity other detected obstacles are not marked. Focusing on the white bounding box (label 1, Fig. 8b), which represents a car, it can be seen that the bounding box is empty at time $t + 1$ and $t + 2$. This means that no correspondence was found in the corresponding frames. The reason is that the car is driving behind bushes (Fig. 8a) and thus no depth data are measured during this period. But at time $t+3$ the tracking process recovers and the correct assignment for the car is found.

The 3D position of an object can be predicted based on observations in the past. It is assumed that the object moves in a continuous way during the observed period. This assumption is valid in our case, because cars do not change motion abruptly and the range images are captured with a high frame rate by our sensor. To determine the predicted object position we use weighted regression as prediction procedure; $k = 4$ observations are taken into account for the computation of the predicted value. The observations are not required to occur

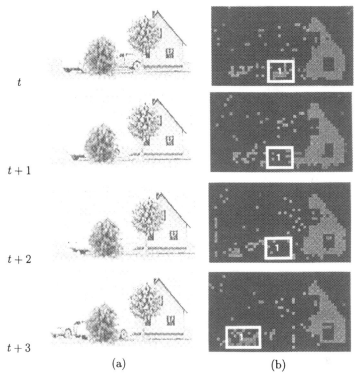

(a)   (b)

Fig. 8. Example for aging mechanism

in consecutive frames. Due to aging it is possible that measurements are missing in between. Using weighted regression for prediction, observations that were made longer ago are less weighted than observations from the near past.

The global merging process is based on the assumption that hypotheses which move similarly over time and which are adjacent to each other in space probably belong to the same obstacle. Thus at the beginning of a time point it is checked for a hypothesis if adjacent hypotheses have similar motion characteristics. Using range data as input, it is not possible to prevent static objects that are very close to each other from being merged together, because, due to egomotion, static objects will have a similar motion characteristic. But this kind of error is not critical for collision avoidance applications, since obstacle detection and not object recognition is the task to be solved. An example of the global merging of hypotheses is shown in Fig. 9. Due to occlusion by a tree, the house is split into two parts (time t, label 1 and 2). By means of global merging, the hypotheses corresponding to the two house parts are merged into one hypothesis and tracked successfully as one hypothesis along the time axis (time $t + 3$, label 1).

Matching features in consecutive frames is a well-known method for object tracking. To be robust against splitting and merging of objects, multiple matches have to be realized. Unlike the fact that in general the integration of multiple

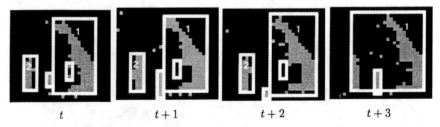

$$t \qquad\qquad t+1 \qquad\qquad t+2 \qquad\qquad t+3$$

Fig. 9. Example for global merge of obstacle hypotheses

matches increases the computational costs significantly, range imagery allows to restrict the search for correspondences to a small depth slice. Thus only a few possible matches have to be evaluated to select the best match. Furthermore temporally global processes such as aging, prediction and global merge allow robust long-term tracking of obstacles. In addition, it is important to emphasize that scene objects that are located in the same depth slice and move in a similar way may be erroneously merged and tracked as one obstacle. More details about our tracking scheme based on feature correspondences can be found in [16].

Kalman Filtering An optimal estimator is implemented by an algorithm that processes measurements to deduce a minimum error estimate of a system by utilizing knowledge of system and measurement dynamics, assumed statistics of system noise and measurement errors, and initial condition information. The optimal estimator for a quadratic error function is the Kalman filter [7,9].

For driver assistance and traffic surveillance systems the Kalman filter is an already known approach to deal with system and measurement noise. An extended Kalman filter can be used to perform quantitative tracking [23]. Three different models are constructed to describe the motion of a moving car and are incorporated into a Kalman filter. The authors in [8] propose two methods to combine the estimates of two tracking systems that provide motion parameters: the first method applies a Kalman filter, while the second method assigns weights to the individual estimates based on the covariance matrix.

The model of a linear dynamic process is defined by

$$x_{t+1} = \Phi_t x_t + w_t, \quad w_t \sim N(0, Q_t) \tag{5}$$

where the *transition matrix* Φ_t models the evolution of the state vector x_t at time t and the measurement model

$$z_t = H_t x_t + v_t, \quad v_t \sim N(0, R_t) \tag{6}$$

determines the measurements z_t as a function of the state x_t. H_t is called the *measurement sensitivity matrix*. The system noise w_t and measurement noise v_t are zero-mean Gaussian sequences with given covariance-matrices Q_t and R_t, respectively. These equations, shown in the dashed-line box in Fig. 10, are simply a mathematical abstraction – a model of what we think the system and

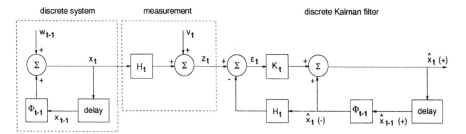

Fig. 10. System model and discrete Kalman Filter

measurement processes are – and the Kalman filter is based upon this model. In the linear, discrete Kalman filter, calculations at the covariance level ultimately serve to provide K_t, which is then used to determine the estimated state $\hat{x}_t$. K_t is called the *Kalman gain matrix*. The estimated state before the measurement at time t is known as the *a priori* estimate $\hat{x}_t(-)$ and the estimated state after the measurement as the *a posteriori* estimate $\hat{x}_t(+)$. This is illustrated in Fig. 10, which is essentially a simplified flow diagram of the discrete Kalman filter. The basic steps of the computational procedure for the discrete-time Kalman estimator are as follows:

1. a priori state estimate extrapolation:
$$\hat{x}_t(-) = \Phi_{t-1}\hat{x}_{t-1}(+)$$
2. error covariance extrapolation:
$$P_t(-) = \Phi_{t-1}P_{t-1}(+)\Phi_{t-1}^T + Q_{t-1}$$
3. Kalman gain matrix :
$$K_t = P_t(-)H_t^T[H_tP_t(-)H_t^T + R_t]^{-1}$$
4. error covariance update:
$$P_t(+) = [I - K_tH_t]P_t(-)$$
5. a posteriori state estimate observational update:
$$\hat{x}_t(+) = \hat{x}_t(-) + K_t[z_t - H_t\hat{x}_t(-)]$$

We employ Kalman filtering to provide most likely estimates of the state of each obstacle from measurement data corrupted by noise. In our system the state vector of each obstacle contains the depth d_t, the relative velocity $\dot{d}_t$, the horizontal angle ψ_t to the obstacle and its change $\dot{\psi}_t$, its vertical angle η_t and the corresponding change $\dot{\eta}_t$, where all values are taken at time t:

$$x_t^T = [d_t \ \dot{d}_t \ \psi_t \ \dot{\psi}_t \ \eta_t \ \dot{\eta}_t]. \tag{7}$$

The a priori state predictions are mapped with the measurements (depth, horizontal and vertical angle to an obstacle) from the segmentation step. We search for the detected object whose position is closest to the a priori estimate from the Kalman filter and whose distance is inside a search area.

Obstacles can disappear in some frames due to the coarseness of the data. This problem is addressed by employing an aging process as described before.

Parameters are updated with the estimated values from the Kalman filter over several frames. The obstacle can thus be recognized if it reappears within an certain time. Otherwise it will be removed. Furthermore, new obstacles are added if not all detected objects of the segmentation could be matched.

As extension a global merging process — as described before — can be added. Adjacent obstacles which move similarly over time are probably part of the same object and can therefore be merged.

Kalman filtering makes the prediction of important state information more robust by employing noise models. It provides the best possible estimation under the assumption of Gaussian noise and a quadratic error function. Regrettably, the initialization of a Kalman filter is difficult and computational cost is not small.

3.3   Results

Since the range sensor is still under development, range image sequences of real traffic scenes are not yet available. Therefore, we use on the one hand simulated range image sequences and on the other hand scaled range images recorded by an ABW range scanner (ABW GmbH, Germany) to develop and test the segmentation and target tracking algorithms.

First, the simulated range images are generated by a graphical simulation software package. An example is illustrated in Fig. 11. From left to right, we see the simulated traffic sequence and the acquired range images. The range data are represented by different greylevels; black represents undefined pixels where no meaningful measurement was obtained. The range sensor has a relatively narrow field of view whereas the driver has a wider field of view. Thus the car driving ahead appears very large in the range images. The scene contains two cars, one driving ahead and the other in the oncoming traffic. During the course of the sequence the oncoming car leaves the sensor's field of view. This 100-frame sequence has a frame rate of 25 images per second, an image size of 16×64 pixels and a field of view of $2.38° \times 10.0°$. In the simulation, the range sensor is installed next to one of the front lights. Secondly, in Fig. 12 an example of a range image sequence of a scaled toy traffic scenario acquired with a structured light sensor is shown. The scene contains one car driving in the oncoming traffic. Several trees are in the background. This sequence has a length of 25 images and a frame rate of 25 images per second. The image size is 65×144 pixels and the field of view is $6.04° \times 14.1°$.

All combinations between the segmentation and tracking approaches are possible. In this paper, however, we focus on the results of two.

Results of obstacle detection and tracking for the simulated traffic sequence of Fig. 11 are shown in Fig. 13. In Fig. 13a obstacles are detected using the radial slope approach. Tracking is performed by finding feature correspondences. Both cars (label 1 and 2) are correctly detected as obstacles and tracked successfully along the time axis. In Fig. 13b we separate image features in the segmentation approach and use a Kalman filter for the tracking. The detected obstacles are represented by their bounding boxes, thus the state vector has to be augmented

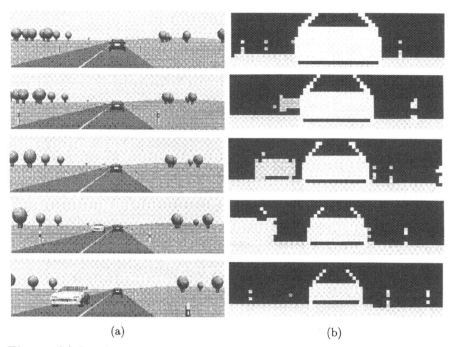

(a)   (b)

Fig. 11. (a) Simulated traffic sequence and (b) corresponding computed range images, Sequence 22

by the width and height of each obstacle. The size and location of the bounding boxes are the estimated state values provided by the Kalman filter. The results show only obstacles which are new or tracked within this segmentation step, updated obstacles are not included. As the results show, the separation of image features in combination with a Kalman filter works well. The oncoming bright car (label 2) which is partially occluded is correctly detected and tracked although the position and size change considerably. By comparing the results in Fig. 13 we can observe differences between the two approaches. As the radial slope approach was used in the left column, single pixels (see for example the fifth image from the top) can not be identified as an obstacle. This segmentation scheme needs at least two valid depth measurements in neighboring rows to detect an obstacle. In contrast we see that the separation of image features which is shown in the right column loses image pixels lying below the separation plane.

Examples for tracking a car in a toy traffic scene is illustrated in Fig. 14. The corresponding toy traffic scene is shown in Fig. 12. In Fig. 14a obstacles are once more detected using the radial slope approach and tracked by finding feature correspondences. As can be seen the car (label 1) is successfully detected as obstacle and tracked over the entire range image sequence. The result of separating image features in the segmentation step and use of a Kalman filter for the tracking are shown in Fig. 14b. Even if a tracker loses its obstacle (label 1 → 4) the system continues without any difficulties by a new initialization of

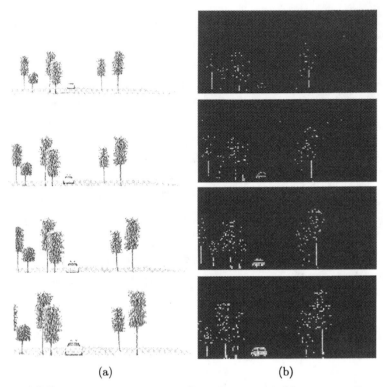

(a)   (b)

Fig. 12. (a) Sensor's view at a toy traffic sequence and (b) corresponding range images recorded with a structured light sensor, Sequence 10

the Kalman filter. In this sequence, only obstacles larger than three pixels are shown. Looking at the results in Fig. 14 the dissimilar object sizes are conspicuous. As already explained for the previous result sequence, this is the effect of the different obstacle detection schemes and the various region groupings. Furthermore, differences between the two tracking methods are illustrated. While the matching scheme has problems to track trees in their entire size, Kalman filtering loses the moving car.

4   Conclusion

In this article several approaches for obstacle detection and tracking in low resolution range image sequences were proposed. We focused on an automotive application. The availability of accurate 3D data is found to be advantageous.

Concerning collision avoidance in traffic scenarios, it was shown that robust obstacle detection is possible by checking traversability in the sensor's field of view. Two approaches were presented based on vertical slope evaluation and computation of a separating plane. Robust long-term tracking of obstacles implies

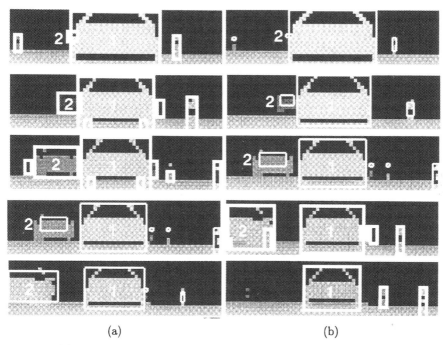

(a)   (b)

Fig. 13. (a) results of the matching scheme, (b) result of the Kalman filtering

the handling of problems such as occlusion, new appearance and disappearance of scene objects, and furthermore, due to the low resolution of our data, the visibility of narrow obstacles. Our findings reveal that the use of 3D data allows us to treat these problems since information about the geometrical arrangement is available. Thus the search for correspondences can be restricted to a small depth slice and in case of temporal disappearance of obstacles, internal parameters are updated by estimated values. As obstacle detection is performed at every discrete point of time, newly appearing obstacles and even obstacles that were partly occluded previously, are detected and tracked in their entirety. Our investigations focused on two tracking schemes. In a first approach a matching scheme was realized using temporally local and global processes. In a second approach Kalman filtering was employed to track obstacles.

The described methods allow real time processing. This is possible due to the small image size, the minimization of the object management and the restriction of the search space. The issue of how to combine the different approaches in order to improve robustness of an overall system, deserves further study and will be subject of our future work.

Our results imply that range imagery is a key technology for many computer vision applications since many ambiguities of interpretation arising between object boundaries and inhomogeneities of intensity, texture and color can thus be trivially solved.

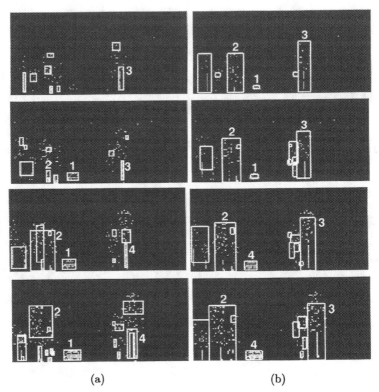

(a)   (b)

Fig. 14. (a) results of the matching scheme, (b) result of the Kalman filtering

Acknowledgments

This research was partially supported by MINORA, a project of the Swiss priority program OPTIQUE II, funded by the ETH Council.

References

1. M. Asada, M. Kimura, Y. Taniguchi, and Y. Shirai. Dynamic integration of height maps into 3D world representation from range image sequences. *Internation Journal of Computer Vision*, 9(1):31–53, 1992.

2. J. Bernasch and W. von Seelen. Visual attention control in complex dynamic image sequences. In *9th Israeli Symposium on Artificial Intelligence and Computer Vision*, Ramat Gan, Israel, 28-29 December 1992.

3. M. Bertozzi, A. Broggi, D. Colla, and A. Fascioli. Sensing of automotive environments using stereo vision. In *30th International Symposium on Automotive Technology and Automation (ISATA), Special Session on Machine Vision and Intelligent Vehicles and Autonomous Robots*, pages 187–193, Florence, Italy, 16th-19th June 1997.

4. M. Bertozzi, A. Broggi, G. Conte, and A. Fasciol. Obstacle and lane detection on argo. In *IEEE Conference on Intelligent Transportation Systems*, pages 1010–1015, 1997.
5. M. J. Daily, J. G. Harris, and K. Reiser. Detecting obstacles in range imagery. In *Image Understanding Workshop*, pages 87–97, February 1987.
6. E. D. Dickmanns and B. D. Mysliwetz. Recursive 3-d road and relative ego-state recognition. *IEEE Transaction on Pattern Analysis and Machine Intelligence*, 14(2):199–213, 1992.
7. A. Gelb. *Applied Optimal Estimation*. MIT Press, 1996.
8. S. Gil, R. Minanese, and T. Pun. Combining multiple motion estimates for vehicle tracking. In *European Conference on Computer Vision*, volume 2, pages 307–320, 1996.
9. M. S. Grewal and A. P. Andrews. *Kalman Filtering Theory and Practice*. Prentice-Hall, Inc., 1993.
10. Q.-T. Luong, J. Weber, D. Koller, and J. Malik. An integrated stereo-based approach to automatic vehicle guidance. In *5th International Conference on Computer Vision*, pages 52–57, 1995.
11. E. B. Meier and F. Ade. Tracking cars in range image sequences. In *IEEE Conference on Intelligent Transportation Systems (ITS)*, Boston,Massachusetts, 9-12 November 1997.
12. E. B. Meier and F. Ade. Object detection and tracking in range image sequences by separation of image features. In *IEEE International Conference on Intelligent Vehicles*, 1998.
13. J. Roberts and D. Charnley. Attentive visual tracking. In *British Machine Vision Conference*, pages 459–468, 1993.
14. S. R. Ruocco. *Robot Sensors and Transducers*. Wiley, New York, 1990.
15. R. Schneider, G. Wanielik, and J. Wenger. Millimeter-wave imaging of traffic scenarios. In *Proceedings of the 1996 IEEE Intelligent Vehicles Symposium*, pages 327–332, September 19-20 1996.
16. K. Sobottka and H. Bunke. Employing range imagery for vision-based driver assistance. In *SPIE conference: Enhanced and Synthetic Vision*, Orlando, Florida, USA, 13-17 April 1998.
17. K. Sobottka and H. Bunke. Obstacle detection in range image sequences using radial slope. In *3rd IFAC Symposium on Intelligent Autonomous Vehicles*, pages 535–540, Madrid, Spain, 25-27 March 1998.
18. T. Spirig, M. Marle, and P. Seitz. The multi-tap lock-in CCD with offset subtraction. *IEEE Transactions on Electron Devices*, 44(10):1643–1647, 1997.
19. T. Spirig, P. Seitz, O. Vietze, and F. Heitger. The lock-in CCD — two-dimensional synchronous detection of light. *IEEE Journal of Quantum Electronics*, 31(9):1705–1708, 1995.
20. C. Thorpe, M. Hebert, T. Kanade, and S. Shafer. The new generation system for the CMU Navlab. In I. Masaki, editor, *Vision-based vehicle guidance*, pages 30–82, 1992.
21. S. Tokoro. Automotive application systems of a millimeter-wave radar. In *Proceedings of the 1996 IEEE Intelligent Vehicles Symposium*, pages 260–265, September 19-20 1996.
22. P. Veatch and L. Davis. Efficient algorithms for obstacle detection using range data. In *Computer Vision, Graphics, and Image Processing 50*, pages 50–74, 1990.
23. L. Zhao and C. Thorpe. Qualitative and quantitative car tracking from a range image sequence. In *Computer Vision and Pattern Recognition*, pages 496–501, 1998.

Active Control of Resolution for Stable Visual Tracking

Nicola Ferrier

University of Wisconsin, Madison WI 53706, USA,
ferrier@engr.wisc.edu,
http://mechatron.me.wisc.edu

Abstract. Success of visual tracking typically relies on the ability to process visual information sufficiently fast. Often a dynamic system model of target motion is used to estimate the target location within the image and a region of interest (ROI) is used to reduce the amount of image data processing. This has proven effective, provided the ROI is sufficiently large to detect the target and sufficiently small to be processed quick enough. Here we formally consider the size of the ROI and the resolution of the ROI to ensure that tracking is stable. Dynamic system formulation of visual tracking usually requires specification of the dynamics of the target. We analyze motions which can be described by linear time-invariant dynamical systems (although the image motion may be highly non-linear). One can successfully analyze the required ROI size and resolution to ensure stable tracking.

1 Motivation

Planning and modeling of sensor utilization in intelligent systems typically covers topics such as how to fuse multiple sensor inputs, or how to utilize a particular sensor for a task such as navigation or map building. For visual control of motion, sensor planning typically involves determining which algorithms to use to process the visual data. A key competency required in order to use vision for motion control (grasping or manipulation) is to be able to locate and track a target. Conventional visual servo control requires continuous tracking of a target, at great computational expense. We are looking to relax the continuous tracking requirement.

Previous systems that use vision for control of manipulation or navigation typically must track the object of interest. Processing closely sampled images enables the use of temporal coherence to facilitate tracking. For advanced systems, the ability to simultaneously track multiple objects (such as a driver moving his eyes to objects/points on either side of the road) either forces the use of multiple visual sensors *or* one must relax the temporal coherence requirement. We are looking to do the later. The first step towards this is to be able to understand the relationship between the target dynamics, system and measurement noise, and time delay and their impact on successful tracking. This chapter develops a tracking model, based on traditional Kalman filter tracking systems,

Christensen et al. (Eds.): Sensor Based Intelligent Robots, LNAI 1724, pp. 140–159, 1999.

that can successfully track given non-fixed time delays between measurements. Initially we assume that the delay is based on computation time (although the delay could be due to other reasons such as "time sharing" the sensor between two or more targets). If there is noise in the system model (as there will be in any practical system), as delay time increases, the positional uncertainty of the target will also increase. The region in the image that must be searched to ensure successful tracking will also increase (thus more computation is required to process the data), and we have a negative feedback system - tracking could become unstable (or effectively non-observable as the delay between observations grows). This chapter presents an algorithm to actively control the resolution of the processed image window to avoid this growth in computation time. While the idea of using variable sized search windows has been used before, we develop the underlying relationships between the system parameters and the required window size and resolution. We show via a simulation and a simple system how active control of the resolution of the image data processed can lead to stable tracking.

2   Background on Visual Tracking

Visual tracking, that is, extracting geometric information of the motion of one or many objects, is a central problem in vision-based control and there has been considerable effort to produce real-time algorithms for tracking moving objects. Typically tracking seeks to determine the image position of a target as the target moves through the camera's field of view. Successive image frames (usually taken at closely spaced intervals) are processed to locate the features or target region. Difficulties in target recognition arise due to the variation in images from illumination, full or partial occlusion of the target, effects of pose on the projected image. Given that the target can be recognized, tracking difficulties arise due to noise (modeling uncertainty) and the time delay required to process the image data.

Visual control systems must take into account the delays associated with the visual computation. Tracking stability can only be achieved if the sensing delays are sufficiently small, or, if the dynamical model is sufficiently accurate. Various approaches have been taken to compensate for the delay of the visual system in visual servo control. Many the tracking systems use *dynamic* models of target motion, and use the Kalman-Bucy filter for improved tracking performance (see e.g. [4,12,16,17,22]). Successful visual tracking for servo control has been achieved by balancing the trade off between the accuracy of the sensing and the visual data processing time. Roberts *et al* [20] show that for systems for which the dynamical model is known with sufficiently accuracy, the search area (window size or resolution) can be kept small enough such that the processing time is sufficiently small for accurate tracking. Clark & Ferrier [3,6] utilize feed-forward compensation for the processing delay. The dynamical model must be known with sufficient accuracy to produce stable tracking. Schnackertz & Grupen [21] also utilize a feed-forward term. Similar feed-forward, or predictive, techniques are employed in visual tracking techniques which incorporate the computation delay (expressed as the number of frames 'dropped' during the compute cycle) within the dynamical model to predict the tracker motion [2]. Nelson *et. al.* [15] assume commensurate delays and model the delay d within the system model and a new control law is developed to account for this delay. One feature uniform to these methods is that, after estimating the approximate image location of the feature, a region of the image must be searched in the next frame to locate the target (either to match the model for wire frame and/or solid models, or locate the image feature point(s), etc.). Given the uncertainty of the estimated position (the covariance), the size of the search window can be modified to reduce computation time. For sufficiently small delays, tracking is successful, however none of these systems have *quantitatively* analyzed the bounds on the accuracy of the performance, the computational delay and the system noise.

Assuming a recursive estimation scheme for tracking, this paper demonstrates that even for simple linear dynamic models, the tracking system can become effectively *unobservable* for particular dynamics, noise models, or processing algorithms. As the size of the search window grows, the measurement sample time increases. This time delay can cause the tracking system to become unstable. Olivier [18] develops some mathematical underpinnings for this analysis, however his work is primarily restricted to scalar stochastic systems.

In this paper we concentrate on the ability of Kalman-Bucy filter based tracking systems to successful track objects using a *search window based* algorithm. We will show that for *search window based* tracking: 1) If computation time is proportional to the estimated covariance then the sequence of measurement times $(t_1, t_2, \ldots, t_k, \ldots)$ can diverge (becomes unstable), and 2) variable resolution can be used to stabilize the tracking (bound the time delay). The dynamic system analysis for tracking commonly used is presented in section 3. We analyze search window based methods giving analytic and experiment results demonstrating our claim.

3   Dynamics System Formulation of Tracking

The ability to describe, or estimate, the motion of the object with respect to the camera enables the tracking system to restrict the search to a local region centered on the estimated position. Point based methods typically assume closely spaced image sequences in order to use correlation methods (and often dynamics are not used under the assumption that the images are closely enough spaced that the point motion is Brownian, so the search window is centered on the current location). The motion of the points, lines or region is typically described using a stochastic dynamic system model. The *state*, $x(t)$, depends on the representation used. For example, for point based tracking x may be a vector of point coordinates, while for b-spline based tracking x may be a vector of control points, and for region based tracking x may be a vector of coordinates of the centroid of the region.

For continuous linear systems we describe the evolution of the state with the equation

$$dx(t) = A(t)x(t)dt + \Gamma(t)dw_t, \quad t \geq t_o \tag{1}$$

where the state $x(t)$ is an n-vector, A is an $n \times n$ continuous matrix time function, Γ is an $n \times r$ continuous matrix time function, w_t is a r-vector Brownian motion process with expected value $\mathcal{E}[dw_t dw_t^T] = Q(t)dt$. At time instants t_k discrete linear measurements are taken:

$$y(t_k) = C(t_k)x_k + v_k; \quad k = 1, 2, \ldots; \quad t_{k+1} > t_k \geq t_o \tag{2}$$

where y_k is the m-vector observation and C is an $m \times n$ nonrandom bounded matrix function. v_k is a m-vector white, Gaussian sequence, $v_k \sim N(0, R_k)$, $R_k > 0$. The initial state is assumed to be a Gaussian random variable, $x_{t_o} \sim N(x_{t_o}^*, P_{t_o})$. The random variables x_{t_o}, v_k and w_t are assumed independent. The covariance of the state is given by $P(t) = \mathcal{E}[(x(t) - x^*(t))(x(t) - x^*(t))^T]$. The actual (true) state is denoted $x^*(t)$. Here we are using a continuous-discrete system. The system (motion of an object) is a continuous process, whereas measurements are taken at discrete times. Jazwinski [13] points out that analysis of the continuous-continuous system and the continuous-discrete system are essentially equivalent.

Under these assumptions, estimation is often performed using the Kalman-Bucy filter (see e.g. [4,12,16,22]). Under the assumed statistical assumptions, the Kalman-Bucy filter provides a state estimate that is an optimal trade-off between the extrapolated state under the system dynamical model and the measured state.

Most tracking systems assume *equally spaced* measurements. Here we *do not* make that assumption in order to evaluate the effect of the search size on the tracking performance. Measurements are taken at discrete instances $(t_1, t_2, \ldots t_k, t_{k+1}, \ldots)$. The time of the measurements effects tracking stability. Previous work has considered the optimal measurement schedule where the growth of the covariance is used to determine *when* to take measurements [14]. Due to the time taken for the processing of image data, the problem with visual tracking is not one of determining *how long* to wait before taking a measurement, but determining how to measure *often enough* to ensure stability. Here we assume one finishes processing previous measurements before taking the next measurement. If the time between measurements $|t_{k+1} - t_k|$ grows, then tracking fails (we call this situation unstable tracking).

For linear time invariant systems, the estimate of the state can be found from the state transition matrix on the interval $[t_k, t_{k+1}]$,

$$\hat{x}(t_{k+1}) = \Phi(t_{k+1}, t_k)x(t_k) = e^{A(t_{k+1} - t_k)} \hat{x}(t_k) \tag{3}$$

and the covariance is determined (before measurement) from the Lyapanov equation [1]:

$$P(t_{k+1}^-) = e^{A(t_{k+1} - t_k)} P(t_k^+) e^{A^T(t_{k+1} - t_k)} + Q(t_{k+1}, t_k) \tag{4}$$

[1] We adopt the notation $P(t_k^-)$ to denote the covariance at time t prior to any measurement. In other notation this may be denoted $P(t_k|t_{k-1})$ and $P(t_k^+)$ denoted $P(t_k|t_k)$

where

$$P(t_k^+) = \left(P^{-1}(t_k^-) + C^T(t_k)R_k^{-1}C(t_k)\right)^{-1} \qquad (5)$$

and

$$Q(t_{k+1}, t_k) = \int_{t_k}^{t_{k+1}} e^{A(t_{k+1}-\tau)}Q(\tau)e^{A^T(t_{k+1}-\tau)}d\tau$$

is the integrated noise over the interval (t_k, t_{k+1}). The covariance immediately before a measurement at time t_{k+1} is given by:

$$P(t_{k+1}^-) = e^{A(t_{k+1}-t_k)}\left(P^{-1}(t_k^-) + C^T(t_k)R_k^{-1}C(t_k)\right)^{-1}e^{A^T(t_{k+1}-t_k)}$$
$$+ Q(t_{k+1}, t_k) \qquad (6)$$

We will utilize this expression later in determining the computational delay for tracking. The ability to determine the growth of the covariance requires solving this Riccati-equation. For certain cases the non-linear coupled differential equations in the elements of P can be solved (see e.g. Gelb [10], chapter 4, example 1 presents four methods to solve the Riccati equation for a second-order integrator with no system noise).

4   Computation Time and Covariance

In this section we explore the relationship between the covariance and the computation time.

Suppose the location of a target is a pixel location within a frame (e.g. point coordinate or centroid of a region). If the entire image frame is searched, the computation time is a function of image size. Often one utilizes a model of object motion then a search window or region of interest (ROI) can be centered on the predicted target location. The covariance of the estimated target location can be used to determine the size of the search window. Although there have been suggestions to utilize the covariance to determine ROI size [2], in practice the ROI is a fixed size or a limited set of sizes [19,20]). If the ROI is chosen to be proportional to the covariance (typically a 2σ or 3σ window size) then under the assumed Gaussian noise models, with high probability the target should be located within the given ROI. If we let the ROI size vary then the size of the ROI affects the processing time. If the algorithm must "visit" each pixel a certain number of times (assumed at least once), then the processing time is proportional to the ROI size:

$$t_{k+1} = t_k + \beta\|P(t_k^-)\|$$

where the *size* of the covariance, denoted $\|P(t_k^-)\|$, represents the number of pixels to be processed. If the ROI size (and hence the covariance) grows, so does the processing time and hence the system can become unstable. Figure 1 graphically depicts this idea. The solid line represents the covariance in a target position *before* a measurement has been taken. The size of this region determines the amount of data to be processed and hence effects the time of the next possible measurement. The dashed ellipse represents the uncertainty in a target position *after* a measurement has been taken. The size of this ellipse grows over the time interval before the next measurement can be taken.

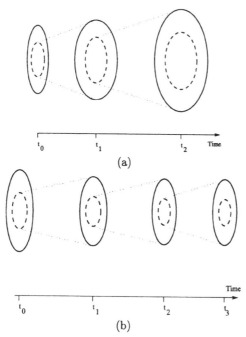

Fig. 1. Graphic representation of growth in covariance. The solid line represents the covariance before a measurement, $P(t_k^-)$. The dashed line represents the covariance after the measurement has been processed, $P(t_k^+)$. The time between measurements depends on the size of the covariance. (a) The covariance shown here increases in size and so the time delay $t_{k+1} - t_k$ increases. This tracking becomes effectively unobservable. (b) In this instance, the covariance decreases in size to a fixed size and so the time delay $t_{k+1} - t_k$ converges to a fixed interval.

If the covariance is expressed with respect to image coordinates (for image based systems) then the covariance can be used to describe an ellipsoidal uncertainty region. For example a 2σ ellipse centered about $(\hat{x}, \hat{y})$ is given by:

$$\left\{ (x,y) \mid [x - \hat{x}, y - \hat{y}] \ P^{-1} \begin{bmatrix} x - \hat{x} \\ y - \hat{y} \end{bmatrix} \le 2 \right\}$$

In order to ensure that the target is found, the entire ellipse must be searched. The exact elliptical region could be searched, for example, fast region filling algorithms from computer graphics [8,11] can be adapted to extract the elliptical region from the image. Alternatively, a bounding rectangle of the uncertainty

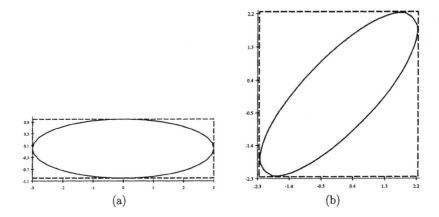

Fig. 2. The bounding rectangles (aligned with the image coordinate frame). In (a) the ellipse is optimally aligned with the image axes. In (b) the worst case alignment occurs with the ellipse rotated 45^0

region can be used. Based on current technology with linear scanning of CCDs images using frame-grabbers, the latter is often used. The major and minor axes of the ellipse are determined by the eigenvalues of the covariance matrix and the desired search scale (2σ). For an ellipse with major/minor axis lengths of a and b, in the worst case, the bounding rectangle will have area $2(a^2 + b^2)$. The bounding rectangle in the the optimal case (the major and minor axes are aligned with the image CCD/frame-grabber axes) has area $4ab$ (and note that the ellipse has area πab). The major and minor axes of the ellipse are given by the maximum and minimum eigenvalues of the covariance. For a σ search region the area of the bounding box for the elliptical ROI is given by

$$4(\lambda_{max})(\lambda_{min}) \leq Area(ROI) \leq 2\left(\lambda_{max}^2 + \lambda_{min}^2\right)$$

where λ denotes the eigenvalues of P. Thus we can obtain bounds for the "size-of" operator, $\|P(t_k^-)\|$, and hence on the computational delay. In general the axes of the uncertainty ellipse will not be aligned with the axes of the image (the lower bound).

5   Covariance Growth and Resolution Control

Tracking with a variable sized ROI will succeed or fail based on the size of the ROI and the computation performed on the ROI. Here the ROI size depends on the growth of the covariance $P(t_k^-)$. P evolves according to the Riccati equation given in equation (6). We noted earlier that the growth of the covariance depends on the dynamics, A, the time interval $(t_{k+1} - t_k)$, and the noise models and that solving the Riccati equation analytically can only be performed in special cases.

We note that the measurement resolution, R, will be determined by the chosen search procedure (algorithm). Centroids can be located to sub-pixel accuracy, thus R will typically be on the order of a pixel squared. This accuracy is not unrealistic for window search approaches. For SSD tracking [1,19] give estimates for the uncertainty of the computed target position.

If the covariance is too large, one can decrease the resolution of the processing within the image by sub-sampling the ROI or using image pyramid algorithms [15,23]. Modifying the resolution of the processing will decrease the accuracy of the measurement but it will also decrease the time delay. A scalar system analysis (from Olivier [18]) is summarized in the appendix. Below we consider a simple dynamical system. We will demonstrate the growth of the covariance for this simple dynamical system and show how modification of the measurement resolution can lead to stable tracking, at the expense of the accuracy of the target estimate. While others (e.g. Nelson [15]) have used multi-scale approaches, no analysis was given to appropriately determine the correct resolution or window-size. In these works the window-size and resolution were determined *a priori* and thus were fixed parameters in the system. Only targets with particular dynamics or noise models would be track-able with fixed window and resolution sizes. Here we show that procedures that control (modify) resolution can indeed stabilize the tracking system. The intuition in previous tracking systems with multi-resolution searches can be formalized by considering both the covariance size and the resolution of search.

If the ROI is sub-sampled then the measurement resolution R is inversely proportional to this sub-sampling factor. As mentioned above many window based routines can achieve sub-pixel accuracy. The resolution will grow with the sub-sampling factor. In the scalar case (see appendix), one can explicitly derive the relationship between the resolution factor, the time delay and the system dynamics. For higher dimensions close-formed analysis requires solution of equation (6), hence we analyze some simple cases, then resort to simulation and experimentation to demonstrate the utility of resolution control.

5.1 Tracking with Simple Linear Dynamics

To demonstrate the potential instability in tracking and subsequent improvement with resolution control we consider a simple system with constant velocity motion. We analyze a first order integrator with measurement of the position (which is a frequently encountered system in tracking). The continuous-discrete linear system has state $\mathbf{x}^T = \begin{bmatrix} \theta, \dot{\theta} \end{bmatrix}$, and the state dynamics is given by

$$\dot{\mathbf{x}} = \begin{bmatrix} 0 & 1 \\ 0 & 0 \end{bmatrix} \mathbf{x} + \begin{bmatrix} 0 \\ w \end{bmatrix} \tag{7}$$

where $w \sim N[0, q]$ is the system noise and the discrete measurement process is

$$y(t_k) = \begin{bmatrix} 1 & 0 \end{bmatrix} \mathbf{x}(t_k) + v_k, \quad v_k \sim N[0, r] \quad t_k \in [t_o, t_1, \ldots] \quad \text{where } t_{i+1} > t_i$$

The transition matrix for the system matrix given above is

$$e^{A(t_{k+1}-t_k)} = \begin{bmatrix} 1 & (t_{k+1}-t_k) \\ 0 & 1 \end{bmatrix} = \begin{bmatrix} 1 & \delta_k \\ 0 & 1 \end{bmatrix}$$

where $\delta_k = (t_{k+1}-t_k)$, and

$$C^T R^{-1} C = \begin{bmatrix} \frac{1}{r_k} & 0 \\ 0 & 0 \end{bmatrix}.$$

The covariances immediately before measurements at times t_k and t_{k+1} are denoted

$$P(t_k^-) = \begin{bmatrix} p_{11}(t_k^-) & p_{12}(t_k^-) \\ - & p_{22}(t_k^-) \end{bmatrix} = \begin{bmatrix} p_{11_k} & p_{12_k} \\ - & p_{22_k} \end{bmatrix}$$

and

$$P(t_{k+1}^-) = \begin{bmatrix} p_{11}(t_{k+1}^-) & p_{12}(t_{k+1}^-) \\ - & p_{22}(t_{k+1}^-) \end{bmatrix} = \begin{bmatrix} p_{11_{k+1}} & p_{12_{k+1}} \\ - & p_{22_{k+1}} \end{bmatrix}.$$

With this system equation (5) becomes

$$P(t_k^+) = \frac{1}{p_{11_k}+r_k} \begin{bmatrix} r_k p_{11_k} & r_k p_{12_k} \\ - & p_{22_k}(p_{11_k}+r_k) - p_{12_k}^2 \end{bmatrix}$$

Using equation (6) and the notation above we solve for $P(t_{k+1}^-)$ and obtain three non-linear difference equations:

$$p_{11_{k+1}} = \frac{1}{p_{11_k}+r_k} \left(r_k p_{11_k} + 2\delta_k r_k p_{12_k} + \delta_k^2 (p_{22_k}(p_{11_k}+r_k) - p_{12_k}^2) \right) + q_{11_k}$$

$$p_{12_{k+1}} = \frac{1}{p_{11_k}+r_k} \left(r_k p_{12_k} + \delta_k \left(p_{22_k}(p_{11_k}+r_k) - p_{12_k}^2 \right) \right) + q_{12_k} \qquad (8)$$

$$p_{22_{k+1}} = \frac{1}{p_{11_k}+r_k} \left(p_{22_k}(p_{11_k}+r_k) - p_{12_k}^2 \right) + q_{22_k}$$

where

$$Q(t_{k+1},t_k) = \int_{t_k}^{t_{k+1}} \begin{bmatrix} (t_{k+1}-\tau)^2 & (t_{k+1}-\tau) \\ (t_{k+1}-\tau) & 1 \end{bmatrix} q(\tau)d\tau = \begin{bmatrix} \frac{1}{3}\delta_k^3 q & \frac{1}{2}\delta_k^2 q \\ - & \delta_k q \end{bmatrix}$$

for constant system noise $dw_\tau^2 = q(\tau) = q$. Note that the evolution of the covariance depends on the initial conditions $P(0)$, the time delay δ_k, the noise covariances q_{ij} and r. The dynamics enter via the state transition matrix. The initial speed and/or position do not enter into the above equations.

To reduce the number of subscripts, we will write $a_k = p_{11_k}$, $b_k = p_{12_k}$, and $c_k = p_{22_k}$ in the following.

5.2 Variable Time and Fixed Resolution Tracking

Here we consider search methods which search at a fixed resolution within the ROI. The size of the ROI is assumed to be proportional to the target positional uncertainty (i.e. δ_k is a function of the size of p_{11}, r_k is constant). We take $\delta_k = 4 * \beta \sqrt{p_{11}}$. The positional covariance is obtained from the 11 element of the covariance matrix[2], $p_{11} = \sigma_p$. To search at a fixed resolution within a 2-σ_p region of the estimated position we must process $4\sqrt{p_{11}}$ pixels. Thus $4\sqrt{p_{11}}$ is the number of pixels to be processed and β is the processing time per pixel giving a time delay due to computation of δ_k.

The difference equations (equations (8)) for this time delay become

$$a_{k+1} = \frac{1}{a_k + r} \left(r a_k + 2\beta a_k^{1/2} r b_k + \beta^2 a_k (c_k(a_k + r) - b_k^2) \right) + q_{11_k}$$

$$b_{k+1} = \frac{1}{a_k + r} \left((r b_k + \beta a_k^{1/2} (c_k(a_k + r) - b_k^2)) \right) + q_{12_k}$$

$$c_{k+1} = \frac{1}{a_k + r} \left(c_k(a_k + r) - b_k^2 \right) + q_{22_k}$$

We solve these numerically. Figure 3 uses default values of $a_o = 25$ (i.e. initial uncertainty in object position is 25 pixels squared), and $b_o = 0$, $c_o = 4$, $r = 1\text{pixel}^2$, $q = 1(\text{pixel}/s)^2$ and $\beta = 0.01 s/\text{pixel}$. These plots show that for various combinations of parameters, the time delay, δ_k, can increase with k (and hence the tracking system fails). In figure 3(b) the values of β are varied. For $\beta > 0.127$, the time delay diverges (and hence only the first few measurements for $\beta = 0.13$ were plotted). The interaction of the various parameters is difficult to observe. From Figure 3 one may be led to believe that, say, increasing the system noise cannot cause the system to fail. In Figure 4 the same parameters are used except that β is increased from $\beta = 0.01$ second/pixel to $\beta = 0.025$ second/pixel. The system diverges for much smaller system noise values. (In figure 3(d), had we increased q sufficiently, the time delay diverges).

5.3 Fixed Time Delay with Resolution Control

A more typical case assumes a fixed delay (typically a multiple of the camera frame rate). In order to have a fixed delay, the covariance must be kept small or, as we demonstrate here, the resolution must be adjusted.

Using the same notation as above, δ_k is constant but r_k depends on the size of p_{11}. If we take β to be the processing time per pixel, and $\delta_k = \delta$ is the processing time, then δ/β is the number of pixels that can be processed in the given time interval. If $4\sqrt{p_{11}}$ is the window size (max number of pixels) then we can only

---

[2] Note that p_{22} represents the velocity variance, and p_{12} is the covariance term between position and velocity. The search window depends only on the current value of the positional uncertainty p_{11}. The other terms will affect the growth of this term, however in determining the search window only the value of p_{11} is considered.

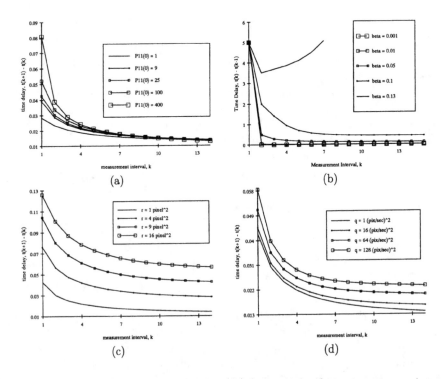

Fig. 3. Graphs showing the time delay (δ_k) for variation in system parameters. The plots show the effect of increasing (a) initial positional covariance, P_o, (b) computation time, β, (c) measurement noise, r, and (d) system noise, q. The default values for parameters used are $P_{11}(0) = 25 pixel^2$, $R = 1 pixel^2$, $Q = 1(pixel/sec)^2$, and $\beta = 0.01 sec/pixel$.

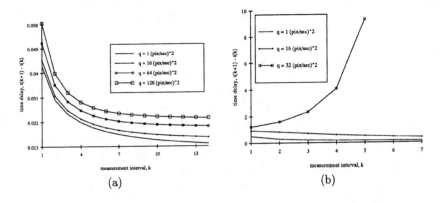

Fig. 4. The time delay (δ_k) diverges at much lower system noise levels when the computation time increases. (a) For $\beta = 0.01$, the time delay did not diverge for $q < 512(pix/s)^2$ (b) $\beta = 0.025$, the time delay diverges for much smaller q.

check every pixel if

$$4\sqrt{p_{11}} < \frac{\delta}{\beta}$$

If this condition does not hold then we must process at a lower resolution (sub-sample). We find n such that

$$\frac{1}{2^n} 4\sqrt{p_{11}} < \frac{\delta}{\beta} \tag{9}$$

i.e.

$$n = \lceil \log_2 \frac{4\beta\sqrt{p_{11}}}{\delta} \rceil \tag{10}$$

where $\lceil\ \rceil$ denotes the ceiling function and we adjust r accordingly (will increase by a factor of 2^n). Note that we could have used an arbitrary scaling factor instead of $\frac{1}{2^n}$, however vision hardware often supports sub-sampling by factors of 2.

Using the above criteria (equation (9)) to set the value of r_k, one must then solve the difference equations (8):

$$a_{k+1} = \frac{1}{a_k + r_k} \left(r_k a_k + 2\delta r_k b_k + \delta^2 (c_k(a_k + r_k) - b_k^2) \right) + q_{11_k}$$

$$b_{k+1} = \frac{1}{a_k + r_k} \left(r_k b_k + \delta \left(c_k(a_k + r_k) - b_k^2 \right) \right) + q_{12_k}$$

$$c_{k+1} = \frac{1}{a_k + r_k} \left(c_k(a_k + r_k) - b_k^2 \right) + q_{22_k}$$

Figure 3 showed various divergent behavior for certain combinations of parameters. Given a fixed delay of $0.05s$, these unstable systems can be stabilized by controlling the resolution of the ROI. The positional covariance (and hence the time delay) in figure 3 (b) and figure 4 (b) diverged. Figure 5 shows these same dynamical system parameters, however the resolution is determined as described above. The positional covariance converges to steady state.

6   Linear Dynamics - Simple Experiments

In this section rather than solve the difference equations for the positional covariance, we perform resolution control during tracking: that is, we actively modify the resolution of the image processed to ensure that the time delay does not diverge. To utilize the methods developed, we consider simple systems that can again be described by linear time-invariant dynamical systems. Suppose we have a circular track on the ground plane described by the path

$$p(\theta(t)) = (r\cos(\theta(t)), r\sin(\theta(t))) = (r\cos(\omega t + \theta_o), r\sin(\omega t + \theta_o)$$

with $p_o = (r, 0)$ and where $\theta(t) = \omega t + \theta_o$.

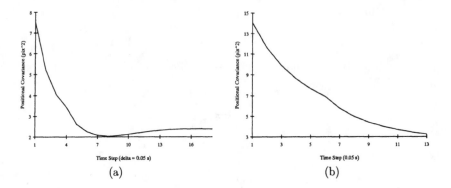

Fig. 5. (a) The parameters used in figure 4(b) caused the covariance (and hence the time delay) to diverge. For fixed time step, using the same parameters but incorporating resolution control we show a convergent positional covariance. The resolution (not shown) is on the order of 8-16 $pixel^2$. (b) For $\beta = 0.13$, tracking failed for the default parameters and fixed resolution. Here we use these same system parameters with resolution control and show a convergent positional covariance.

If we are viewing the motion with an **overhead camera**, the image of the track will also be a circular path[3]. We can view all of the track within the field of view and the path in the image is a circle centered at pixel (u_c, v_c) with radius ρ. The image path is described by

$$(u(t), v(t)) = (u_c + \rho \cos \phi(t), v_c + \rho \sin \phi(t))$$

where

$$\theta(t) = \phi(t) + \phi_o$$

(the angular measure in the image and on the circular path may be offset by a fixed amount ϕ_o). The target centroid is located at pixel $(\hat{u}, \hat{v})$ Assuming that the actual target location is on the path, and we project this detected location to the closest path point, we have a measurement of ϕ of the form:

$$\hat{\phi} = \tan^{-1}\left(\frac{\hat{v} - v_c}{\hat{u} - u_c}\right)$$

The accuracy of the estimated angle depends on the accuracy of the calibrated terms (center of path), and on the accuracy of the extracted target location $(\hat{u}, \hat{v})$ A sensitivity analysis of the computation of $\hat{\phi}$ yields

$$\delta\hat{\phi} = \frac{(\hat{u} - u_c)d(\hat{v} - v_c) - (\hat{v} - v_c)d(\hat{u} - u_c)}{(\hat{v} - v_c)^2 + (\hat{u} - u_c)^2}$$

---

[3] We comment later on how to relax this assumption

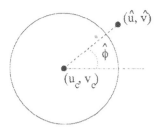

Fig. 6. Measurement of target position in the image. Example of object tracked while moving along a circular path.

and assuming.that the center has been calibrated with high accuracy

$$\delta\hat{\phi} = \frac{1}{(\hat{v} - v_c)^2 + (\hat{u} - u_c)^2} \left((\hat{u} - u_c)\delta\hat{v} - (\hat{v} - v_c)\delta\hat{u} \right).$$

If the target is close to the path then $(\hat{v} - v_c)^2 + (\hat{u} - u_c)^2 \approx \rho^2$ so we have

$$\delta\hat{\phi} = \frac{1}{\rho^2} \left((\hat{u} - u_c)\delta\hat{v} - (\hat{v} - v_c)\delta\hat{u} \right)$$

Assuming a Gaussian distribution for the target location with $\mathcal{E}[\hat{u}] = u_T$, $\mathcal{E}[\hat{v}] = v_T$ where (u_T, v_T) is the actual target location in the image. $\mathcal{E}[\delta\hat{u}^2] = \sigma_u^2$ and $\mathcal{E}[\delta\hat{v}^2] = \sigma_v^2$ depend on the target geometry and image resolution. For symmetric targets and image sampling we can assume $\sigma_u^2 = \sigma_v^2 = \sigma^2$ (i.e. the uncertainty in image target centroid computation) and thus the covariance of the measurement is given by

$$\sigma_\phi^2 = \mathcal{E}[\delta\hat{\phi}^2] = \frac{1}{\rho^4} \left((u_T - u_c)^2\sigma_v^2 + (v_T - v_c)^2\sigma_u^2 \right)$$

$$= \frac{1}{\rho^4} \left((u_T - u_c)^2 + (v_T - v_c)^2 \right) \sigma^2 = \frac{\sigma^2}{\rho^2}$$

From this we see that it is desirable to have the image of the circular path as large as possible in the field of view (maximize ρ). Note, in practice, ϕ is function of the ratio of two Gaussian random variables. The true distribution will be a Cauchy distribution [7,9]. The Gaussian assumption for errors in ϕ is only approximate.

The examples given in figure 7 show a system with slow processing. The time delays increase and eventually the tracking is unstable. Figure 8 shows tracking of the same dynamic system using fixed time intervals and resolution control. By sub-sampling the ROI the resolution of processing is controlled to stabilize the tracking. Thus in order to process the images as they arrive (say at 30 Hz) the either the ROI size must be adjusted (risking tracking failure) or the ROI must be sub-sampled. Figure 9 shows the resulting positional covariance for two different measurement resolutions. Lower resolution can facilitate or stabilize tracking, at the loss of accuracy of the tracking system.

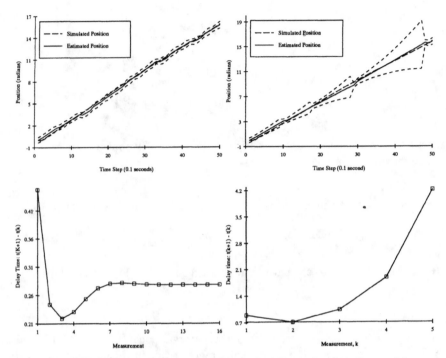

Fig. 7. Estimated position of target with dashed lines showing the position co-variance. Right hand side shows doubled processing time. (Initial speed is 33 RPM, system noise is $N[0, 0.04rad/s^2]$).

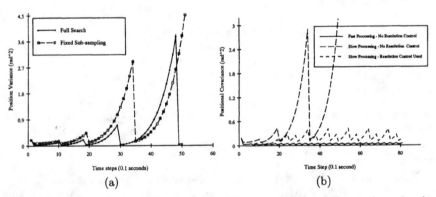

Fig. 8. (a) The covariance for target position shown for full search and sub-sampling. Even with sub-sampling the covariance can grow monotonically. (b) Graph comparing target position covariance for a fast algorithm, a slow algorithm with no resolution control and a slow algorithm using resolution control. Notice that the slow algorithm with resolution control reaches a "steady state" (it oscillates between two covariance values). The "steady-state" covariance is larger than the equivalent dynamic system with fast processing.

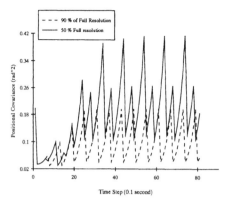

Fig. 9. Comparison of resolution control for two different sub-sampling factors. The dashed line shows tracking at close to full resolution, the solid line shows tracking with 50% resolution

In steady state in both figure 8 and 9, the positional covariance oscillates between two values. The size of the covariance determines the resolution to use. For particular values of system parameters, the resolution factor n in equation (10) can jump between two values.

6.1 Extensions to other Motion

The examples presented were simple to demonstrate feasibility of actively controlling resolution. There are some simple extensions to the dynamic systems presented.

It is simple to extend the circular motion to a non-fronto parallel camera. If the camera is not directly overhead, but views the ground plane at an oblique angle, the image of the circular path will be an ellipse centered at pixel (u_c, v_c) with major and minor axes of length a and b. On rotated coordinates we have

$$(u'(t), v'(t)) = (u'_c + a \cos \phi(t), v'_c + b \sin \phi(t))$$

which in the image are

$$\begin{bmatrix} u(t) \\ v(t) \end{bmatrix} = \begin{bmatrix} \cos \alpha & -\sin \alpha \\ \sin \alpha & \cos \alpha \end{bmatrix} \begin{bmatrix} u'(t) \\ v'(t) \end{bmatrix}$$

The target centroid is located at pixel $(\hat{u}, \hat{v})$ Assuming that the actual target location is on the path, and we project this detected location to the closest path point, we have a measurement of ϕ of the form:

$$\hat{\phi} = \tan^{-1} \left(\frac{a(\hat{v} - v_c)}{b(\hat{u} - u_c)} \right)$$

and
$$\theta(t) = \phi(t) + \phi_o$$

The measured angle provides an estimate of the actual angle given that the path has been calibrated within the image.

Viewing motion on known paths For object motion along known paths, often a linear time-invariant dynamical system can be used to describe the motion (for example, vehicle motion along roads). The path can be determined *a priori* by calibration or from known "maps" (for motion along roads). A special case is motion on the ground plane. The mapping of the ground plane to the image plane is a collineation (or projectivity) In homogeneous coordinates, points on the ground plane $(x, y, 1)$ are mapped to points on the image (u, v, s) by a 3x3 matrix, known up to a scale [5]. The coefficients of this matrix can be determined using calibration techniques. The path can be described by a curve (parameterized by arc-length) $p(s) = [x(s), y(s)]^T$, we can formulate the *state* as $S(t) = [s(t), \dot{s}(t), \ddot{s}(t)]^T$ and then dynamics (up to 2nd order) along the path are thus described by the simple linear system used earlier (2nd order integrator).

Thus if the path geometry is known, or available from prior computation or calibration, the dynamic system formulation of motion along that path can still be described by a linear system. The measurement, however, involves a non-linear mapping from the path to the image.

7   Discussion

While we have demonstrated that active control of resolution can yield stable tracking performance, the situations where we have applied it are very simple. Generally some form of calibration must be performed to enable the motion to be described by a linear system. We monitor the size of the covariance and/or the delay time to assess the performance of the tracking system (and determine whether to employ resolution control). Other metrics, such as the Fisher Information matrix are being considered for possible use in assessing tracking performance. Without control of the ROI size (and resolution of processing within the ROI), one cannot guarantee stability of the tracking systems. Previous tracking systems have intuitively followed these bounds. We have tried to quantify the relationships between the system parameters and tracking performance. Future work aims to provide a mechanism to evaluate tracking performance online. Monitoring either the condition number of the information matrix or the sequence of delay times between measurements allows the tracker to perform on-line modification of resolution to maintain stability. Currently the positional covariance size is used to instantiate resolution control. This method works as long as the modified resolution does not effect the target recognition process. We implicitly assumed that the target was still recognizable at the computed resolution. If the resolution employed is "larger" than the target, tracking will fail.

We have presented dynamical systems which can be described with simple linear models. Certain examples, (e.g. motion of vehicles along curvilinear paths) produce motion which is highly non-linear in the image, yet the tracking can be formulated with a linear system. While the linear dynamic system may seem restrictive, the ability to separate path geometry from path dynamics enables tracking of complex motion such as vehicles on known roads and landmark tracking for navigation. Many tracking applications can fit within this framework.

References

1. P. Anandan. A computational framework and an algorithm for measurement of visual motion. *International Journal of Computer Vision*, 2:283–310, 1989.
2. A. Blake, R. Curwen, and A. Zisserman. A framework for spatio-temporal control in the tracking of visual contours. *International Journal of Computer Vision*, 1993.
3. J. Clark and N. Ferrier. Modal control of visual attention. In *Proc. of the Int'l Conf. on Computer Vision*, pages 514–531, Tarpon Springs, Florida, 1988.
4. A. Cretual and F. Chaumette. Image-based visual servoing by integration of dynamic measurements. In *International Conf. on Robotics and Automation*, pages 1994–2001, 1998.
5. O. Faugeras. *Three Dimensional Vision*. MIT Press, 1993.
6. N. Ferrier and J. Clark. The Harvard Binocular Head. *International Journal of Pattern Recognition and AI*, pages 9–32, March 1993.
7. E. Fieller. The distribution of the index in a normal bivariate population. *Biometrika*, 24:428–440, 1932.
8. J. D. Foley and A. van Dam. *Fundamentals of interactive computer graphics*. Addison-Wesley, 1982.
9. R. Geary. The frequency distribution of the quotient of two normal variables. *Royal Statistical Society Series A*, 93:442–446, 1930.
10. A. Gelb and the Technical Staff, Analytic Sciences Corporation. *Applied optimal estimation*. MIT Press, Cambridge, MA, 1974.
11. A. S. Glassner, editor. *Graphics gems (Volumes I-V)*. Academic Press, 1990-1995.
12. C. Harris and C. Stennett. Rapid – a video-rate object tracker. In *Proc. 1st British Machine Vision Conference*, pages 73–78, 1990.
13. A. Jazwinski. *Stochastic processes and filtering theory*. Academic Press, 1970.
14. R. Mehra. Optimization of measurement schedules and sensor designs for linear dynam ic systems. *IEEE Transactions on Automatic Control*, 21(1):55–64, 1976.
15. B. Nelson, N. Papanikolopoulos, and P. Khosla. Visual servoing for robotic assembly. In *Visual Servoing-Real-Time Control of Robot Manipulators Based on Visual Sensory Feedback*, pages 139–164. World Scientific Press, 1993.
16. K. Nichols and S. Hutchinson. Weighting observations: the use of kinematic models in object tracking. In *Proc. 1998 IEEE International Conf. on Robotics and Automation*, 1998.
17. D. Okhotsimksy, A.K.Platonov, I. Belousov, A. Boguslavsky, S. Emelianov, V. Sazonov, and S. Sokolov. Real time hand-eye system: Interaction with moving objects. In *International Conf. on Robotics and Automation*, pages 1683–1688, 1998.
18. C. Olivier. Real-time observatiblity of targets with constrained processing power. *IEEE Transactions on Automatic Control*, 41(5):689–701, 1996.

19. N. Papnikolopoulos, P. Khosla, and T. Kanade. Visual tracking of a moving target by a camera mounted on a robot: A comp bination of control and vision. *IEEE Transactions on Robotics and Automation*, 9(1), 1993.
20. J. Roberts and D. Charnley. Parallel attentive visual tracking. *Engineering Applications of Artificial Intelligence*, 7(2):205–215, 1994.
21. T. Schnackertz and R. Grupen. A control basis for visual servoing tasks. In *Proc. 1995 IEEE Conf. on Robotics and Automation*, Nagoya, Japan, 1995.
22. G. Sullivan. Visual interpretation of known objects in constrained scenes. *Phil. Trans. R. Soc. Lond. B.*, B(337):109–118, 1992.
23. M. Vincze and C. Weiman. On optimising tracking performance for visual servoing. In *International Conf. on Robotics and Automation*, pages 2856–2861, 1997.

A. 1D Analytic Results

We present a 1D example analytically. The following example follows closely from that given in [18]). The one dimensional linear system:

$$dx = axdt + dw \; ; \; E[dw^2] = \sigma^2 dt \tag{11}$$

$$y(t_k) = x(t_k) + v \; ; \; v(t_k) \sim N[0, R_k] \tag{12}$$

For this scalar system equation 6 is

$$P(t_{k+1}^-) = e^{2a(t_{k+1}-t_k)}\frac{P(t_k^-)R_k}{P(t_k^-) + R_k} + \frac{\sigma^2}{2a}\left(e^{2a(t_{k+1}-t_k)} - 1\right) \tag{13}$$

This general form of $P(t_{k+1}^-)$ can be shown to grow for particular dynamics and measurement noise. Note that if

$$t_{k+1} - t_k > \frac{1}{2a}\log\left(\frac{P+R}{R}\right)$$

(where we abbreviate $P = P(t_k^+)$ and $R = R(t_k)$) then

$$\begin{aligned}
P(t_{k+1}^-) &> \left(\frac{P+R}{R}\right)\frac{PR}{P+R} + \frac{\sigma^2}{2a}\left(\frac{PR}{R} - 1\right) \\
&= P(t_k^-)\left(1 + \frac{\sigma^2}{2aR}\right) - \frac{\sigma^2}{2a} \\
&= P(t_k^-)\left(1 + \frac{\sigma^2}{4aR}\right) + P(t_k^-)\frac{\sigma^2}{4aR} - \frac{\sigma^2}{2a} \\
&> P(t_k^-)\left(1 + \frac{\sigma^2}{4aR}\right) \quad \text{for } P(t_k^-) > \max[2R, a]
\end{aligned}$$

Hence the covariance grows at a rate greater than unity.

If we choose the resolution to be some fraction of the covariance, $R = \alpha P$, where $\alpha << 1$ then equation 13 becomes

$$P(t_{k+1}^-) = e^{2a(t_{k+1}-t_k)}\frac{\alpha}{1+\alpha}P(t_k^-) + \frac{\sigma^2}{2a}\left(e^{2a(t_{k+1}-t_k)} - 1\right)$$

and the covariance evolves as a geometric sequence

$$P(t_{k+1}^-) = \gamma P(t_k^-) + \zeta.$$

For $\gamma < 1$ the covariance decreases and hence the target can be tracked. The expression

$$\gamma < 1 \qquad \text{or} \qquad e^{2a(t_{k+1}-t_k)}\frac{\alpha}{1+\alpha} \; < \; 1$$

makes explicit the trade off between the sample interval $(t_{k+1}-t_k)$, the resolution sub-sampling α, and the system dynamics, a. If the system is sufficiently slow for the time scale used then we can pick a resolution to ensure tracking.

Near-Optimal Sensor-Based Navigation in an Environment Cluttered with Simple Shapes

Hiroshi Noborio and Kenji Urakawa

Division of Information and Computer Science, Graduate School of Engineering
Osaka Electro-Communication University
Hatsu-Cho 18-8, Neyagawa, Osaka 572-8530, Japan

Abstract. For the last decade, many sensor-based path-planning algorithms have been proposed. In all the algorithms, the convergence of a mobile robot to a destination is theoretically ensured if a path to the destination exists. However, due to no information of obstacle shape and location, a mobile robot basically takes a very long path to its destination. To overcome this drawback in this paper, we consider how a mobile robot selects its direction to follow when encountering an obstacle. Then, the result is as follows: a mobile robot should select a tangential direction to avoid a circular obstacle, whose absolute angle against the goal directions is smaller. The obstacle avoidance procedure is very simple. Therefore, by adding it into good classic sensor-based path-planning algorithms $Class1$ and $Bug2$, we get near-optimal algorithms $Simple(Class1)$ and $Simple(Bug2)$. In this paper, we firstly ascertain that $Simple(Class1)$ and $Simple(Bug2)$ always select slightly longer paths than the path generated by the optimal (model-based) path-planning algorithm. Secondly, we describe that $Simple(Class1)$ and $Simple(Bug2)$ always select extremely shorter paths than paths generated by $Class1$ and $Bug2$. The near-optimality and the superiority are given by theoretical proofs in an uncertain 2-D environment with circular obstacles. Thirdly, we theoretically investigate whether the near-optimality and the superiority are still kept or not in an uncertain environment with square, rectangular, elliptic, or triangular obstacles. Finally based on simulation and experiment results, we conclude that $Simple(Class1)$ and $Simple(Bug2)$ keep the near-optimality and the superiority in an environment with simple shapes as obstacles.

1 Introduction

In order for a mobile robot to select a near-optimal deadlock-free path to its destination in an unknown 2-D environment with simple shapes as obstacles, we propose new sensor-based navigation algorithms in this paper. From the viewpoint of path length, we ascertain a near-optimality for these new sensor-based navigation algorithms by theoretical proofs and simulation and experimental results.

In the sensor-based navigation, a mobile robot aims at its destination by position-feedback control while avoiding uncertain obstacles by sensor-feedback

Christensen et al. (Eds.): Sensor Based Intelligent Robots, LNAI 1724, pp. 160–179, 1999.
© Springer-Verlag Berlin Heidelberg 1999

control. The convergence of a mobile robot to its destination is exactly ensured by metric, geometric, or topologic characteristics [1]. Because of no information of obstacle shape and location, all the classic sensor-based navigation algorithms sometimes select very long paths until the destination is reached. To overcome this, we are recently developing two types of sensor-based navigation algorithms running in unknown environments with simple and complex shapes as obstacles. In an uncertain environment with complicated shapes as obstacles, if a mobile robot avoids an encountered obstacle by changing its direction alternatively, the robot can select a shorter deadlock-free path until a destination is reached [2]. The superiority is evaluated in terms of probabilities. Needless to say, the convergence is ensured by the metric characteristic.

As contrasted with this, in this paper, we propose another sensor-based navigation algorithm for selecting a near-optimal (very short) path to a destination in a 2-D environment with simple shaped obstacles, e.g. circle, square, rectangle, ellipse, and so on. A new algorithm is obtained if we add a new rule concerning obstacle avoidance into its classic algorithm. The rule is as follows: a robot selects a tangential direction V_T at a hit point H_i, whose angle against its goal direction V_G is smaller. In Fig.1, a mobile robot follows an encountered obstacle in the counter-clockwise order by the rule. The rule is very simple, but is very powerful to pick up a near-optimal (very short) path from an unknown 2-D environment with simple shapes as obstacles. In addition, because of the simplicity, the rule can be easily added into all kinds of sensor-based navigation algorithms, i.e., metric, geometric, and topologic algorithms [1]. The metric algorithms supervise a mobile robot directly by the Euclidean distance toward a destination. For this reason, they select relatively shorter paths. Especially, the algorithms $Class1$ and $Bug2$ select shorter paths within an unknown 2-D environment [3],[4]. Therefore, we designed two new algorithms $Simple(Class1)$ and $Simple(Bug2)$ by adding the new rule into the classic algorithms $Class1$ and $Bug2$.

Firstly, we determine a competitive ratio r_1=(Length of a path selected by $Simple(Class1)$)/(The shortest path length selected by the model-based path-planning), and also determine a worst ratio r_2=(Path length selected by $Class1$)/(Path length selected by $Simple(Class1)$). Also, we determine a competitive ratio r_1=(Path length selected by $Simple(Bug2)$)/(The shortest path length selected by the model-based path-planning), and also determine a worst ratio r_2=(Path length selected by $Bug2$)/(Path length selected by $Simple(Bug2)$). Because a competitive ratio r_1 is mathematically bounded by a small finite number, the new algorithms are regarded as near-optimal algorithms. On the other hand, since the worst ratio r_2 is mathematically given by infinite or a large finite number, the new algorithms $Simple(Class1)$ and $Simple(Bug2)$ are extremely improved from the classic algorithms $Class1$ and $Bug2$.

Secondly, we discuss whether the near-optimality of $Simple(Bug2)$ against the model-based path-planning algorithm is kept or not for an uncertain environment with square, rectangular, elliptic, or triangular obstacles. In addition to this, we check whether the superiority of $Simple(Bug2)$ against the clas-

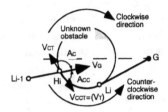

Fig. 1. A mobile robot selects a tangential direction V_T whose angle against its goal direction V_G is smaller.

sic algorithm $Bug2$ is maintained or not for the environment. Thirdly, we address simulation results so as to see the near-optimality in $Simple(Class1)$ and also simulation and experiment results in order to conclude the superiority in $Simple(Class1)$ on the average.

Finally, two sensor-based navigation algorithms for selecting a near-optimal path were recently proposed [5],[6]. In the former paper, a proposed algorithm is designed under the Hamilton-Jacob-Bellman (HJB) theory, which is similar to the artificial potential method. The authors said the HJB algorithm is to be an on-line navigation algorithm. However in section IV of the paper, the algorithm recursively changes a part of a deadlock-free path among multiple obstacles. This means that a robot knows the shape and location of such obstacles locally beforehand. Furthermore in the HJB algorithm, a selected path is regarded as a near-optimal path only in an environment with circular obstacles. Nevertheless, a competitive ratio between the optimal and selected paths was not exactly evaluated. The reason is that a path is implicitly selected by the HJB equation. In the latter paper, the authors propose the same rule to select a direction following an encountered obstacle. But it is just an idea, and they also did not evaluate a competitive ratio against the optimal (shortest) path.

The present paper is organized as follows: Section 2 defines a mobile robot, a sensor, and obstacles in an uncertain 2-D environment. In section 3, we explain near-optimal sensor-based path-planning algorithms $Simple(Class1)$ and $Simple(Bug2)$. In sections 4 and 5, we theoretically describe how much the proposed algorithms are worse than the optimal algorithm. Also, we theoretically show how much the proposed algorithms are better than the classic algorithms. In section 6, we discuss whether the same property can be maintained or not for an uncertain environment with square, rectangular, elliptic, or triangular obstacles. In section 7, we show the same tendency by averaging several simulation and experiment results. Finally, section 8 presents a few concluding remarks.

2   A Mobile Robot and Its 2-D Unknown Environment

In this section, we describe a robot with a sensor in an unknown 2-D environment. Firstly, a robot expressed as a point, is able to move omnidirectionally in the environment. A mobile robot can recognize its position exactly by GPS (Global Positioning System), and therefore aims at its destination by position-feedback control. Also, a mobile robot can identify the normal vector of its surrounding obstacle, and consequently is able to follow a tangential direction around an encountered obstacle by sensor-feedback control. In the environment, all the uncertain obstacles are cluttered (Fig.2). The number of obstacles and the perimeter of each obstacle are finite.

Fig. 2. An example of a cluttered uncertain 2-D environment.

3   Sensor-Based Algorithms Simple(Class1) and Simple(Bug2)

In this section, we propose sensor-based navigation algorithms $Simple(Class1)$ and $Simple(Bug2)$ for selecting near-optimal paths in an uncertain 2-D environment. They are designed from their classic algorithms $Class1$ and $Bug2$ [3],[4].

The algorithm $Simple(Class1)$ is described as follows:

1. A mobile robot R goes straight to its destination G until one of the following occurs:

(**1a**) If R arrives at G, the algorithm ends.

(**1b**) If R encounters an obstacle, R memorizes its present position as a hit point H_i and a closest point C_i to G, and then we move to step 2.

2. R senses clockwise and counter-clockwise tangential vectors (V_{CT} and V_{CCT}) at H_i. Then, R calculates two absolute angles A_C and A_{CC} from V_{CT} and V_{CCT} to the goal vector V_G, respectively (Fig.1). Finally by considering a smaller angle from two absolute angles A_C and A_{CC}, R selects its direction. In Fig.1, R selects the counter-clockwise order. Then according to the direction, R goes around an encountered obstacle while R seeks a closest point C_i to G until one of the following occurs:

(**2a**) If R arrives at G, the algorithm ends.

(**2b**) If the goal vector V_G is not obstructed by any obstacle and if a present position is closer to G than the closest point C_i, R memorizes its present position as a leave point L_i and we return to step 1.

(**2c**) If R returns to H_i, the algorithm finishes because no path to G exists.

To design the algorithm $Simple(Bug2)$, we replace the procedure **(2b)** by the following:

(2b) If V_G is not obstructed by any obstacle, if R is closer to G than C_i, and if R is located on the segment SG, R memorizes its position as a leave point L_i and we return to step 1.

Both algorithms have the metric condition, i.e., L_i is always closer to G than H_i. Needless to say, if only circular obstacles exist, the property is automatically maintained without the metric condition [7]. As a result, an inequality $|H_iG| > |L_iG|$ is kept. In addition, R always goes straight from L_{i-1} to H_i. Thus, an inequality $|L_{i-1}G| > |H_iG|$ is naturally kept. By these inequalities, the convergence of a mobile robot to its destination is ensured (e.g., $|S(=L_0)G| > |H_1G| > |L_1G| > ... > |L_{i-1}G| > |H_iG| > |L_iG| > ... > |L_nG|$) [3]. Consequently, R directly arrives at G after it leaves the final leave point L_n.

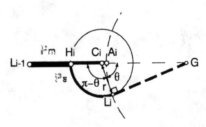

Fig. 3. In an interval between successive leave points L_{i-1} and L_i, a mobile robot decreases a lower boundary of its shortest path P_m by $LP_{mi}(=|L_{i-1}G| - |L_iG|)$, and synchronously decreases its path P_s by the sum LP_{si} of distances of segment $L_{i-1}H_i$ and arc H_iL_i.

4   A Near-Optimality of Simple(Class1)

In this section, for every triplet of start point, goal point, and uncertain environment with circular obstacles, we can evaluate how much a path P_s selected by the new algorithm $Simple(Class1)$ is longer than the optimal path P_m generated in the model-based algorithm. Moreover, we can evaluate how much a path P_n selected by the new algorithm $Simple(Class1)$ is shorter than a path P_c made in the classic algorithm $Class1$.

For this purpose, we can calculate a competitive ratio $r_1=$(Length LP_s of a path P_s selected by $Simple(Class1)$) / (Length LP_m of the shortest path P_m selected by the model-based path-planning), and also we can calculate a worst ratio $r_2=$(Length LP_c of a path P_c selected by $Class1$)/(Length LP_n of a path P_n made in $Simple(Class1)$). Finally, because r_1 is kept small, we see $Simple(Class1)$ has a near-optimal property in a 2-D unknown environment with a simple shape. On the other hand, because r_2 is kept large or is unbounded, we see that $Simple(Class1)$ becomes better than $Class1$.

4.1   A Competitive Ratio $r_1 = LP_s/LP_m$

First of all, we define LP_m as $D(= |SG|)$. This is the lower boundary of the shortest path P_m between S and G. As mentioned previously, R goes straight from L_{i-1} to H_i and avoids an encountered obstacle from H_i to L_i. Therefore, a path P_s is generated by switching the two behaviors in the sensor-based navigation. For this reason, we can focus on each interval between L_{i-1} and L_i. In the interval, length $LP_m(= D)$ of the lower boundary of a path P_m decreases by $LP_{mi}=|L_{i-1}G| - |L_iG|$. On the other hand, length LP_s of a path P_s decreases by the sum LP_{si} of distances of segment $L_{i-1}H_i$ and arc H_iL_i. Therefore in the interval between L_{i-1} and L_i, a competitive ratio $r_{1i} = (|L_{i-1}H_i| +$ arc length $H_iL_i)$ / $(|L_{i-1}G| - |L_iG|) = (|L_{i-1}H_i| +$ arc length $H_iL_i)$ / $(|L_{i-1}H_i| + |H_iA_i|)$ $\leq$ (arc length H_iL_i) / (segment length H_iA_i) $= r(\pi-\theta)/r(1+\frac{1}{cos\theta} - \frac{sin\theta}{cos\theta})$ (Fig.3). The point A_i is defined by the intersection between the segment H_iG and the circle whose radius is L_iG and its center is at G. The competitive ratio r_{1i} is bounded by 1.66 whose θ is 41.80. The detail is described in [8].

As long as a mobile robot goes straight to its destination, the length $|L_{i-1}H_i|$ becomes larger and consequently a competitive ratio $r_{1i} = LP_{si}/LP_{mi}$ becomes smaller. Needless to say, we can locate all sizes of circles in an uncertain environment, and consequently the length $|L_{i-1}H_i|$ can be kept small enough. For this reason, every unknown environment with all sizes of circles can be evaluated by the above discussion.

Moreover, the total ratio r_1 is calculated by $\Sigma_i LP_{si}/\Sigma_i LP_{mi}(\Sigma_i LP_{mi} = D)$. Also, the fractional expression is bounded by the maximum of all ratios $r_{1i}(i = 1, ..., n)$ if and only if each ratio r_{1i} is greater than one and each numerator LP_{si} and denominator LP_{mi} are larger than zero. For this reason, the proposed algorithm $Simple(Class1)$ generates at the most a 1.66 times longer path than the optimal (shortest) path made in the model-based navigation algorithm. Even if D, r, or location of start, goal, and circular obstacles is flexibly changed, a competitive ratio r_1 is fixed by a small constant number of 1.66. As a result, we regard $Simple(Class1)$ as a near-optimal sensor-based (on-line) navigation algorithm.

4.2   A Worst Ratio $r_2 = LP_c/LP_n$

In this paragraph, we show an environment whose worst ratio r_2 becomes continuously larger (Fig.4). In this case, a robot supervised by $Simple(Class1)$ always selects a counter-clockwise direction to avoid an obstacle by the new rule, on the other hand, a robot supervised by $Class1$ sometimes selects a clockwise direction to avoid the obstacle in advance. In this case, $r_2 = LP_c/LP_n$ is bounded by $(D - \epsilon + 2\pi r)/(D + \epsilon)$ because LP_c is bounded by $D - \epsilon + 2\pi r$ and LP_n is defined by $D + \epsilon$. Here, ϵ is defined as the difference between shorter arc H_iL_i and segment H_iL_i. Therefore, even if a radius r of a circular obstacle is constant, the ratio r_2 approaches an infinite number as D converges to zero. On the other hand, even if the distance D is constant, the ratio r_2 approaches an infinite number as r approaches an infinite number.

Fig. 4. A case whose ratio r_2 is maximized.

As a result, a worst ratio r_2 is defined as infinite or a large finite number, which is determined by D and r. For this reason, we can conclude that the new algorithm $Simple(Class1)$ is extremely improved from the classic algorithm $Class1$.

5   A Near-Optimality of Simple(Bug2)

In this section, for every triplet of start point, goal point, and uncertain environment with circular obstacles, we can evaluate how much a path P_s selected by the new algorithm $Simple(Bug2)$ is longer than the optimal path P_m generated in the model-based algorithm. Moreover, we can evaluate how much a path P_n selected by the new algorithm $Simple(Bug2)$ is shorter than a path P_c made in the classic algorithm $Bug2$.

For this purpose, we can calculate a competitive ratio r_1=(Length LP_s of a path P_s selected by $Simple(Bug2)$) / (Length LP_m of the shortest path P_m selected by the model-based path-planning), and also calculate a worst ratio r_2=(Length LP_c of a path P_c selected by $Bug2$)/(Length LP_n of a path P_n made in $Simple(Bug2)$). Finally, because r_1 is kept small, we see $Simple(Bug2)$ has a near-optimal property in a 2-D unknown environment with simple shapes as obstacles. On the other hand, because r_2 is kept large or is unbounded, we see that $Simple(Bug2)$ becomes better than $Bug2$.

5.1   A Competitive Ratio $r_1 = LP_s/LP_m$

First of all, we define LP_m as $D(= |SG|)$. This is the lower boundary of the shortest path P_m between S and G. Secondly, the algorithm $Simple(Bug2)$ has the segment condition as follows: a mobile robot hits and leaves uncertain obstacles only on the segment SG. By mixing the segment condition and the new rule, a mobile robot supervised by $Simple(Bug2)$ avoids its encountered obstacle by at the most $r\pi/2$ while the robot supervised by the model-based navigation decreases the lower boundary D by r. As a result, we indicate a case whose competitive ratio r_1 is maximized (Fig.5).

In this case, all the obstacles are arranged in a straight line on SG. Firstly, between A and B, every competitive ratio $r_{1i} = LP_{si}/LP_{mi} = (r\pi/2)/r(i =$

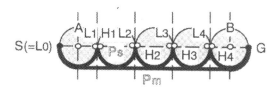

Fig. 5. A case whose ratio r_1 is maximized.

$2, ..., n - 1$) is denoted by $\pi/2$. Secondly, between S and A or B and G, the competitive ratio r_{11} or r_{1n} is one. Here, length of the sum of two intervals SA and BG decreases gradually in the length $D(= |SG|)$ as r decreases. Moreover, the total ratio r_1 is calculated by $\Sigma_i LP_{si}/\Sigma_i LP_{mi}(\Sigma_i LP_{mi} = D)$. Also, the fractional expression is bounded by the maximum of all ratios $r_{1i}(i = 1, ..., n)$ if and only if each ratio r_{1i} is greater than one and each numerator LP_{si} and denominator LP_{mi} are larger than zero. As a result, the competitive ratio r_1 is strictly bounded by $\pi/2$. The ratio r_1 is independent of the radius r of an obstacle and the location of start, goal, and circular obstacles.

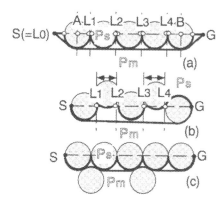

Fig. 6. (a) If a free space in SG exists, the ratio r_1 decreases. (b) If obstacles do not form a straight line, the ratio r_1 decreases. (c) The straight sequence of obstacles is pushed by other obstacles, the ratio r_1 decreases.

If a free space in SG exists, length of the sum of two intervals SA and BG increases in the length $D(= |SG|)$. Consequently, the ratio r_1 decreases (Fig.6(a)). Furthermore in the case where the obstacles do not form a straight line, there are intervals whose r_{1i} are extremely smaller than $\pi/2$ (Fig.6(b)). Moreover in case that the straight sequence of obstacles is pushed by other obstacles, only a denominator of r_1 increases and consequently the ratio r_1 decreases (Fig.6(c)). For this reason, we need not consider the other cases.

As a result, the proposed algorithm $Simple(Bug2)$ generates at the most a $\pi/2$ times longer path than the optimal (shortest) path made in the model-based navigation algorithm. For this reason, we regard $Simple(Bug2)$ as a near-optimal sensor-based (on-line) navigation algorithm.

5.2   A Worst Ratio $r_2 = LP_c/LP_n$

In this paragraph, we will show an environment whose worst ratio r_2 becomes larger continuously (Fig.7). In this case, a robot supervised by $Simple(Bug2)$ always selects a counter-clockwise direction to avoid an obstacle by the new rule, on the other hand, a robot supervised by $Bug2$ sometimes selects a clockwise direction to avoid the obstacle in advance. In this case, $r_2 = LP_c/LP_n$ is bounded by $(D - \epsilon + 2\pi r)/(D + \epsilon)$ because LP_c is bounded by $D - \epsilon + 2\pi r$ and LP_n is defined by $D + \epsilon$. Here, ϵ is defined as the difference between the shorter arc $H_i L_i$ and segment $H_i L_i$. Therefore, even if a radius r of a circular obstacle is constant, the ratio r_2 approaches an infinite number as D converges to zero. On the other hand, even if the distance D is constant, the ratio r_2 approaches an infinite number as r approaches an infinite number.

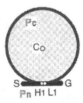

Fig. 7. A case whose ratio r_2 is maximized.

As a result, a worst ratio r_2 is defined as infinite or a large finite number, which is determined by D and r. For this reason, we see that the new algorithm $Simple(Bug2)$ is extremely improved from the classic algorithm $Bug2$.

6   Discussion for Several Kinds of Obstacle Shapes

In this section, we consider what kind of obstacle shape keeps the near-optimality and the superiority explained in the last section. For this purpose, we test an uncertain environment with square, rectangular, elliptic, or triangular obstacles in the algorithm $Simple(Bug2)$.

First of all, if we pick up a pair of start and goal points in an uncertain environment with square obstacles, we can find cases whose r_1 and r_2 are maximized

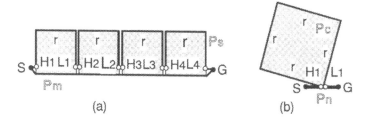

Fig. 8. (a) A case whose ratio r_1 is maximized in an uncertain environment with square obstacles. (b) A case whose ratio r_2 is maximized in an uncertain environment with square obstacles.

in Fig.8(a) and (b), respectively. If an edge length is denoted by r, a competitive ratio $r_1 = LP_s/LP_m < 3r*(D/r)/D$ is given by 3 (Fig.8(a)). As contrasted with this, a worst ratio $r_2 = LP_c/LP_n$ is bounded by $(4r+D-\epsilon)/(D+\epsilon) < 1+(4r/D)$ (Fig.8(b)). Here, ϵ is defined as the difference between the turn H_1L_1 and segment H_1L_1.

Even though D, r, or location of start, goal, and obstacles is flexibly changed, a competitive ratio r_1 is fixed by three. As a result, for every triplet of start point, goal point, and uncertain environment with square obstacles, the proposed algorithm $Simple(Bug2)$ generates at the most a 3 times longer path than the optimal (shortest) path made in the model-based navigation algorithm. For this reason, we can conclude that the optimality of $Simple(Bug2)$ against the optimal (model-based) navigation algorithm is maintained. As contrasted with this, if the inequality $D \leq r$ is satisfied, a worst ratio r_2 is always larger than one and consequently the superiority of $Simple(Bug2)$ against $Bug2$ is kept (Fig.9(a)). Especially, if D converges to zero or r approaches an infinite number, r_2 increases infinitely. As a result, for a triplet of start point, goal point, and uncertain environment with square obstacles, the new algorithm $Simple(Bug2)$ always selects an extremely shorter path than a path made in the classic algorithm $Bug2$. For this reason, we can conclude that the superiority of $Simple(Bug2)$ against $Bug2$ is maintained. Otherwise, that is, if the inequality $D > r$ is satisfied, r_2 is sometimes smaller than one and consequently $Simple(Bug2)$ sometimes selects a longer path than a path made in the classic algorithm $Bug2$ (Fig.9(a)). As a result, the characteristics described in the previous section are partially kept.

Secondly, if we select a pair of start and goal points in an uncertain environment with rectangular obstacles, we can find cases whose r_1 and r_2 are maximized in Fig.10(a) and (b), respectively. If lengths of short and long sides are denoted by l_1 and l_2, a competitive ratio $r_1 = LP_s/LP_m < (2l_2+l_1)*(D/l_1)/D$ is bounded by $1 + 2l_2/l_1$ (Fig.10(a)). As contrasted with this, a worst ratio $r_2 = LP_c/LP_n$ is bounded as the maximum of $(2(l_1+l_2)+D-\epsilon)/(D+\epsilon) < 1+2(l_1+l_2)/D$ and $1 + 2l_2/l_1$ (Fig.10(b)). Here, ϵ is defined as the difference between the turn H_iL_i and segment H_iL_i.

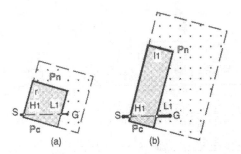

Fig. 9. (a) If $D > r$ is satisfied, r_2 is not always larger than one, otherwise, it is always larger than one around a square obstacle. (b) If $D > l_1$ is satisfied, r_2 is not always larger than one, otherwise, it is always larger than one around a rectangular obstacle.

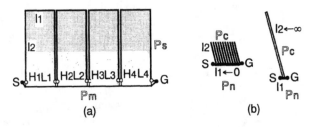

Fig. 10. (a) A case whose ratio r_1 is maximized in an uncertain environment with rectangular obstacles. (b) A case whose ratio r_2 is maximized in an uncertain environment with rectangular obstacles.

Therefore, if l_2/l_1 is constant, r_1 is determined as a small number and consequently the optimality of $Simple(Bug2)$ against the optimal (model-based) navigation algorithm is maintained. Otherwise, for example, if l_2 approaches an infinite number or l_1 converges to zero, r_1 approaches an infinite number and consequently the optimality is lost. On the other hand, if the inequality $D \leq l_1$ is satisfied, r_2 is always larger than one (Fig.9(b)). Especially if D converges to zero or l_1 approaches an infinite number, the worst ratio r_2 increases infinitely. Therefore, the superiority of $Simple(Bug2)$ against $Bug2$ is maintained. Otherwise, that is, if the inequality $D > l_1$ is satisfied, r_2 is sometimes less than one, i.e., $0 < r_2 < 1$, and consequently the superiority of $Simple(Bug2)$ against $Bug2$ is lost (Fig.9(b)). As a result, the characteristics described in the previous section are partially kept.

Thirdly, if we pick up a pair of start and goal points in an uncertain environment with elliptic obstacles, we can find cases whose r_1 and r_2 are maximized in Fig.11(a) and (b), respectively. The perimeter of each ellipse is denoted by P, and lengths of minor and major axes are denoted by l_1 and l_2. P is determined as a

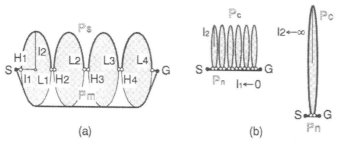

(a)   (b)

Fig. 11. (a) A case whose ratio r_1 is maximized in an uncertain environment with elliptic obstacles. (b) A case whose ratio r_2 is maximized in an uncertain environment with elliptic obstacles.

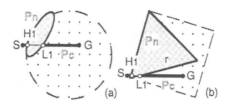

Fig. 12. (a) The superiority of $Simple(Bug2)$ against $Bug2$ is not always maintained around an elliptic obstacle $(0 < r_2 < 1)$. (b) That is not always kept around a triangular obstacle $(0 < r_2 < 1)$.

function of l_1 and l_2. In this case, a competitive ratio $r_1 = LP_s/LP_m$ is bounded by $(P/2)(D/2l_1)/D = P/4l_1$ (Fig.11(a)). As contrasted with this, a worst ratio $r_2 = LP_c/LP_n$ is bounded by the maximum of $(P+D-\epsilon)/(D+\epsilon) < 1+P/D$ and $P(D/2l_1)/D = P/2l_1$ (Fig.11(b)). Here, ϵ is defined as the difference between the shorter arc H_iL_i and segment H_iL_i.

Therefore, if P/l_1 is constant, r_1 is determined as a small number and consequently the optimality of $Simple(Bug2)$ against the optimal (model-based) navigation algorithm is maintained. Otherwise, for example, if l_2 approaches an infinite number or l_1 converges to zero, r_1 approaches an infinite number and consequently the optimality is not kept. On the other hand, r_2 is not always larger than one, i.e., $0 < r_2 < 1$, on any condition concerning to D, l_1, l_2, or location of start, goal, and obstacles (Fig.12(a)). In this figure, in contrast to a circular obstacle, $Simple(Bug2)$ sometimes selects a longer path than a path made in $Bug2$ around an elliptic obstacle. Consequently, the superiority of $Simple(Bug2)$ against $Bug2$ is not kept. As a result, the characteristics described in the previous section are almost lost.

Finally, if we determine a pair of start and goal points in an uncertain environment with triangular obstacles, we can find a case whose r_1 and r_2 are

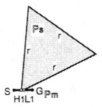

Fig. 13. A case whose r_1 and r_2 are maximized in an uncertain environment with triangular obstacles.

Table 1. A comparison of the near-optimality and the superiority in an uncertain environment with circle, square, rectangle, ellipse, or triangle obstacles (α and β are denoted as small finite numbers).

Circle	Square	Rectangle	Ellipse	Triangle
$1.58 > r_1 > 1$	$3 > r_1 > 1$	$Partly\alpha > r_1 > 1$	$Par.\beta > r_1 > 1$	$\infty > r_1 > 1$
$\infty > r_2 > 1$	$Par.\infty > r_2 > 1$	$Par.\infty > r_2 > 1$	$\infty > r_2 > 0$	$\infty > r_2 > 0$

maximized in Fig.13. In this case, $r_1 = LP_s/LP_m$ and $r_2 = LP_c/LP_n$ are simultaneously bounded by $(3r + D - \epsilon)/(D + \epsilon) < 1 + 3r/D$. Here, ϵ is defined as the difference between the turn H_1L_1 and segment H_1L_1. The smaller the distance D is or the larger the edge length r is, the larger both r_1 and r_2 are. For this reason, the optimality of $Simple(Bug2)$ against the optimal (model-based) navigation is completely lost. On the other hand, r_2 is not always larger than one, i.e., $0 < r_2 < 1$, on any condition concerning to D, r, or location of start, goal, and obstacles (Fig.12(b)). In this figure, in contrast to a square obstacle, $Simple(Bug2)$ frequently selects a longer path than a path made in $Bug2$ around a triangular obstacle. Consequently, the superiority of $Simple(Bug2)$ against $Bug2$ is not maintained. As a result, the characteristics described in the previous section are completely lost.

In conclusion, whether the near-optimality and the superiority are maintained or not in $Simple(Bug2)$ is described in Table 1. To revive the characteristics especially in an unknown environment with elliptic or triangular obstacles, we will try two solutions in this paper. One is an absolute solution which does not depend on size of an obstacle, the other is a relative solution which depends on the size.

In the former solution, a mobile robot changes its following direction alternatively and drives two times longer from a middle point (a hit point H_i), e.g., 1, 2, 4, 8, ..., from a hit point H_i (Fig.14). If the solution is adopted between leave points L_{i-1} and L_i, a mobile robot regards a competitive ratio r_{1i} as 9 in order to find a leave vertex L_i closer to H_i. The calculation detail is described in the

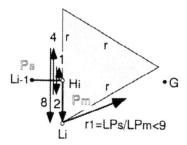

Fig. 14. An absolute solution whose ratio r_1 is bounded by 9 in an uncertain environment with triangular obstacles.

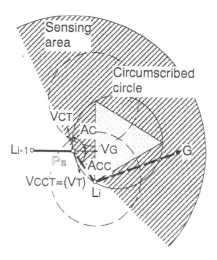

Fig. 15. A relative solution whose r_1 is bounded by a finite number in an uncertain environment with triangular obstacles.

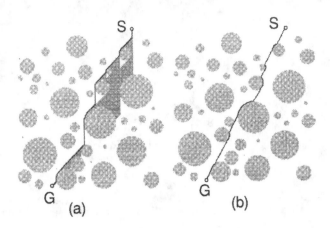

Fig. 16. (a) The (optimal) shortest path and an investigated area in a digital map [512,512] with uncertain circular obstacles by the A^* algorithm. (b) A near-optimal (extremely shorter) path and an investigated area in the map by the $Simple(Class1)$ algorithm.

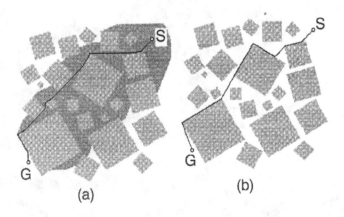

Fig. 17. (a) The (optimal) shortest path and an investigated area in a digital map [512,512] with uncertain square obstacles by the A^* algorithm. (b) A near-optimal (extremely shorter) path and an investigated area in the map by the $Simple(Class1)$ algorithm.

Fig. 18. The Nomad 200 equipped with a sonar ranging system (the Sensus 200) and a laser ranging system (the Sensus 500) runs in an unknown 2-D environment with board fences.

Appendix. This is equivalent to the case where there are rectangular obstacles whose l_2/l_1 is 4. For this reason, we can conclude that the solution is not so bad.

In the latter solution, we assume that a mobile robot has a sensing area to identify neighbor unknown obstacles, which is sufficiently (about 2 or 3 times) larger than the circumscribed circle of an unknown obstacle (Fig.15). Before entering into the circumscribed circle, a mobile robot with a sensing area finds a leave vertex L_i closer to the hit point H_i. This means that a mobile robot avoids an unknown obstacle with inadequate shape by avoiding implicitly its circumscribed circle. As a result, a mobile robot supervised by this solution always avoids an obstacle with inadequate shape by an adequate direction.

7   Simulation and Experiment

In this section, we ascertain two characteristics of our proposed navigation algorithms. One is the near-optimality of $Simple(Class1)$ and $Simple(Bug2)$ against the optimal (model-based) navigation algorithm, and the other is the superiority of $Simple(Class1)$ and $Simple(Bug2)$ against $Class1$ and $Bug2$.

Firstly, as compared with the optimal (model-based) navigation algorithm, we will check the performance of our algorithm $Simple(Class1)$. For this purpose, we investigate the difference between the optimal path selected by the A^* algorithm and a path selected by $Simple(Class1)$ in an environment cluttered with simple shapes. Both paths are measured by the $piecewiseL_2$ metric in a 2-D digital map [512,512]. If the digital map includes a lot of uncertain circular obstacles, both paths are written as sequences of black cells in Fig.16(a) and (b). The lengths are 590 and 615, respectively. The unit is defined as the length of an edge of a cell. In Fig.16(a) and (b), numbers of cells investigated by A^* and $Simple(Class1)$ are 11070 and 506, respectively, and cells investigated by A^* and $Simple(Class1)$ are described as dark gray cells. Furthermore, if the digital map includes uncertain rectangular obstacles, both paths are shown as sequences

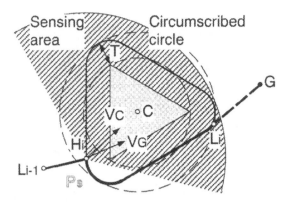

Fig. 19. An extended obstacle avoidance of a relative solution.

of black cells in Fig.17(a) and (b). The lengths are 672 and 731, respectively. In Fig.17(a) and (b), numbers of cells investigated by A^* and $Simple(Class1)$ are 57989 and 591, respectively, and cells investigated by A^* and $Simple(Class1)$ are described as dark gray cells. From these results, we can conclude the following: (1) $Simple(Class1)$ selects a near-optimal path to a destination on the average in an uncertain map with simple shapes as obstacles. (2) $Simple(Class1)$ is regarded as the on-line and sensor-based path-planning algorithm because the search space (the set of investigated cells) is small enough.

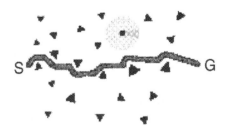

Fig. 20. In the simulation, the Nomad200 selects a path until a destination in another uncertain 2-D environment.

Secondly, we check feasibility of the relative solution addressed in the last section. To check the feasibility, we use a mobile robot (The Nomad 200, Nomadic Tech. Inc.) with an ultrasonic ring (the Sensus 200) and a laser range finder (the Sensus 500) (Fig.18). In order to use the relative solution in a real environment with triangular obstacles, we extend it as follows: Firstly, the Nomad200 goes straight to its destination until the Sensus200 omnidirectionally identifies a triangular obstacle whose Euclidean distance is less than or equal to 30 inches. It is a margin for absorbing the radius (8.8 inches) of the Nomad 200 and errors (about 10 inches) of dead-reckoning and sensing. Secondly, the Nomad200 turns its upper part including the Sensus500 to identify at least one edge of an encountered obstacle. Then on the assumption that each obstacle is an equilateral triangle, the Nomad200 calculates an uncertain triangle and its center of gravity C. Thirdly, the Nomad200 calculates the outer product $V_G \times V_C$, and if it is negative or zero [positive], the Nomad200 selects the clockwise [counter-clockwise] order. According to the selected direction, the Nomad200 traces the obstacle while keeping 30 inches until it is able to go straight to a destination (Fig.19).

To ascertain feasibility of the extended approach, we give simulation and experiment results described in Fig.20 and 21. In the graphics simulation developed in Nomadic Tech. Inc., twenty or more triangular obstacles whose edge lengths are larger than 20 inches and less than 40 inches exist (Fig.20). In the unknown environment, a mobile robot identifies each obstacle by an ultrasonic ring and detects its edge by a laser range finder and then avoids it by the extended approach (Fig.20). In the practical experiment, an uncertain environment is built as a wide room whose floor is roughly flat, and obstacles whose number is less than ten are constructed by a set of board fences (Fig.21). Here, two types of triangular obstacles whose edge lengths are 24 and 36 inches exist. In this case, the Sensus200 precisely identifies an uncertain obstacle omnidirectionally whose distance is within 40 inches, and also the Sensus500 exactly detects it for front 30 degrees within 80 inches. By using the extended approach based on two kinds of sensors, the Nomad200 smart selects a near-optimal path to its destination while avoiding a lot of triangular uncertain obstacles (Fig.20 and Fig.21).

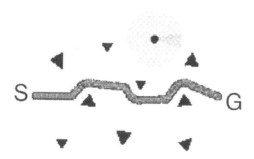

Fig. 21. In the experiment, the Nomad200 selects a path until a destination in an unknown 2-D environment with triangular obstacles constructed by board fences.

8   Conclusions

In this paper, we proposed two near-optimal sensor-based navigation algorithms $Simple(Class1)$ and $Simple(Bug2)$ in an uncertain 2-D environment with simple shapes as obstacles. Because of theoretical proofs and simulation results and experimental results, the near-optimality of the proposed algorithms and the superiority of them against the original algorithms are ascertained. Since the near-optimality can be acquired by a small investigation cost, our new algorithms are categorized into the sensor-based navigation.

In future, we will design a good algorithm running in a general uncertain environment with concave obstacles such as a maze.

Appendix

In this appendix, we calculate a competitive ratio r_{1i} between L_{i-1} and L_i for a mobile robot supervised by the absolute solution. By this solution, the robot changes its following direction alternatively and drives two times longer from a hit point H_i. Therefore, before reaching a leave point L_i closer to H_i, a mobile robot successively runs distances d_0, d_1, ..., and d_{k-1} from the middle point H_i to positions x_0, x_1, ..., and x_{k-1} (Fig.22). Because of the rule, the equation $d_j = 2d_{j-1}$ is always maintained. In this case, we are able to calculate a competitive ratio r_{1i} by the following.

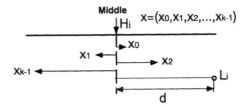

Fig. 22. A mobile robot changes its following direction alternatively and drives two times longer from the hit point H_i.

Because of $d \geq d_{k-2}$,

$$r_{1i} = \frac{d + 2\sum_{j=0}^{k-1} d_j}{d} \leq 1 + \frac{2\sum_{j=0}^{k-1} d_j}{d_{k-2}}, d_j > 0$$

Then, we define an upper limit as follows:

$$C \equiv \underset{k \geq 0}{\overset{sup}{}} \left(1 + 2\frac{\sum_{j=0}^{k-1} d_j}{d_{k-2}} \right)$$

$$1 + 2\frac{\sum_{j=0}^{k-1} d_j}{d_{k-2}} = C$$

$$\sum_{j=0}^{k-1} d_j = \left(\frac{C-1}{2}\right) d_{k-2} \tag{1}$$

$$\sum_{j=0}^{k-2} d_j = \left(\frac{C-1}{2}\right) d_{k-3} \tag{2}$$

Because of (1) - (2), we get the following equation.

$$d_{k-1} = cd_{k-2} - cd_{k-3}, c = \frac{C-1}{2}$$

$$d_{k-1} - cd_{k-2} + cd_{k-3} = 0$$

By solving $4\lambda - 2c\lambda + c\lambda = 0$ ($\lambda = d_{k-3} > 0$), we get $c = 4$ and then obtain a competitive ratio $r_{1i} \leq C = 1 + 2c = 9$.

Acknowledgments

This research is supported in part by 1997 Grants-in-aid for Scientific Research from the Ministry of Education, Science and Culture, Japan (No.09650300). The authors would also like to thank Dr. Sven Schuierer (University of Freiburg) for his helpful comment concerning to the Appendix.

References

1. H.Noborio, "On a sensor-based navigation for a mobile robot," *Journal of Robotics Mechatronics*, Vol.8, No.1, pp.2-14, 1996.
2. H.Noborio, T.Yoshioka, and S.Tominaga, "On the sensor-based navigation by changing a direction to follow an encountered obstacle," *Proc. of the IEEE/RSJ Int. Conf. on Intelligent Robots and Systems*, pp.510-517, September 1997.
3. H.Noborio,"A path-planning algorithm for generation of an intuitively reasonable path in an uncertain 2D workspace", *Proc. of the Japan-USA Symposium on Flexible Automation*, Vol.2, pp.477-480, July 1990.
4. V.J.Lumelsky and A.A.Stepanov, "Path-planning strategies for a point mobile automaton moving amidst unknown obstacles of arbitrary shape," *Algorithmica*, Vol.2, pp.403-430, 1987.
5. S.Sundar and Z.Shiller, "Optimal obstacle avoidance based on the hamilton-jacobi-bellman equation," *IEEE Trans. on Robotics and Automation*, Vol.13, No.2, pp.305-310, 1997.
6. I.Kamon and E.Rivlin, "Sensory-based motion planning with global proofs," *IEEE Trans. on Robotics and Automation*, Vol.13, No.6, pp.814-822, 1997.
7. H.Noborio, "A relation between workspace topology and deadlock occurrence in the simplest path-planning algorithm," *Proc. of the 1992 Second Int. Conf. on Automation, Robotics and Computer Vision*, pp.RO-10.1.1-RO.10.1.5, 1992.
8. H.Noborio and K.Urakawa, "On the near-optimality of sensor-based navigation in a 2-D unknown environment with simple shape" *Proc. of the IEEE Int. Conf. on Robotics and Automation*, to appear, 1999.

Realistic Environment Models
and Their Impact on the Exact Solution
of the Motion Planning Problem

A. Frank van der Stappen

Department of Computer Science, Utrecht University,
P.O.Box 80089, 3508 TB Utrecht, the Netherlands.

Abstract. The ability of exact motion planning algorithms to handle
all possible environments often leads to high worst-case running times.
These high running times are an important reason for the fact that exact algorithms are hardly used in practice. It has been noted that certain
weak assumptions on the size of the robot and the distribution of the
obstacles in the workspace lead to a drastic reduction of the complexity
of the motion planning problem, and, in some cases, to an equally-drastic
improvement of the performance of exact motion planning algorithms.
We review recent extensions of the known result for one robot in a low-
density environment with stationary obstacles to one robot amidst moving obstacles and to two and three robots among stationary obstacles.
In addition we consider the consequences of further weakening of the
assumptions on the distribution of the obstacles.

1   Introduction

Research on the motion planning problem within the field of computational geometry focusses on the exact solution of this problem. Exact motion-planning
algorithms must find a path for the robot whenever one exists and may only report failure if no path exists. It is well known that the maximum complexity of
the free part of the configuration space, or free space, of a robot with q degrees of
freedom moving in a scene consisting of n disjoint obstacles can be $\Omega(n^q)$. As the
performance of exact motion-planning algorithms heavily depends on the complexity of the free space, these algorithms often have a worst-case running time
of at least the same order of magnitude. This is probably one of the reasons that
most of the exact algorithms were never implemented. One exception is Bañon's
implementation [6] of the $O(n^5)$ algorithm of Schwartz and Sharir [28] for a ladder moving in a two-dimensional workspace, which performs surprisingly well,
and much better than the worst-case theoretical analysis predicts. The reason is
that the running time of the algorithm is sensitive to the actual complexity of
the free space, and this is in practice far less than the $\Theta(n^q)$ worst-case bound.
    These observations inspired research [1,2,3,8,9,10,11,14,15,16,17,19,20,21,23]
[24,25,26,32,33,36,37,38,39,40,41,43] where geometric problems are studied under
certain (hopefully realistic) assumptions on the input—in the case of motion

Christensen et al. (Eds.): Sensor Based Intelligent Robots, LNAI 1724, pp. 180–199, 1999.

planning: the environment in which the robot is moving. The goal of this line of research is to be able to predict better the practical performance of algorithms. For instance, van der Stappen et al. [36] studied the free-space complexity for not-too-large robots moving in environments consisting of *fat obstacles*. They showed that in this restricted type of environments the worst-case free-space complexity is only $\Theta(n)$, which is more in line with the experimental results of Bañon. Van der Stappen and Overmars [39,38] used this result to obtain an efficient algorithm for robot motion planning amidst fat obstacles. These results were extended to the more general setting of *low-density* environments by van der Stappen et al. [41] (see also [40]).

In this paper we review the extensions on the results from [40,41] reported in a sequence of recent papers [3,10,11]. These extensions concern motion planning in the even more general setting of an uncluttered environment [10], motion planning in a low-density environment with moving obstacles [11], and motion planning for two and three robots in a low-density environment [3].

The result from [10] that we discuss in Section 4.2 says that the linear upper bound on the complexity of the free space does not extend to motion planning in uncluttered environments. Instead we find that the free-space complexity in such environments is $\Theta(n^{q/2})$ when the workspace is two-dimensional and $\Theta(n^{2q/3})$ when the workspace is three-dimensional. These bounds fit between the $\Theta(n)$ bound for low-density environments and the $\Theta(n^q)$ bound for general environments.

The motion planning problem among moving obstacles is significantly more complex than the same problem among stationary obstacles. Reif and Sharir [27] showed that, when obstacles in a 3-dimensional workspace are allowed to rotate, the motion planning problem is PSPACE-hard if the velocity modulus is bounded, and NP-hard otherwise. (A similar result was obtained by Sutner and Maass [42].) Canny and Reif [12] showed that dynamic motion planning for a point in the plane, with a bounded velocity modulus and an arbitrary number of polygonal obstacles, is NP-hard, even when the obstacles are convex and translate at constant linear velocities. The result from [11] discussed in Section 5 shows that the circumstances change dramatically if we assume that the environment satisfies the low-density assumption, the robot is not too large, and the obstacles move along polyline trajectories, at constant speed per segment. In the case of a two-dimensional workspace with polygonal obstacles, the motion planning problem can then be solved in $O(n^2 \alpha(n) \log^3 n)$ time regardless of the number of degrees of freedom of the robot[1]. This bound is close to optimal as it turns out that in certain situations the robot may have to make $\Omega(n^2)$ simple motions to reach its goal.

The common approach to the exact solution of the (coordinated) motion planning for $m > 1$ robots $\mathcal{R}_i$ $(1 \leqslant i \leqslant m)$ is *centralized planning*; centralized planning regards the m robots as a single (multi-body) robot with $q = q_1 + \ldots + q_m$ degrees of freedom, where q_i is the number of degrees of freedom of $\mathcal{R}_i$. Techniques by Basu et al. [5] and—under certain general position assumptions—

---

[1] $\alpha(n)$ is the extremely slowly-growing inverse of the Ackermann function.

by Canny [13] result in $O(n^{q+1})$ and $O(n^q \log n)$ algorithms. Better bounds can only be obtained in specific cases. The result from [3] discussed in Section 6 shows that when two or three not-too-large robots move in the same low-density environment the motion planning problem for these robots can be solved in $O(n \log n)$ time, regardless of the number of degrees of freedom of the robots, and despite the fact that the complexity of the free space is $\Omega(n^2)$ in the case of two robots and $\Omega(n^3)$ in the case of three robots. Apparently it suffices to restrict the search for free paths to a subspace of the free space with lower complexity.

Before we discuss the above extensions, we review the environment models that play a role in this paper—fatness, low density, and unclutteredness—and determine the relations among these models. In Section 3, we repeat the known consequences of low density (and fatness) for the complexity and the solution of the motion planning problem. Section 4.1 shows how these ideas can be used to solve the motion planning problem for a single robot in almost-linear time.

2   Environment Models

2.1   Historical Perspective

The study of geometric problems under certain assumptions on the input—in the case of motion planning, the environment in which the robot moves—commenced in the early 1990s with a considerable number of papers focussing on fat objects. An object is fat if it has no long and thin parts. Among the first papers in this research direction are those by Alt et al. [2] and Matoušek et al. [23]. Alt et al. considered the influence of the length-to-width ratio—essentially a measure for its fatness—of a rectangular robot on the complexity of the motion planning problem for this robot. Matoušek et al. studied the complexity of the union of n triangles in the plane—a problem with applications in motion planning. They showed that if the triangles are fat, i.e., if each of the angles has at least a certain magnitude δ, for some constant $\delta > 0$, then the complexity of their union is at most roughly linear rather than quadratic in the number of triangles. Among the other achievements of the study of fatness are near-linear bounds on the complexity of the union of fat figures (e.g. wedges) in the plane [14,21], a linear bound on the complexity of the free space for motion planning amidst fat obstacles [36], and efficient algorithms for computing depth orders on certain fat objects [1], hidden surface removal for fat horizontal triangles [19], and range searching and point location among fat objects [25,26].

More recent papers have turned their attention to different input models such as low density, unclutteredness, and simple-cover complexity. Definitions of the first two notions are given in Section 2.2. Van der Stappen et al. [41] extended earlier results on motion planning amidst fat obstacles [38,39] to environments that satisfy the low density assumption (see Section 3 for details). Schwarzkopf and Vleugels [33] considered the range searching and point location problems in low density scenes. De Berg [8] proved that binary space partitions of size linear in the number of objects exist for scenes that satisfy the unclutteredness assumption. Finally, Mitchell et al. [24] introduced simple-cover complexity to measure the complexity of a scene with objects by the number of balls, each intersecting no more than a constant δ of the objects, required to cover the entire scene. The authors gave an algorithm for solving ray-shooting queries, i.e., for finding the first object hit by a ray emanating from a given point in a given direction, in time proportional to the number of balls intersected by the query ray.

2.2 Models and Model Hierarchy

Below we provide formal definitions of the three environment models that will play a central role throughout this paper. Moreover, we establish relations among the models. Each of the models assigns a value to a workspace and the obstacles in it. We shall denote the volume of an object $\mathcal{P}$ by $\mathrm{vol}(\mathcal{P})$ and assume that the bounding box $\mathrm{bb}(\mathcal{P})$ of an object $\mathcal{P}$ is the smallest *axis-parallel* rectangloid containing $\mathcal{P}$. Furthermore, we let the size of an object $\mathcal{P}$—denoted by $\mathrm{size}(\mathcal{P})$— be the radius of its minimal enclosing circle or sphere.

The first—and most extensively studied—model we consider is fatness. Intuitively, an object (or obstacle) is called fat if it contains no long and skinny parts. There are many different definitions of fatness [1,2,23,21], which are all more or less equivalent—at least for convex objects. We shall use the definition proposed in [39].

Definition 1. *Let $\mathcal{P} \subseteq \mathbb{E}^d$ be an object and let β be a positive constant. Define $U(\mathcal{P})$ as the set of all balls centered inside $\mathcal{P}$ whose boundary intersects $\mathcal{P}$. We say that the object $\mathcal{P}$ is β-fat if for all balls $B \in U(\mathcal{P})$, $\mathrm{vol}(\mathcal{P} \cap B) \geq \beta \cdot \mathrm{vol}(B)$. The fatness of $\mathcal{P}$ is defined as the maximal β for which $\mathcal{P}$ is β-fat.*

The fatness of a scene of objects is defined as the maximal β for which every individual object is β-fat. We speak of a scene of fat objects if the fatness of the scene is β for some (small) constant β.

The model of low density was introduced by van der Stappen [39] (see also [41]) and refined by Schwarzkopf and Vleugels [33]. It forbids any ball B to be intersected by many objects that are at least as large as B.

Definition 2. *Let $\mathcal{S} := \{\mathcal{P}_1, \ldots, \mathcal{P}_n\}$ be a d-dimensional scene, and let $\lambda \geq 1$ be a parameter. We call $\mathcal{S}$ a λ-low-density scene if for any ball B, the number of objects $\mathcal{P}_i \in \mathcal{S}$ with $\mathrm{size}(\mathcal{P}_i) \geq \mathrm{size}(B)$ that intersect B is at most λ. The density of $\mathcal{S}$ is defined to be the smallest λ for which $\mathcal{S}$ is a λ-low-density scene.*

We say that a scene has *low density* if its density is a small constant.

Clutteredness was introduced by de Berg [8]. It is defined as follows.

Definition 3. *Let $\mathcal{S}$ be a d-dimensional scene, and let $\kappa \geq 1$ be a parameter. We call $\mathcal{S}$ a κ-cluttered scene if any hypercube whose interior does not contain a vertex of one of the bounding boxes of the objects in $\mathcal{S}$ is intersected by at most κ objects in $\mathcal{S}$. The clutter factor of a scene is the smallest κ for which it is κ-cluttered.*

We sometimes call a scene whose clutter factor is a small constant *uncluttered*. Of the models we consider, this is the only one not invariant under rotations, since it uses bounding boxes and axis-parallel hypercubes.

Theorem 1 shows that the three models form a hierarchy in the sense that fatness implies low density, and low density implies unclutteredness. (The result can in fact be extended to simple-cover complexity, but since we leave this model out of consideration, we restrict the theorem to the relations between fatness, low density, and unclutteredness.)

Theorem 1. *[9]*
- *Any d-dimensional scene consisting of β-fat objects has density at most $2^d\beta^{-1}$.*
- *Any d-dimensional λ-low-density scene is at most $\lceil\sqrt{d}\rceil^d\lambda$-cluttered.*

The theorem says that the assumptions imposed by the unclutteredness model are weaker than those imposed by the low-density model, which are in turn weaker than those imposed by the fatness model. In other words, the unclutteredness model is more general than the low-density model, which is more general than the fatness model.

3   Environment Models and Motion Planning

We focus on motion planning problems in a workspace W with a collection $\mathcal{E}$ of n obstacles, each of constant descriptional complexity. The objective is to find a motion for a robot from a specified initial placement Z_0 to a specified final placement Z_1 during which it avoids collision with the obstacles or to report that no such motion exists. The robot $\mathcal{R}$ is assumed to have q degrees-of-freedom and to be of constant descriptional complexity. We assume that the maximum size of the robot is at most a constant multiple of the size of the smallest obstacle.

The motion planning problem is commonly modelled and solved in the *configuration space* C, which is the space of parametric representations of robot placements. The dimension of C equals the number of degrees of freedom q of the robot $\mathcal{R}$. A point in C (representing a robot placement) is referred to as a configuration. Although there is a subtle difference between a placement and a configuration, we will use both terms interchangeably. The *free space* FP is the subspace of C consisting of points that represent placements of the robot in which it does not intersect any obstacle in $\mathcal{E}$. The free space can be regarded as the union of certain cells—the free cells—in the arrangement[2] of constraint hypersurfaces. A constraint hypersurface $f_{\phi,\Phi}$ is the set of placements in which a robot feature ϕ, i.e., a basic part of the boundary like a vertex, edge, or face, touches an obstacle feature Φ of appropriate dimension. The constraint hypersurfaces are assumed to be algebraic and to have bounded degree. A collision-free *path* or *motion* for a robot from an initial placement to a final placement is a

---

[2] The arrangement of a set of geometric objects is the subdivision of space into connected pieces of any dimension induced by these objects.

continuous curve in the free space FP connecting the points representing these two placements. The effort that is required to find such a curve clearly depends on the complexity of FP.

Exact motion planning algorithms process the free space into a query structure that allows for the efficient solution of one or more path-finding queries. Although there essentially exist two different approaches to exact motion planning (cell decomposition and retraction), the time spent in processing the free space and the size of the resulting query structure clearly depend on the complexity of the free space. In our case, the complexity of the free space, in turn, depends on the number of different q-fold contacts of the robot with the obstacles. We shall use this fact in several complexity analyses. The cell decomposition approach [18,22,28,29,30,31,34] that we follow below solves the motion planning problem by decomposing the free space into simple (constant-complexity) subcells. The construction of a graph on the subcells, connecting pairs of adjacent subcells by edges, then reduces the motion planning problem to a graph search problem.

Van der Stappen [39] considered configuration spaces C with a projective subspace B that can be decomposed into (constant-complexity) regions R that satisfy

$$|\{f_{\phi,\Phi}|\phi \in_f \mathcal{R} \wedge \Phi \in_f \mathcal{E} \wedge f_{\phi,\Phi} \cap (R \times D) \neq \emptyset\}| = O(1), \tag{1}$$

where D is such that $C = B \times D$, and $\in_f$ should be read as 'is a feature of'. The equality says that the cylinder $R \times D$ obtained by lifting R back into the configuration space is intersected by only a constant number of constraint hypersurfaces. Figure 1 shows a three-dimensional configuration space C, a two-dimensional subspace B, a region $R \subset B$ and a constraint hypersurface intersecting the cylinder $R \times D$. We refer to a subspace B of C with the above property as a *base space*. The property tells us two things.

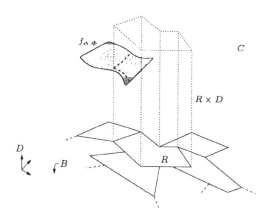

Fig. 1. A three-dimensional configuration space C, a two-dimensional base space B, a regions $R \subset B$, and its lifting $R \times D$ back into C.

Firstly, the cylinders $R \times D$ obtained by lifting the regions R subdivide the free space into constant-complexity subcells and thus form an appropriate cell decomposition of the free space. The constant number of constraint hypersurfaces that intersect $R \times D$ define a constant-complexity arrangement inside $R \times D$. Since $R \times D$ itself has constant complexity as well, the arrangement subdivides $R \times D$ into a constant number of constant-complexity free and forbidden cells. The free cells of all cylinders form a cell decomposition of the free space.

Secondly, we observe that the cell decomposition of the free part of C is obtained by decomposing a lower-dimensional subspace B of C, subject to certain constraints. Thus, we have transformed the problem of finding a decomposition of the free part of C into the problem of finding a constrained decomposition of the lower-dimensional subspace B

The above approach will only be succesful if the problem of finding the constrained decomposition is easier than the original problem. At first sight, this seems not to be the case since the constraints are still formulated in the original space C. In each of the applications of the approach we shall strengthen the original constraints to obtain simpler constraints in the base space. It turns out that strengthening the constraints does not harm the ability to find (almost) optimal decompositions.

A set V_B of regions R that decompose B alone is insufficient to solve the motion planning problem. The preceding paragraphs show that we need the constraint hypersurfaces intersecting the cylinder obtained by lifting R into C. In each of the applications we shall do this by computing (a constant-size superset of) the constant-size set of obstacles E for which at least one constraint hypersurface $f_{\phi,\Phi}$ with $\Phi \in_f E$ intersects $R \times D$. Moreover, we need to establish which pairs of free subcells are adjacent. Using the fact that adjacent free subcells either lie inside the same cylinder or in cylinders obtained by lifting adjacent regions R and R' in B, we find that it is sufficient to report the set E_B of all pairs of adjacent regions in V_B. Provided that the necessary information is available, it is easy to transform the decomposition of B given by V_B and E_B into a decomposition of the free space, in time proportional to the sizes of V_B and E_B (see [39,40,41] for details). The time $T(n)$ to find the decomposition of B along with the required additional information thus determines the time to compute a cell decomposition of the free space.

Theorem 2. *[39,41] Let B be a base space in the configuration space C of a motion planning problem, and let $T(n)$ be the time required to compute a base partition in B. Then the motion planning can be solved in $O(T(n))$ time.*

Van der Stappen showed [39,41] that in the case of a single free-flying robot moving in a low-density scene, the workspace W of the robot can be chosen as a base space. This initally [39,40] resulted in a (nearly-optimal) almost-linear algorithm for motion planning in a 2D workspace, and in quadratic and cubic algorithms for motion planning in 3D workspaces with polyhedral and arbitrary obstacles respectively. Although this represents a considerable improvement over the general bound of $O(n^q)$, there remains a gap in the 3D case since there is a linear upper bound on the complexity of the free space. We close this gap in

Section 4.1. Our other challenges in this paper concern extensions of the result of the previous paragraph to weaker environments and to different instances of the motion planning problem. In Section 4.2 we will try to generalize the result for one free-flying not-too-large robot to uncluttered scenes. In Sections 5 and 6 we study motion planning problems among moving obstacles and for more than one robot.

4   Motion Planning for a Single Robot

We first show that the motion planning problem for a single robot in a low-density environment can be solved in $O(n \log n)$ time, regardless of the dimension of the workspace, thereby improving the bounds for 3D workspaces quoted in the previous section. This bound is close to optimal as the complexity of the free space for motion planning problems in low density environments is $\Theta(n)$. Moreover, we will see that the $\Theta(n)$ bound does not extend to environments that only satisfy the weaker unclutteredness assumption.

4.1   Low Density Environments

We assume that our robot moves in a low-density environment. In the case of a single free-flying robot, the (low-density) workspace is a base space in the configuration space. As a result, we can obtain a cell decomposition of the free space—and thus solve the motion planning problem—by decomposing the workspace subject to the constraints formulated in Section 3.

Our first step is to replace the configuration-space constraints by slightly stronger workspace constraints. Let $G(E)$ be obtained by growing (or dilating) an obstacle E in all directions by the size of the robot. The key observation here is that a robot can only touch an obstacle E if it moves in the proximity of E, or, more precisely, in $G(E)$. In configuration space, it says that a constraint hypersurface $f_{\phi,\Phi}$ induced by E (i.e., $\Phi \in_f E$) can only intersect a cylinder $R \times D$ if R intersects $G(E)$. As a consequence, a decomposition of the workspace into constant-complexity regions R each intersecting a constant number of grown obstacles suits our purposes.

It turns out that a very simple quadtree-like decomposition leads to a partition of the workspace of size linear in the number of obstacles. Let $Cov(R)$ be the set of obstacles whose grown version intersects R. (Thus, our aim is to decompose the workspace into regions R with $|Cov(R)| = O(1)$). Let Σ be the set of the $2^d n$ vertices of the axis-parallel bounding boxes of all grown obstacles. For simplicity, we enclose these vertices by an axis-parallel square or cube C. Let $\Sigma(R)$ be the subset of vertices of Σ contained in the interior of a region R. We subdivide C recursively until all resulting regions R satisfy $\Sigma(R) = \emptyset$. Let C be a (square or) cube with $\Sigma(C) \neq \emptyset$ and let $C_1, \ldots, C_{2^d}$ be the equally-sized sub-cubes resulting from a 2^d-tree split of C, i.e., from cutting C with the d planes perpendicular to and bisecting its edges. Note that every sub-cube C_i shares exactly one corner with C. A 2^d-tree split is called *useless* if all vertices of

$\Sigma(C)$ lie in a single sub-cube C_i of C, and called *useful* otherwise. The cube C is subdivided in one of the following ways.

- If the 2^d-tree split of C is useful, then perform the 2^d-tree split.
- If the 2^d-tree split of C is useless, then replace the single sub-cube C_i containing all vertices of $\Sigma(C)$ by the smallest cube that shares a corner with C and still contains all vertices of $\Sigma(C)$. The resulting cube $C' \subseteq C$ has one of the vertices of $\Sigma(C)$ on its boundary. The 'L-shaped' complement $C \setminus C'$ satisfies $\Sigma(C \setminus C') \neq \emptyset$.

Figure 2 gives two-dimensional examples of both types of splits.

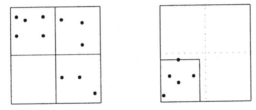

Fig. 2. A useful and useless 2^d-tree split. In the latter case the square is subdivided into an L-shaped region and a square.

Application of the recursive splits to the workspace with the grown obstacles leads to a decomposition into $O(n)$ cubic and L-shaped regions that have no vertices from Σ in their interiors. Using the fact that the workspace with the grown obstacles still has low density and the result from Theorem 1 that any low density scene is also uncluttered, yields by Definition 3 that each cubic region is intersected by only $O(1)$ grown obstacles. If we cover the L-shapes by $2^d - 1$ cubes then it is clear that the same is also true for these regions. As a result, the cubic and L-shaped regions decompose the workspace into regions R with $|Cov(R)| = O(1)$. Let V be the set of cubic and L-shaped regions, and observe that the set of all constraint hypersurfaces induced by obstacles in $Cov(R)$ form a constant-size superset of the constraint hypersurfaces that really intersect $R \times D$. The computation of V, along with all sets $Cov(R)$, which aid in computing the free subcells (see Section 3) takes $O(n \log n)$ time. Furthermore, it can be shown (see [41]) that the set of pairs of adjacent regions has size $O(n)$ and can—with the use of a point location structure—be computed in $O(n \log n)$ time as well. By Theorem 2, we obtain the following result.

Theorem 3. *[41] The motion planning problem for a single not-too-large robot in a low-density environment can be solved in $O(n \log n)$ time.*

The $O(n \log n)$ bound also applies to a robot moving amidst fat obstacles by Theorem 1.

4.2   Uncluttered Environments

The linear upper bound on the complexity of the free space no longer holds when we move one step in the model hierarchy—to uncluttered environments. We construct a motion planning problem for a robot in a two-dimensional uncluttered environment for which the free space complexity is more than linear in the number of obstacles. We recall that the complexity of the free space depends on the number of q-fold contacts for the robot.

The robot $\mathcal{R}$ in our lower bound example consists of q links, which are all attached to the origin. The links have length $1 + \varepsilon$, for a sufficiently small $\varepsilon > 0$. Obviously $\mathcal{R}$ has q degrees of freedom.

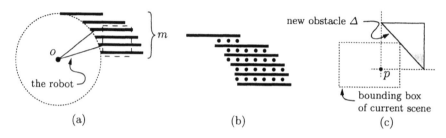

Fig. 3. (a) Part of the lower bound construction. (b,c) Adding bounding-box vertices to make the scene uncluttered.

The set of n obstacles for the case of a 2-cluttered planar scene is defined as follows. (Later we adapt the construction to get the bound for κ-cluttered scenes for larger κ.) Fix an integer parameter m; it will turn out later that the appropriate value for m is roughly $\sqrt{n}$. For a given integer i, let E_i be the horizontal segment of length 1 whose y-coordinate is i/m and whose left endpoint lies on the unit circle—see Figure 3(a) for an example. Let $\mathcal{E}_1 = \{E_i | 1 \leqslant i \leqslant m\}$; this set forms a subset of the set of all obstacles. The remaining obstacles, which we describe later, are only needed to turn the environment into an uncluttered environment.

Consider any subset of q segments from $\mathcal{E}_1$. It is easy to see that there is a (semi-)free placement of $\mathcal{R}$ such that each segment in the subset is touched by a link of $\mathcal{R}$. Hence, the free-space complexity is $\Omega(m^q)$. When m is large, however, the set $\mathcal{E}_1$ does not form an uncluttered enviroment: the dashed square in Figure 3(a) for instance, intersects $\Omega(m)$ obstacles without having a bounding-box vertex of one of the obstacles in its interior. This problem would disappear if between every pair of adjacent horizontal segments there would be a collection of $O(m)$ equal-spaced bounding-box vertices, as in Figure 3(b). Notice that in total we need $\Theta(m^2)$ bounding-box vertices for this. We cannot add tiny obstacles between the segments to achieve this, because such obstacles would be much smaller than the robot, so the robot would no longer have bounded size. There is no need, however, to add obstacles between the segments; we can also create bounding-box vertices there by adding large—for example, triangular—obstacles outside the current scene (see Figure 3(c)). The point p is a bounding-box vertex of Δ, and Δ is disjoint from the current set of obstacles. By iteratively adding obstacles that generate the necessary bounding-box vertices between the segments in $\mathcal{E}_1$ we transform the cluttered environment into an uncluttered one. It is not difficult to see that we can even obtain a 2-cluttered environment in this manner, if we choose the distance between adjacent bounding-box vertices to be $1/m$.

We now have a collection of $\Theta(m^2)$ obstacles forming a 2-cluttered scene such that the free-space complexity is $\Omega(m^q)$. By choosing a suitable value for m (in the order of $\sqrt{n}$), we obtain a collection of n obstacles such that the free-space complexity is $\Omega(n^{q/2})$.

To get the general bound we replace each of the m segments in the set $\mathcal{E}_1$ by $\lfloor \kappa/2 \rfloor$ segments of length 1 that are quite close together. The left endpoints of these segments still lie on the unit circle. Since the original scene was 2-cluttered, the new scene is κ-cluttered. The number of q-fold contacts has increased to $\Omega(\kappa^q m^q)$. By again choosing m to be roughly $\sqrt{n}$ and assuming κ to be a constant we get the final result.

A similar construction exists for 3D workspaces, leading to a somewhat different bound. Theorem 4 summarizes both worst-case lower bounds.

Theorem 4. *[10] The free-space complexity for a not-too-large robot with q degrees of freedom moving in an uncluttered scene consisting of n obstacles can be*

- $\Omega(n^{q/2})$ *if the workspace is 2D, and*
- $\Omega(n^{2q/3})$ *if the workspace is 3D.*

Even though we have seen that there exist 2D and 3D uncluttered scenes that lead to $\Omega(n^{q/2})$ and $\Omega(n^{2q/3})$ free space complexities it is also possible (see [10] for details) to show that the free space complexity of an uncluttered scene cannot be worse than these bounds. The proof uses a decomposition that resembles to a certain extent the decomposition in Section 4.1. Again we grow the obstacles by an amount proportional to the size of the robot. Then we apply a similar quadtree-like decomposition, but stop subdividing squares one their size becomes proportional to the size of the robot. The resulting decomposition consists of $O(n)$ regions. Next, we count the number of q-fold contacts of the robot when placed inside any of these regions. Summing the contacts over all regions yields the bounds of Theorem 5, assuming again that κ is a constant.

Theorem 5. *[10] The free-space complexity for a not-too-large robot with q degrees of freedom moving in an uncluttered scene consisting of n obstacles is*

- $O(n^{q/2})$ *if the workspace is 2D, and*
- $O(n^{2q/3})$ *if the workspace is 3D.*

We observe that the bound of Theorem 5 fits nicely between the $O(n)$ bound for low-density scenes and the $O(n^q)$ bound for general scenes.

5   Motion Planning Among Moving Obstacles

Let us now see what happens if we allow the obstacles in a low-density space to move. We focus on two-dimensional workspaces and assume that the obstacles are polygonal and translate along piecewise-linear paths, at constant speed per segment. At every moment in time, the workspace with the obstacles satisfies the low-density assumption. The robot leaves its initial configuration Z_0 at time t_0 and must reach its final configuration Z_1 at t_1.

Motion planning problems among moving obstacles are normally solved in the configuration-time space, which is the Cartesian product of the configuration space of the stationary version of the problem—that is, the version in which the obstacles do not move—and time. It turns out [11] that in this configuration-time space $CT = C \times T$, the subspace $W \times T$ can be decomposed into regions R satisfying $|\{f_{\phi,\varPhi} | \phi \in_f \mathcal{R} \wedge \varPhi \in_f \mathcal{E} \wedge f_{\phi,\varPhi} \cap (R \times D) \neq \emptyset\}| = O(1)$, where D is such that $C = W \times D$. Hence, the space $W \times T$ is a base space. As in earlier cases, our objective is to transform the constraint into a constraint in the base space, in this case the work-time space.

The approach we follow is similar to the approach taken in Section 4.1. We grow the obstacles by the size of the robot, but since a future decomposition requires polyhedral imput, we immediately replace the grown obstacles by polygonal outer approximations. In our base space—the work-time space—we see these polygonal approximations move along linear trajectories, resulting in a polyhedral arrangement. The complexity of this arrangement is roughly quadratic in the number of obstacles, and the cylinders resulting from lifting the three-dimensional cells of the arrangement into CT are intersected by only a constant number of constrain hypersurfaces. The single remaining problem that prevents these cells from being an appropriate decomposition of $W \times T$ is that the three-dimensional cells may have more than constant complexity. We resolve this by applying (a modified version of) the vertical decomposition algorithm by De Berg et al. [7]. It results in $O(n^2 \alpha(n) \log^2 n)$ regions, each with a constant number of neighbors. The partition is computable in $O(n^2 \alpha(n) \log^2 n)$ time. As a consequence, we can obtain a cell decomposition of the free part of the configuration-time space of size $O(n^2 \alpha(n) \log^2 n)$ in $O(n^2 \alpha(n) \log^2 n)$ time.

It remains to find the actual free path for the robot, or in other words, to find a curve connecting the configuration-time space points (Z_0, t_0) and (Z_1, t) that is completely contained in the union of the free subcells. This is not as simple as in the case of a motion planning problem involving stationary obstacles: as the robot cannot travel back in time, the curve in the free part of the configuration-time space must be time-monotone. To facilitate the search for such a path we cut the free subcells into time-monotone parts, by slicing them with planes orthogonal to the time axis at appropriate points (see again [11] for details). Each of the resulting subcells has the property that if one point inside it is reachable by a time-monotone path then all points inside it with a larger time-coordinate are reachable by a time-monotone path.

To find a path we sweep the refined cell decomposition with a plane orthogonal to the time axis, from $t = t_0$ to $t = t_1$, in order to label each of the subcells with the first moment in time at which it can be reached. The sweep requires the use of a priority queue for efficient event handling [11]. The event handling increases the computation time by a $\log n$-factor. If the subcell containing the point (Z_1, t_1) has received a label during the sweep then it is reachable by a time-monotone path. An actual path can be determined by tracing back the sequence of subcells to (Z_0, t_0). If the subcell containing (Z_1, t_1) has received no label then no path exists. We get the following result.

Theorem 6. *[11] The motion planning problem for a single not-too-large robot in a two-dimensional low-density environment with obstacles translating along piecewise-linear paths can be solved in $O(n^2 \alpha(n) \log^3 n)$ time.*

The result of Theorem 6 is close to optimal as there exist situations in which a robot has to make $\Omega(n^2)$ simple moves to reach its goal. This number provides a worst-case lower bound on the complexity of a single connected component of the free space, and thus on the complexity of the free space itself. The complexity of the free space, in turn, provides a lower bound on the number of subcells in a cell decomposition. Consider the workspace in Figure 4. The grey rectangular robot

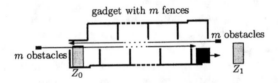

Fig. 4. A quadratic lower bound construction.

must translate from position Z_0 to Z_1. The gadget in the middle forces the robot to make $\Omega(m)$ moves to move from left to right. It can easily be constructed from $O(m)$ stationary obstacles. The big black obstacle at the bottom right moves slowly to the right: it takes a long time before the robot can actually get out of the gadget to move to its goal. Now a small obstacle moves from the left to the right, through the gaps in the middle of the gadget. This forces the robot to go to the right as well. Only there can it move slightly further up to let the obstacle pass. But then a new obstacle comes from the right through the gaps, forcing the robot to move to the left of the gadget to let the obstacle pass above it. This is repeated m times after which the big obstacle is finally gone and the robot can move to its goal. The robot has to move $2m$ times through the gadget, each time making $\Omega(m)$ moves, leading to a total of $\Omega(m^2)$ moves. As $m = \Omega(n)$, the total number of moves is $\Omega(n^2)$. It can be verified that at any moment, the low obstacle density property is satisfied.

6   Motion Planning for Two and Three Robots

Our problem in this section is to plan motions for two or three not-too-large robots (operating in the same low-density environment) from their respective initial placements to their respective final placements during which they avoid collision with the stationary obstacles and with each other.

The common approach the exact solution of the (coordinated) motion planning for $m > 1$ robots $\mathcal{R}_i$ ($1 \leqslant i \leqslant m$)—centralized planning—regards these robots $\mathcal{R}_i$ as one (multi-body) robot $\mathcal{R}$. The configuration space C for this composite robot $\mathcal{R}$ equals the Cartesian product of the configuration spaces C_i of the robots $\mathcal{R}_i$. As each of the original robots $\mathcal{R}_i$ is free to move, the composite robot $\mathcal{R}$ does not have bounded size. As a consequence, we may not expect the free space of the composite robot to have linear complexity. In fact, it is quite easy to see that the free-space complexity can be $\Omega(n^m)$: each of the m original robots $\mathcal{R}_i$ can be placed in contact with any of the n obstacles. This gives us n^m different m-fold contacts for the composite robot $\mathcal{R}$. As a result, the complexity of the free space can be $\Omega(n^m)$. Processing of this free space for motion planning queries would inevitably lead to an algorithm with at least $\Omega(n^m)$ running time. Surprisingly however, it suffices to restrict the search for a free path to the union of a few lower-dimensional subspaces of the configuration space C, leading to a drastic reduction of the amount of time required to find a path.

We focus on the case of two robots $\mathcal{R}_1$ and $\mathcal{R}_2$ with d_1 and d_2 degrees of freedom respectively. The motion planning problem for the composite robot consisting of $\mathcal{R}_1$ and $\mathcal{R}_2$ can be solved in the $d_1 + d_2$-dimensional configuration space $C = C_1 \times C_2$. The objective is to reduce the dimension of the space we have to consider. To this end we limit the possible multi-configurations— combinations of configurations for the two robots—that we allow. Of course, we have to guarantee that a feasible multi-path continues to exist.

The multi-configurations that we allow—we call them *permissible multi-configurations*—are as follows.

- When $\mathcal{R}_1$ is at its start or goal configuration, we allow any configuration of $\mathcal{R}_2$.
- When $\mathcal{R}_2$ is at its start or goal configuration, we allow any configuration of $\mathcal{R}_1$.
- When neither $\mathcal{R}_1$ nor $\mathcal{R}_2$ is at its start or goal configuration, we only allow configurations where $\mathcal{R}_1$ and $\mathcal{R}_2$ touch each other.

Consider the situation depicted in Fig. 5, where we have two disk-shaped robots moving amidst polygonal obstacles in the plane. The initial and final (or start and goal) configurations of the robots are indicated in Figure 5(a); start configurations are solid and goal configurations are dotted. A feasible multi-path for this problem that uses permissible multi-configurations is indicated in Figure 5(b)–(d): first $\mathcal{R}_2$ moves towards $\mathcal{R}_1$ until it touches it, then $\mathcal{R}_2$ and $\mathcal{R}_1$ together move until $\mathcal{R}_2$ is at its end configuration, and finally $\mathcal{R}_1$ breaks off its contact with $\mathcal{R}_2$ and moves to its own goal configuration.

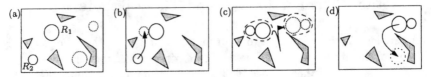

Fig. 5. A feasible multi-path using permissible multi-configurations.

At first sight, it may seem severe to restrict oneself to permissible multi-configurations. Nevertheless, it turns out that solutions using only permissible multi-configurations always exist, provided a solution exists at all. Let π_1 and π_2 be a coordinated feasible paths for our robots. (Hence, the robots do not collide with the obstacles or each other along these paths.) We define the *coordination diagram* for π_1 and π_2 as follows. Let U be the unit square. We call the edges of U incident to the origin the *axes* of U. The horizontal axis, or t_1-*axis*, of U represents the configuration of $\mathcal{R}_1$ along π_1; the vertical axis, or t_2-*axis*, represents the configuration of $\mathcal{R}_2$ along π_2. Thus a point $(t_1, t_2) \in U$ corresponds to placing $\mathcal{R}_1$ and $\mathcal{R}_2$ at configurations $\pi_1(t_1)$ and $\pi_2(t_2)$ respectively. Observe that the left edge of U corresponds to multi-configurations where $\mathcal{R}_1$ is at its start configuration, the top edge of U corresponds to configurations where $\mathcal{R}_2$ is at its goal configuration, and so on. A point $(t_1, t_2) \in U$ is called *forbidden* if $\mathcal{R}_1$ at $\pi_1(t_1)$ intersects $\mathcal{R}_2$ at $\pi_2(t_2)$; otherwise it is called *free*—see Fig. 6. The coordination diagram for π_1 and π_2 is the subdivision of U into free and forbidden regions.

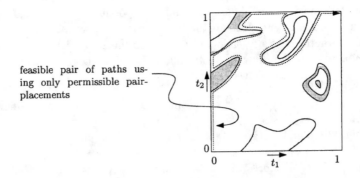

feasible pair of paths using only permissible pair-placements

Fig. 6. The coordination diagram; forbidden regions are shaded and a feasible multi-path using only permissible multi-configurations is shown dotted.

We call a path in U from $(0, 0)$ to $(1, 1)$ a 01-path. Since π_1 and π_2 are feasible paths, $\mathcal{R}_1$ does not intersect any obstacle along π_1 and $\mathcal{R}_1$ does not intersect any obstacle along π_2. Hence, a 01-path that lies in the free region corresponds to a pair of coordinated feasible paths; we call this a feasible 01-path. Notice that the diagonal from $(0, 0)$ to $(1, 1)$ is by definition a feasible 01-path. This means that

$(0,0)$ and $(1,1)$ lie in the same component of the free region. Since they both lie on the boundary of U, they must lie in the same component of the boundary of the free region as well. Hence, there is a feasible 01-path along the boundary of the free region, as illustrated in Fig. 6. Any point on such a 01-path corresponds to a permissible multi-configuration: the point either lies on the boundary of U, in which case one of the robots is at its start or goal configuration, or it lies on the boundary of a forbidden region, in which case the robots touch each other.

This result—formulated in Lemma 1—does not rely on the low-density property of the environment: it applies to arbitrary environments. The interested reader is referred to [3] for details on the implications for motion planning in general environments; in the sequel we shall only focus on the consequences for low-density motion planning.

Lemma 1. *[3] Let $\mathcal{R}_1$ and $\mathcal{R}_2$ be two robots operating in the same workspace. If there is a feasible multi-path for given initial and final configurations, there is also a feasible multi-path for those initial and final configurations that only uses permissible multi-configurations.*

We now know that we can solve the motion-planning problem by looking at only a subspace of the composite configuration space $C_1 \times C_2$. This subspace consists of five *configuration-space slices*, or *slices* for short.

- In the first slice, $\mathcal{R}_2$ is free to move and $\mathcal{R}_1$ is stationary at its initial configuration; here we can simply consider $\mathcal{R}_1$ as an additional obstacle. We denote this configuration-space slice by $C_{1,s}$; its dimension is d_2.
- In $C_{1,g}$, the second slice, $\mathcal{R}_2$ is again free to move and $\mathcal{R}_1$ is stationary, but this time $\mathcal{R}_1$ is at its final configuration. Again, $\mathcal{R}_1$ is an additional obstacle, and the dimension of the slice is d_2.
- The third slice, $C_{2,s}$, is defined analogously to $C_{1,s}$, with the roles of $\mathcal{R}_1$ and $\mathcal{R}_2$ reversed. Its dimension is d_1.
- The fourth slice, $C_{2,g}$, is defined analogously to $C_{1,g}$, with the roles of $\mathcal{R}_1$ and $\mathcal{R}_2$ reversed. Its dimension is d_1.
- The fifth slice, C_c, is a configuration space for the *contact robot* $\mathcal{R}_1\mathcal{R}_2$, which is the robot composed of $\mathcal{R}_1$ and $\mathcal{R}_2$ where $\mathcal{R}_1$ and $\mathcal{R}_2$ are required to touch each other. The requirement that $\mathcal{R}_1$ and $\mathcal{R}_2$ should touch implies that the number of degrees of freedom is one less than the composite robot would normally have, so the dimension of C_c is $d_1 + d_2 - 1$.

Let us consider the subproblems corresponding to each of the slices in more detail. We recall that the free-space complexity for a not-too-large robot in a low-density environment is is $O(n)$, when n is the number of obstacles, and that a cell decomposition of the free space can be obtained in $O(n \log n)$ time (see Sections 3 and 4.1). Clearly, when one robot is fixed at its initial or final configuration and the other robot moves, the fixed robot may be regarded as one additional obstacle for the moving robot. This additional obstacle cannot increase the asymptotic complexity of the free space and the cell decomposition, so we can obtain cell decompositions of $O(n)$ size of the free parts of $C_{1,s}$, $C_{1,g}$,

$C_{2,s}$, and $C_{2,g}$ in $O(n \log n)$ time. The remaining type of multi-configuration is where the two robots form a contact robot. The size of a contact robot is bounded by the size of one of the constituent robots plus twice the size of the other constituent robot. As a result, it is possible to compute a cell decomposition of the free part of C_c of size $O(n)$ in $O(n \log n)$ time.

Of course we cannot treat the five slices completely separately; a feasible path using only permissible multi-configurations will in general switch between slices a number of times. In the example of Figure 5, for instance, the first part of the path lies in $C_{1,s}$, then (when $\mathcal{R}_2$ reaches $\mathcal{R}_1$) a switch is made to C_c, and finally a switch is made to $C_{2,g}$. We have to connect the slices to make such switches possible. We do this by identifying certain *transition points* in each slice. These points correspond to configurations that are represented by a point in one of the other slices as well. For a given transition point in a slice, we call the point in another slice that corresponds to the same configuration its *twin* in the other slice. Thus if we travel along a curve in one slice and reach a transition point, we may continue in the other slice from its twin. It suffices to identify $O(n)$ transition points (see [3] for details on the choice of these points). Their twins can be determined in $O(\log n)$ time per transition point using a simple point-location structure.

The results for two robots can be extended to three robots. The permissable multi-configurations are now defined as follows.

> There are k, for some $k = 0, 1, 2$, robots placed at either their start or goal configuration, and $2 - k$ pairs of robots that are in contact.

There are several different ways to achieve a permissible multi-configuration. We mention a few of the possibilities. One type of permissible multi-configuration is that $\mathcal{R}_1$ is at its start configuration and $\mathcal{R}_2$ and $\mathcal{R}_3$ form a contact robot. Another type is that $\mathcal{R}_1$ is at its goal configuration, $\mathcal{R}_2$ touches $\mathcal{R}_1$, and $\mathcal{R}_3$ is unconstrained. Although the number of types of permissible multi-configurations is fairly large, it is a constant. Each type of permissible multi-configuration gives rise to a configuration space slice, as in the case of two robots. Again it is possible to show—though the proof is considerably more complex (see [3])— that a solution consisting of only permissable multi-configurations exists provided that a solution exists at all. Moreover, it turns out that a cell decomposition of the (linear-complexity) free part of each of the slices can be obtained in $O(n \log n)$ time. Subsequent linking of the slices takes again $O(n \log n)$ time.

Theorem 7 summarizes the results obtained in this section. Again, these results also apply to motion planning amidst fat obstacles.

Theorem 7. *The motion planning problem for two or three not-too-large robots moving in a low-density environment can be solved in $O(n \log n)$ time.*

7   Conclusion

Exact algorithms often suffer from high worst-case running times, which prevent them from being used in practice. It has been noted that weak assumptions on the size of the robot compared to the obstacles and on the distribution of the obstacles—such as the low-density assumption—lead to a surprising reduction of the complexity of the motion planning problem for a single robot. In addition, these weak assumptions allow for a considerably more efficient approach to the exact solution of the motion planning problem.

We have discussed recent extensions of the result for one robot among stationary obstacles to the case of one robot in a two-dimensional low-density environment with moving obstacles [11] and to the case of two or three robots in a low-density environment with stationary obstacles [3]. We have obtained an almost-optimal $O(n^2\alpha(n)\log^3 n)$ algorithm for the motion planning problem among moving obstacles and almost-optimal $O(n\log n)$ algorithms for the motion planning problems for two or three robots. Both results are remarkable improvements of the bounds for general environments. In addition, we have shown that the complexity result for a single robot does not extend to environments that satisfy the even weaker unclutteredness assumption: the bounds for the complexity of motion planning in uncluttered environments fit between the $\Theta(n)$ bound for low-density environments and the $\Theta(n^f)$ bound for arbitrary environments.

Acknowledgements

Most of the results discussed in this paper are taken from the recent papers [3], [10], [11], and [41]. I thank my co-authors Boris Aronov, Mark de Berg, Robert-Paul Berretty, Matthew Katz, Mark Overmars, Petr Švestka, and Jules Vleugels.

References

1. P.K. AGARWAL, M.J. KATZ, AND M. SHARIR, Computing depth orders for fat objects and related problems, *Computational Geometry: Theory and Applications* **5** (1995), pp. 187-206.

2. H. ALT, R. FLEISCHER, M. KAUFMANN, K. MEHLHORN, S. NÄHER, S. SCHIRRA, AND C. UHRIG, Approximate motion planning and the complexity of the boundary of the union of simple geometric figures, *Algorithmica* **8** (1992), pp. 391-406.

3. B. ARONOV, M. DE BERG, A.F. VAN DER STAPPEN, P. ŠVESTKA, AND J. VLEUGELS, Motion planning for multiple robots, *Proceedings of the 14th Annual ACM Symposium on Computational Geometry* (1998), pp. 374-382.

4. F. AVNAIM, J.-D. BOISSONNAT, AND B. FAVERJON, A practical exact motion planning algorithm for polygonal objects amidst polygonal obstacles, *Proc. Geometry and Robotics Workshop* (J.-D. Boissonnat and J.-P. Laumond Eds.), Lecture Notes in Computer Science **391**, Springer Verlag, Berlin (1988), pp. 67-86.

5. S. BASU, R. POLLACK, AND M.-F. ROY, Computing roadmaps of semi-algebraic sets on a variety, in: *Foundations of Computational Mathematics* (F. Cucker and M. Shub eds.) (1997), pp. 1-15.

6. J. BAÑON, Implementation and extension of the ladder algorithm, *Proc. IEEE Int. Conf. on Robotics and Automation* (1990), pp. 1548-1553.

7. M. DE BERG, L. GUIBAS, AND D. HALPERIN, Vertical decompositions for triangles in 3-space, *Discrete & Computational Geometry* **15** (1996), pp. 35-61.

8. M. DE BERG, Linear size binary space partitions for uncluttered scenes, Technical Report UU-CS-1998-12, Dept. of Computer Science, Utrecht University.

9. M. DE BERG, M. KATZ, A.F. VAN DER STAPPEN, AND J. VLEUGELS, Realistic input models for geometric algorithms, *Proc. 13th Ann. ACM Symp. on Computational Geometry* (1997), pp. 294-303.

10. M. DE BERG, M. KATZ, M. OVERMARS, A.F. VAN DER STAPPEN, AND J. VLEUGELS, Models and motion planning, *Proc. 6th Scandinavian Workshop on Algorithm Theory (SWAT'98)*, Lecture Notes in Computer Science **1432**, Springer Verlag, Berlin (1998), pp. 83-94.

11. R.-P. BERRETTY, M. OVERMARS, AND A.F. VAN DER STAPPEN, Dynamic motion planning in low obstacle density environments, *Computational Geometry: Theory and Applications* **11** (1998), pp. 157-173.

12. J. CANNY AND J. REIF, New lower bound techniques for robot motion planning problems, *Proc. 28th IEEE Symp. on Foundations of Computer Science* (1987), pp. 49-60.

13. J. CANNY, Computing roadmaps of general semi-algebraic sets, *Comput. J.* **36** (1994), pp. 409-418.

14. A. EFRAT, G. ROTE, AND M. SHARIR, On the union of fat wedges and separating a collection of segments by a line, *Computational Geometry: Theory and Applications* **3** (1993), pp. 277-288.

15. A. EFRAT AND M. SHARIR, On the complexity of the union of fat objects in the plane, *Proc. 13th Ann. ACM Symp. on Computational Geometry* (1997), pp. 104-112.

16. A. EFRAT, M.J. KATZ, F. NIELSEN, AND M. SHARIR, Dynamic data structures for fat objects and their applications, *Proc. Workshop on Algorithms and Data Structures (WADS'97)*, Lecture Notes in Computer Science **1272**, Springer Verlag, Berlin (1997), pp. 297-306.

17. A. EFRAT AND M.J. KATZ, On the union of κ-curved objects, *Proc. 14th Ann. ACM Symp. on Computational Geometry* (1998), pp. 206-213.

18. D. HALPERIN, M.H. OVERMARS, AND M. SHARIR, Efficient motion planning for an L-shaped object, *SIAM Journal on Computing* **21** (1992), pp. 1-23.

19. M.J. KATZ, M.H. OVERMARS, AND M. SHARIR, Efficient hidden surface removal for objects with small union size, *Computational Geometry: Theory and Applications* **2** (1992), pp. 223-234.

20. M.J. KATZ 3-D vertical ray shooting and 2-D point enclosure, range searching, and arc shooting amidst convex fat objects, *Computational Geometry: Theory and Applications* **8** (1997), pp. 299-316.

21. M. VAN KREVELD, On fat partitioning, fat covering and the union size of polygons, *Computational Geometry: Theory and Applications* **9** (1998),pp. 197-210.

22. D. LEVEN AND M. SHARIR, An efficient and simple motion planning algorithm for a ladder amidst polygonal barriers, *Journal of Algorithms* **8** (1987), pp. 192-215.

23. J. MATOUŠEK, J. PACH, M. SHARIR, S. SIFRONY, AND E. WELZL, Fat triangles determine linearly many holes, *SIAM Journal on Computing* **23** (1994), pp. 154-169.

24. J.S.B. MITCHELL, D.M. MOUNT, S. SURI Query-sensitive ray shooting, *International Journal on Computational Geometry and Applications* **7** (1997), pp. 317-347.

25. M.H. OVERMARS, Point location in fat subdivisions, *Information Processing Letters* **44** (1992), pp. 261-265.

26. M.H. OVERMARS AND A.F. VAN DER STAPPEN, Range searching and point location among fat objects, *Journal of Algorithms* **21** (1996), pp. 629-656.
27. J. REIF AND M. SHARIR, Motion planning in the presence of moving obstacles, *Journal of the ACM* **41** (1994), pp. 764-790.
28. J.T. SCHWARTZ AND M. SHARIR, On the piano movers' problem: I. The case of a two-dimensional rigid polygonal body moving amidst polygonal boundaries, *Communications on Pure and Applied Mathematics* **36** (1983), pp. 345-398.
29. J.T. SCHWARTZ AND M. SHARIR, On the piano movers' problem: II. General techniques for computing topological properties of real algebraic manifolds, *Advances in Applied Mathematics* **4** (1983), pp. 298-351.
30. J.T. SCHWARTZ AND M. SHARIR, On the piano movers' problem: III. Coordinating the motion of several independent bodies: the special case of circular bodies moving amidst polygonal barriers, *International Journal of Robotics Research* **2** (1983), pp. 46-75.
31. J.T. SCHWARTZ AND M. SHARIR, On the piano movers' problem: V. The case of a rod moving in three-dimensional space amidst polyhedral obstacles, *Communications on Pure and Applied Mathematics* **37** (1984), pp. 815-848.
32. J.T. SCHWARTZ AND M. SHARIR, Efficient motion planning algorithms in environments of bounded local complexity, Report 164, Department of Computer Science, Courant Inst. Math. Sci., New York NY (1985).
33. O. SCHWARZKOPF AND J. VLEUGELS, Range searching in low-density environments, *Information Processing Letters* **60** (1996), pp. 121-127.
34. M. SHARIR AND E. ARIEL-SHEFFI, On the piano movers' problem: IV. Various decomposable two-dimensional motion planning problems, *Communications on Pure and Applied Mathematics* **37** (1984), pp. 479-493.
35. S. SIFRONY AND M. SHARIR, A new efficient motion planning algorithm for a rod in two-dimensional polygonal space, *Algorithmica* **2** (1987), pp. 367-402.
36. A.F. VAN DER STAPPEN, D. HALPERIN, M.H. OVERMARS, The complexity of the free space for a robot moving amidst fat obstacles, *Computational Geometry: Theory and Applications* **3** (1993), pp. 353-373.
37. A.F. VAN DER STAPPEN, The complexity of the free space for motion planning amidst fat obstacles, *Journal of Intelligent and Robotic Systems* **11** (1994), pp. 21-44.
38. A.F. VAN DER STAPPEN AND M.H. OVERMARS, Motion planning amidst fat obstacles, *Proc. 10th Ann. ACM Symp. on Computational Geometry* (1994), pp. 31-40.
39. A.F. VAN DER STAPPEN, *Motion planning amidst fat obstacles*, Ph.D. Thesis, Dept. of Computer Science, Utrecht University (1994).
40. A.F. VAN DER STAPPEN, Efficient exact motion planning in realistic environments, in: *Modelling and Planning for Sensor Based Intelligent Robot Systems* (H. Bunke, H. Noltemeier, T. Kanade Eds.), Series on Machine Perception & Artificial Intelligence, Volume **21**, World Scientific Publ. Co., Singapore (1995), pp. 51-66.
41. A.F. VAN DER STAPPEN, M.H. OVERMARS, M. DE BERG, AND J.VLEUGELS, Motion planning in environments with low obstacle density, *Discrete & Computational Geometry* **20** (1998), pp. 561-587.
42. K. SUTNER AND W. MAASS, Motion planning among time dependent obstacles, *Acta Informatica* **26**, pp. 93-133.
43. J. VLEUGELS, *On fatness and fitness—realistic input models for geometric algorithms*, Ph.D. Thesis, Dept. of Computer Science, Utrecht University (1997).

Robot Localization Using Polygon Distances*

Oliver Karch[1], Hartmut Noltemeier[1], and Thomas Wahl[2]

[1] Department of Computer Science I, University of Würzburg,
Am Hubland, 97074 Würzburg, Germany
{karch,noltemei}@informatik.uni-wuerzburg.de
[2] ATR Interpreting Telecommunications Research Laboratories,
2-2 Hikari-dai, Seika-cho, Soraku-gun, Kyoto 619-0288, Japan
twahl@itl.atr.co.jp

Abstract. We present an approach to the localization problem, for which polygon distances play an important role. In our setting of this problem the robot is only equipped with a map of its environment, a range sensor, and possibly a compass.

To solve this problem, we first study an idealized version of it, where all data is exact and where the robot has a compass. This leads to the pure geometrical problem of fitting a visibility polygon into the map. This problem was solved very efficiently by Guibas, Motwani, and Raghavan. Unfortunately, their method is not applicable for realistic cases, where all the data is noisy.

To overcome the problems we introduce a *distance function*, the *polar coordinate metric*, that models the resemblance between a range scan and the structures of the original method. We show some important properties of the polar coordinate metric and how we can compute it efficiently. Finally, we show how this metric is used in our approach and in our experimental Robot Localization Program RoLoPro.

1 The Localization Problem

We investigate the first stage of the *robot localization problem* [3,12]: an autonomous robot is at an unknown position in an indoor-environment, for example a factory building, and has to do a complete relocalization, that is, determine its position and orientation. The application we have in mind here is a wake-up situation (e.g., after a power failure or maintenance works), where the robot is placed somewhere in its environment, powered on, and then "wants" to know where it is located. Note, that we do not assume any knowledge about previous configurations of the robot (before its shutdown), because the robot might have been moved meanwhile.

In order to perform this task, the robot has a polygonal map of its environment and a range sensor (e.g., a laser radar), which provides the robot with a set of range measurements (usually at equidistant angles). The localization should

---

* This research is supported by the Deutsche Forschungsgemeinschaft (DFG) under project numbers No 88/14-1 and No 88/14-2.

Christensen et al. (Eds.): Sensor Based Intelligent Robots, LNAI 1724, pp. 200–219, 1999.
© Springer-Verlag Berlin Heidelberg 1999

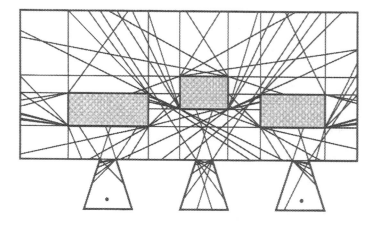

Fig. 1. Polygonal map and its decomposition into visibility cells

be performed using only this minimal equipment. In particular, the robot is not allowed to use landmarks (e.g., marks on the walls or on the floor). This should make it possible to use autonomous robots also in fields of application where it is not allowed or too expensive to change the environment.

The localization process usually consists of two stages. First, the non-moving robot enumerates all hypothetical positions that are *consistent* with its sensor data, i.e., that yield the same visibility polygon. There can very well be several such positions if the map contains identical parts at different places (e.g., buildings with many identical corridors, like hospitals or libraries). All those positions cannot be distinguished by a non-moving robot. Figure 1 shows an example: the marked positions at the bottom of the two outermost niches cannot be distinguished using only their visibility polygons.

If there is more than one hypothetical position, the robot eliminates the wrong hypotheses in the second stage and determines exactly where it is by traveling around in its environment. This is a typical on-line problem, because the robot has to consider the new information that arrives while the robot is exploring its environment. Its task is to find a path as efficient (i.e., short) as possible for eliminating the wrong hypotheses. Dudek et al. [4] have already shown that finding an optimal localization strategy is NP-hard, and described a competitive greedy strategy, the running time of which was recently improved by Schuierer [10].

This paper concentrates on the first stage of the localization process, that is, on generating the possible robot configurations (i.e., positions and orientations), although in our current work we also want to give solutions for the second stage, which can be applied in practice. With the additional assumption that the robot already knows its orientation (i.e., the robot has a *compass*) and all sensors and the map are *exact* (i.e., without any noise), this problem turns into a pure geometric one, stated as follows: for a given map polygon M and a star-shaped polygon V (the visibility polygon of the robot), find all points $p \in M$ that have V as their visibility polygon.

Guibas et al. [5] described a scheme for solving this idealized version of the localization problem efficiently and we will briefly sketch their method in the following section. As this more theoretical method requires exact sensors and an exact map, it is not directly applicable in practice, where the data normally is *noisy*. In Sect. 3 we consider these problems and show in Sections 4 and 5 an approach to avoiding them, which uses *distance functions* to model the resemblance between the noisy range scans (from the sensor) and the structures of the original method (extracted from the possibly inexact map).

2   Solving the Geometric Problem

In the following we sketch the method of Guibas et al., which is the basis for our approach described in Sections 4 and 5. We assume that the robot navigates on a plain surface with mostly vertical walls and obstacles such that the environment can be described by a polygon M, called the *map polygon*. Additionally, we assume that M has *no holes* (i.e., there are no free-standing obstacles in the environment), although the algorithm remains the same for map polygons with holes; the preprocessing costs, however, may be higher in that case.

The (exact) range sensor generates the star-shaped *visibility polygon* V of the robot. As the range sensor is only able to measure the distances relative to its own position, we assume the origin of the coordinate system of V to be the position of the robot. Using the assumption that we have a compass, the geometric problem is then to find all points $p \in M$ such that their visibility polygon V_p is identical with the visibility polygon V of the robot.

The main idea of Guibas et al. [5] to solving this problem is to divide the map into finitely many visibility cells such that a certain structure (the visibility skeleton, which is closely related to the visibility polygon) does not change inside a cell.

For a localization query we then do not search for points where the visibility polygon fits into the map, but instead for points where the corresponding skeleton does. That is, the continuous problem[1] of fitting a visibility polygon into the map is discretized in a natural way by decomposing the map into visibility cells.

2.1   Decomposing the Map into Cells

At preprocessing time the map M is divided into convex *visibility cells* by introducing straight lines forming the boundary of the cells such that the following property holds:

---

[1] "Continuous" in the sense that we cannot find an $\varepsilon > 0$ such that the visibility polygon V_p of a point p moving by at most ε does not change.

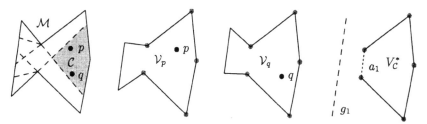

Fig. 2. Decomposition of a map polygon into visibility cells (left), two visibility polygons (middle), and the corresponding skeleton (right)

> The set of visible map vertices does not change when we travel around within a cell.

As the visibility of a vertex only changes if we cross a straight line induced by that vertex and an occluding *reflex* vertex (i.e., having an internal angle $\geq \pi$), the subdivision into visibility cells can be constructed in the following way: we consider all pairs consisting of a vertex v and a reflex vertex v_r that are visible from each other; for each such pair (v, v_r) we introduce the ray into the map that goes along the line through v and v_r, starts at v_r, and is oriented as to move away from v. An example of such a decomposition is depicted in the left part of Fig. 2. The introduced rays are drawn as dashed lines. The points p and q from cell $\mathcal{C}$ see the same set of five map vertices (marked gray in the corresponding visibility polygons in the middle). Figure 1 shows a decomposition for a more complex map with three obstacles (gray), generated with our software RoLoPro described in Sect. 6.

If the map consists of a total number of n vertices, of which r are reflex, the number of introduced rays is in $\Theta(nr)$ in the worst-case. Therefore the worst-case complexity of the decomposition is in $\Theta(n^2r^2)$. For map polygons without holes it can be shown that this complexity is actually in $\Theta(n^2r)$. Moreover, it is easy to give worst-case examples that show that these bounds are tight.

2.2 The Visibility Skeleton

When we compare two visibility polygons of points from the same cell (see Fig. 2), we observe that they are very similar and differ only in certain aspects, namely in the *spurious edges* that are caused by the reflex vertices and are collinear with the viewpoint, and in those map edges that are only *partially visible*. The remaining *full edges* (which are completely visible) are the same in both polygons. This observation can be used to define a structure that does not change inside a visibility cell, the *visibility skeleton*.

For a visibility polygon $\mathcal{V}_p$ with viewpoint p, the corresponding visibility skeleton $\mathcal{V}_p^*$ is constructed by removing the spurious edges, and by substituting the partially visible edges (they lie between two spurious vertices or between a spurious and a full vertex) with an *artificial edge* a_i together with the corresponding line g_i on which the original (partially visible) edge of V_p lies. Thus, we simply ignore the spurious edges and the spurious vertices, as this information continuously depends on the exact position p.

As the skeleton does not change inside a cell, we can define the *cell skeleton* V_C^* as the common skeleton of all visibility polygons of points from cell C. Figure 2 shows an example of the common skeleton V_C^* of two visibility polygons V_p and V_q for viewpoints p and q from the same cell C.

2.3   Performing a Localization Query

When we construct the skeleton V_p^* from the visibility polygon V_p, we "throw away" some information about the spurious edges and the partially visible edges. But this information can be reconstructed using the position of the viewpoint p relative to the skeleton. It is already shown by Guibas et al. that exactly those points q are valid robot positions for a given visibility polygon V_p that have the following two properties:

1. The point q lies in a visibility cell C with $V_C^* = V_p^*$.
2. The position of q relative to V_C^* is the same as the position of p relative to V_p^*.

The consequence is that in order to determine all points in the map that have V_p as their visibility polygon, it suffices to consider the equivalence class of all cells with visibility skeleton V_p^* (first property) and then to determine the subset of cells consisting of those that contain a viewpoint with the same relative position to V_p^* as p (second property). After that, all remaining viewpoints are valid robot positions.

Hence, for performing the localization query, the skeleton V_p^* of the given visibility polygon V_p is computed and the corresponding equivalence class of skeletons is determined. As we know the position of the point p relative to the skeleton V_p^* and as we also know the position of each cell C relative to its cell skeleton V_C^*, we can easily determine for each cell in the equivalence class the corresponding viewpoint of the robot and check whether it lies in C. This test can be performed very efficiently by using a point location structure; in fact, using a sophisticated preprocessing of the visibility cells only a *single point location query* is necessary in order to test the viewpoints of *all cells* in an equivalence class at once.

This way we get a query time of $\mathcal{O}(m + \log n + A)$ where m is the number of vertices of V_p and A denotes the size of the output, that is, the number of all reported robot locations, which can easily be shown to be in $\mathcal{O}(r)$.

The total preprocessing time and space of this approach is in $\mathcal{O}(n^2 r \cdot (n + r^2))$ for map polygons without holes (see [7]).

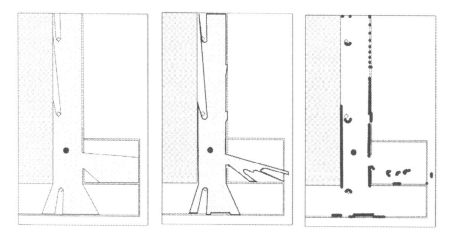

Fig. 3. Exact visibility polygon (left) and approximated visibility polygon (middle) of a noisy scan (right)

3   Problems in Realistic Scenarios

The idealizing assumptions of the method described in Sect. 2 prevent us from using it in realistic scenarios, as we encounter several problems:

- Realistic range sensors do not generate a visibility *polygon* V as assumed for the method, but only a finite sequence S of *scan points* (usually, measured at equidistant angles). Furthermore, these scan points do not lie exactly on the robot's visibility polygon, but are perturbed due to sensor uncertainties. An example is depicted in Fig. 3, which shows in the left part an exact visibility polygon of a robot (based on a map of our department). In the right part of the figure we see a real noisy laser range scan taken at the corresponding position in our department using a SICK LMS 200 laser scanner. Even if we connect the scan points by straight line segments as shown in the middle part of the figure, we only get an approximation V_S of the exact visibility polygon V based on the map.
- For the localization process we assume that we already know the exact orientation of the robot. But in practice this is often not the case, and we only have inexact knowledge or no knowledge at all about the robot's orientation.
- There may be obstacles in the environment that are not considered in the map and which may affect the robot's view. For example, furniture that is too small to be considered for map generation or even dynamic obstacles like people or other robots. Such obstacles can also be recognized in the right part of the example scan of Fig. 3.
- Realistic range sensors have a limited sensing range and obstacles that have a greater distance to the robot cannot be detected.

The consequence is that the (approximated) visibility skeleton V_S^*, which the robot computes from its approximated visibility polygon V_S, usually does not match any of the preprocessed skeletons exactly. That is, the robot is not able to determine the correct equivalence class, and the localization process completely fails.

4   Adaptation to Practice

Our approach to tackling these problems is, for a given range scan S (from the sensor), to search for the preprocessed skeleton that is *most similar* to the scan. For modeling the resemblance between a scan S and a skeleton V^* we use an appropriate *distance function* $d(S, V^*)$. Then, instead of performing an *exact match* query as in the original algorithm, we carry out a *nearest-neighbor* query in the set of skeletons with respect to the chosen distance function $d(S, V^*)$ to find the skeleton with the highest resemblance to the scan.

Depending on the distance function, we then additionally have to apply a local matching algorithm to the scan and the skeleton in order to determine the position of the robot. The reason is that not all methods for determining a distance measure yield an optimal matching (i.e., a translation vector and a rotation angle) as well. Consider, for example, the algorithm for computing the *Arkin metric* [2] for polygons, which, besides the distance measure, only provides the optimal rotation angle and no translation vector. In contrast to this, algorithms for computing the *minimum Hausdorff distance* (under rigid motions) [1] provide both, the distance measure and the corresponding matching.

4.1   Requirements to the Distance Function

In order to be useful in practice, a distance function $d(S, V^*)$ should at least have the following properties:

Continuity The distance function should be continuous in the sense that small changes in the scan (e.g., caused by noisy sensors) or even in the skeleton (e.g., caused by an inexact map) should only result in small changes of the distance. More precisely: Let $d_S(S_1, S_2)$ and $d_{V^*}(V_1^*, V_2^*)$ be functions that measure the resemblance between two scans S_1 and S_2 and between two skeletons V_1^* and V_2^*, respectively. An appropriate reference distance measure for $d_S(S_1, S_2)$ and $d_{V^*}(V_1^*, V_2^*)$ is, for example, the Hausdorff distance (see Sect. 5.1).

The distance $d(S, V^*)$ is said to be *continuous with respect to scans* if

$$\forall_{\varepsilon>0} \exists_{\delta>0} : d_S(S_1, S_2) < \delta \Rightarrow |d(S_1, V^*) - d(S_2, V^*)| < \varepsilon$$

holds, for all scans S_1, S_2 and all skeletons V^*. Analogously, $d(S, V^*)$ is said to be *continuous with respect to skeletons* if

$$\forall_{\varepsilon>0} \exists_{\delta>0} : d_{V^*}(V_1^*, V_2^*) < \delta \Rightarrow |d(S, V_1^*) - d(S, V_2^*)| < \varepsilon$$

holds, for all skeletons V_1^*, V_2^* and all scans S.

The requirement of continuity is also motivated by the fact that particularly the classification of the edges of the visibility polygon into different types (spurious edges, partially visible edges, etc.) makes the original method susceptible to perturbations: even a small translation of a vertex can change the type of an edge which yields a skeleton that does not match any equivalence class. In this sense, the exact match query of the original method can also be interpreted as a discrete distance between a visibility polygon and a skeleton, which, however, strongly violates the continuity requirement, because it takes only two values (e.g., 0 – "match" and 1 – "no match").

Similarity preservation A skeleton V^* that is similar to S should have a small distance value $d(S, V^*)$. Otherwise, the distance would not give any advice for finding a well-matching skeleton and therefore would be useless for the localization algorithm. In particular, if we take a scan S from a point p whose skeleton equals V^*, we want the distance $d(S, V^*)$ to be zero or at least small, depending on the amount of noise and the resolution of the scan.

Translational invariance As the robot has no knowledge about the relative position of the coordinate systems of the scan and the skeleton to each other, a translation of the scan or the skeleton in their local coordinate systems must not influence the distance. Rather finding this position is the goal of the localization algorithm.

Rotational invariance If the robot does not have a compass, the distance must also be invariant under rotations of the scan (or the skeleton, respectively).

Fast computability As the distance $d(S, V^*)$ has to be determined several times for a single localization query (for different skeletons, see Sect. 4.2), the computation costs should not be too high.

4.2   Maintaining the Skeletons

As we do not want to compare a scan with all skeletons to find the skeleton with the highest resemblance (remember that their number can be in $\Omega(n^2 r^2)$, see Sect. 2.1), the skeletons should be stored in an appropriate data structure that we can search through efficiently.

For this purpose we can use the *Monotonous Bisector Tree* [9], a spatial index that allows to partition the set of skeletons hierarchically with respect to a second distance function $D(V_1^*, V_2^*)$ that models the resemblance between two skeletons V_1^* and V_2^*. The set of skeletons is recursively divided into clusters with monotonously decreasing cluster radii in a preprocessing step. This division then represents the similarities of the skeletons among each other.

The distance function $D(V_1^*, V_2^*)$ should be chosen "compatible" to the function $d(S, V^*)$, such that in the nearest-neighbor query not all clusters have to be investigated. That is, at least the *triangle inequality*

$$d(S, V_2^*) \leq d(S, V_1^*) + D(V_1^*, V_2^*)$$

should be satisfied. This way, we can determine lower bounds for the distance values $d(S, V^*)$ of complete clusters, when traversing the tree. Such a cluster can then be rejected and does not have to be examined.

5   Suitable Distances for $d(S, V^*)$ and $D(V_1^*, V_2^*)$

It is hard to find distance functions that have all the properties from Sect. 4.1. Particularly, the fifth requirement (fast computability) is contrary to the remaining ones. Moreover, it is often not possible to simply use existing polygon distances, because in our problem we have to cope with scans and skeletons instead of polygons. Therefore, a careful adaptation of the distance functions is almost always necessary. And of course, it is even more difficult to find for a given scan-skeleton distance $d(S, V^*)$ a compatible distance $D(V_1^*, V_2^*)$, which we need for performing the nearest-neighbor query efficiently.

In the following, we investigate two distance functions, the Hausdorff distance and the polar coordinate metric, and illustrate the occurring problems.

5.1   The Hausdorff Distance

For two point sets $A, B \subset \mathbb{R}^2$, their *Hausdorff distance* $\delta(A, B)$ is defined as

$$\delta(A, B) := \max\{\vec{\delta}(A, B), \vec{\delta}(B, A)\} \; ,$$

where

$$\vec{\delta}(A, B) := \sup_{a \in A} \inf_{b \in B} \|a - b\|$$

is the *directed Hausdorff distance* from A to B, and $\|\cdot\|$ is the Euclidean norm. Accordingly, the term $\vec{\delta}(A, B)$ stands for the maximum distance of a point from A to the set B.

Let $\mathcal{T}$ be the set of all Euclidean transformations (i.e., combinations of translations and rotations). The undirected and directed *minimum Hausdorff distances* with respect to these transformations are defined as

$$\delta_{\min}(A, B) := \inf_{t \in \mathcal{T}} \delta(A, t(B)) \quad \text{and} \quad \vec{\delta}_{\min}(A, B) := \inf_{t \in \mathcal{T}} \vec{\delta}(A, t(B)) \; .$$

It can easily be shown that the minimum Hausdorff distances are continuous and by definition also fulfill the third and forth property of Sect. 4.1. But their computation is very expensive. According to [1], this can be done in time $\mathcal{O}((ms)^4(m + s) \log(m + s))$ if m is the complexity of the scan and s is the complexity of the skeleton. This is surely too expensive in order to be used in our application.

On the other hand, the computation of the Hausdorff distance without minimization over transformation application is relatively cheap [1], namely in $\mathcal{O}((m+s)\log(m+s))$. The property of continuity is also not affected, but we now have to choose a suitable translation vector and a rotation angle by hand.

An obvious choice for such a translation vector for a scan S and a skeleton V^* is the vector that moves the scan origin (i.e., the position of the robot) somewhere into the corresponding visibility cell C_{V^*} (e.g., the center of gravity of C_{V^*}). This is reasonable, because by the definition of the visibility cells, exactly the points in C_{V^*} induce the skeleton V^*. Of course, the consequence of doing so is that all cells with the same skeleton (e.g., the big cells in the two outermost niches in Fig. 1) must be handled separately, because the distance $d(S, V^*)$ now does not only depend on V^*, but also on the visibility cell itself.[2] Besides, their intersection may be empty and we might not find a common translation vector for all cells. Of course, the bigger the cell is that the scan has to be placed into, the bigger is the error of this approach, compared with the minimum Hausdorff distance.

A compromise for computing a good matching, which does have the advantages of the previous algorithms, is using an *approximate matching strategy*, which yields only a pseudo-optimal solution. This means, the algorithm finds a transformation $t \in \mathcal{T}$ with $\delta(A, t(B)) \leq c \cdot \delta_{\min}(A, B)$, for a constant $c \geq 1$. Alt et al. [1] showed that for any constant $c > 1$ an approximate matching with respect to Euclidean transformations can be computed in time $\mathcal{O}(ms \log(ms) \log^*(ms))$ using so-called *reference points*. If we only want an approximate matching with respect to translations instead of Euclidean transformations, the time complexity would even be in $\mathcal{O}((m+s)\log(m+s))$.

Another point to consider is that a skeleton (interpreted as a point set) in general is not bounded, because it includes a straight line for each artificial edge. The result is that the directed distances $\vec{\delta}(V^*, S)$ and $\vec{\delta}_{\min}(V^*, S)$ almost always return an infinite value (except for the trivial case when V^* equals the convex map polygon and has no artificial edges). Therefore, we must either modify the skeletons or we can only use the directed distances $\vec{\delta}(S, V^*)$ and $\vec{\delta}_{\min}(S, V^*)$. Note that if we pursue the second approach, the distance $\vec{\delta}_{\min}(S, V^*)$ is also similarity preserving, provided that the resolution of the scan is high enough such that no edge, in particular, no artificial edge, is missed.

5.2 The Polar Coordinate Metric

A more specialized distance for our problem than the Hausdorff distance is the *polar coordinate metric* (PCM for short) investigated by Wahl [11], which takes

---

[2] In this case, the notation $d(S, V^*)$ is a bit misleading, since there might exist several cells that have the same skeleton V^*. To be correct, we should use the notation $d(S, C_{V^*})$, where the dependence of the distance from the cell is expressed more clearly. But we will use the easier-to-understand expression $d(S, V^*)$.

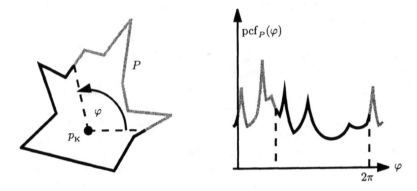

Fig. 4. The polar coordinate function $\mathrm{pcf}_P(\varphi)$ for a star-shaped polygon P

a fundamental property of our problem into account: all occurring polygons are *star-shaped* in the following sense, and we even know a kernel point:

– The approximate visibility polygon V_S (generated from the scan points) is star-shaped by construction with the origin as a kernel point.
– Every skeleton V^* is star-shaped in the sense that from every point in the corresponding visibility cell C_{V^*} all full edges are completely visible, and for each artificial edge a_i a part of the corresponding straight line g_i is visible.

To define the PCM between two (star-shaped) polygons P and Q with kernel points p_K and q_K we first define the value of the *polar coordinate function* (PCF for short)

$$\mathrm{pcf}_P(\varphi) : \mathbb{R} \to \mathbb{R}_{\geq 0}$$

as the distance from the kernel point p_K to the intersection point of a ray starting at p_K in direction φ with the boundary of P. That is, the function $\mathrm{pcf}_P(\varphi)$ corresponds to a description of the polygon P in polar coordinates (with p_K as the origin) and is periodical with a period of 2π. Figure 4 depicts the PCF for a star-shaped polygon as an example. In the same way we define the function $\mathrm{pcf}_Q(\varphi)$ for the polygon Q.

Then, the PCM between the polygons P and Q is the minimum integral norm between the functions pcf_P and pcf_Q in the interval $[0, 2\pi[$ over all horizontal translations between the two graphs (i.e., rotations between the corresponding polygons):

$$\mathrm{pcm}(P,Q) := \min_{t \in [0,2\pi[} \sqrt{\int_0^{2\pi} \left(\mathrm{pcf}_P(\varphi - t) - \mathrm{pcf}_Q(\varphi)\right)^2 \, \mathrm{d}\varphi} \tag{1}$$

Figure 5 shows an example, where the two graphs are already translated in a way such that this integral norm is minimized.

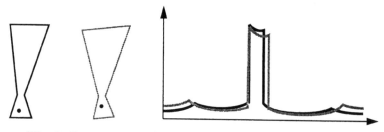

Fig. 5. Computation of the PCM as minimum integral norm

For a fixed kernel point the function pcf_P is continuous in φ except for one special case: when we move a vertex of a polygon edge such that the edge becomes collinear to p_K, the function pcf_P has a discontinuity at the corresponding angle, the height of which represents the length of the collinear edge. Moreover, the PCF is also continuous in the sense of the definitions in Sect. 4.1 with respect to translations of the polygon vertices or translations of the kernel point unless this special case occurs. But as pcf_P and pcf_Q may have only finitely many such discontinuities, the integration makes them continuous with respect to *all* translations of polygon vertices and translations of the kernel points, provided that P and Q remain star-shaped with kernel points p_K and q_K.

It can easily be seen that the PCM fulfills the continuity requirement of Sect. 4.1, if the kernel points are considered as a part of the polygons (i.e., part of the *input* of the PCM). This means that, given two polygons P and Q and an $\varepsilon > 0$, we can find a $\delta > 0$ such that $|\mathrm{pcm}(P,Q) - \mathrm{pcm}(P',Q)| < \varepsilon$, for all polygons P' that are created from P by moving all vertices *and* the kernel point p_K by at most δ. Moreover, if the kernel points are *not* considered as input of the PCM (that is, they are computed from P and Q by the algorithm that computes $\mathrm{pcm}(P,Q)$), the PCM is continuous as well, provided that the kernel points depend continuously on the polygons. For example, the center of gravity of the kernel of a polygon P depends continuously on P and can be used as a kernel point p_K, whereas, for example, the left-most kernel point does *not* depend continuously on the polygon.

Wahl [11] also showed that the function $\mathrm{pcm}(P,Q)$ is a polygon *metric*, provided that the kernel points are invariant under Euclidean transformations. That is, if p'_K denotes the kernel point of a polygon $P' = t(P)$ for a transformations $t \in \mathcal{T}$, the equality $t(p_K) = p'_K$ must hold, for all polygons P and all $t \in \mathcal{T}$. For example, the center of gravity of the kernel of the polygon has this property.

Using the PCM as distance $d(\mathcal{S}, V^*)$ If we want to use the PCM as a distance function $d(\mathcal{S}, V^*)$ we need corresponding star-shaped polygons for $\mathcal{S}$ and V^* that can be used as polygonal representatives for the scan and the skeletons:

- For the scan, we choose the approximated visibility polygon $\mathcal{V}_S$, which is star-shaped by construction. Again, the coordinate origin can be used as a kernel point.
- For generating a polygon from a skeleton V^* (with corresponding cell $\mathcal{C}_{V^*}$), we choose a point c inside the cell $\mathcal{C}_{V^*}$ (e.g., the center of gravity) and determine the visibility polygon $\mathcal{V}_c$ of this point. By construction, c is a kernel point of V_c^*.

In the sequel we will only use $\mathcal{V}_S$ and $\mathcal{V}_c$ for determining the distance measures. Then, our goal is to find the polygon $\mathcal{V}_c$ that is most similar to the approximated visibility polygon $\mathcal{V}_S$ with respect to $\mathrm{pcm}(\mathcal{V}_S, \mathcal{V}_c)$. With this choice we obtain the following theorem about the polar coordinate metric as a distance function:

Theorem 1. *The distance function $d(\mathcal{S}, V^*) := \mathrm{pcm}(\mathcal{V}_S, \mathcal{V}_c)$, with $\mathcal{V}_S$ and $\mathcal{V}_c$ as defined above, fulfills the following requirements from Sect. 4.1: continuity and invariance against translations and rotations.*

Note that the PCM is not similarity preserving: if the point c chosen above for computing a corresponding polygon for a visibility cell $\mathcal{C}$ does not equal the robot's position, the two polygons that are compared by the PCM are different and their distance value cannot be zero. But in practice, the visibility cells usually are not too large. That means, if we take a scan at a position $p \in \mathcal{C}$, the distance from p to the corresponding point $c \in \mathcal{C}$ is not too large. Thus, the approximated visibility polygon $\mathcal{V}_S$ and the visibility polygon $\mathcal{V}_c$ differ not too much, and the value of $\mathrm{pcm}(\mathcal{V}_S, \mathcal{V}_c)$ is small.

Computing the PCM value efficiently The exact computation of the minimum in (1) seems to be difficult and time consuming, since the polar coordinate functions of a polygon with p edges consists of p pieces of functions of the form $c_i / \sin(\varphi + \alpha_i)$. For one fixed translation t (this corresponds to the case, when the robot already knows its exact orientation), the integral

$$\sqrt{\int_0^{2\pi} \left(\mathrm{pcf}_P(\varphi - t) - \mathrm{pcf}_Q(\varphi)\right)^2 \, \mathrm{d}\varphi}$$

can be computed straightforward in linear time $\mathcal{O}(p+q)$, where p and q stand for the complexities of the two polygons. But the global minimum over all possible values of $t \in [0, 2\pi[$ seems to be much harder to determine.

Therefore, we use two different approximative approaches for computing a suitable PCM value. Both approaches use a set of supporting angles/points for each of the two involved polar coordinate functions. For a given polygon its supporting angles are the angles that correspond to a vertex of the polygon plus the angles of the local minima of the inverse sine functions (see the left part of Fig. 6). The ideas of the two approximative approaches are then as follows:

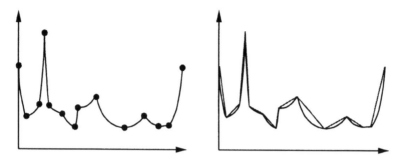

Fig. 6. Two approaches for an approximative PCM value

1. For the first approach we concentrate on the $\mathcal{O}(p)$ (or $\mathcal{O}(q)$, respectively) supporting angles (see the left part of Fig. 6). Namely, we do not minimize over *all rotation angles* of the two polygons, but only over the $\mathcal{O}(pq)$ rotation angles, that place one supporting angle of the first polygon on a supporting angle of the second one. The integral values are then computed exactly in time $\mathcal{O}(p + q)$. Summing up, we need $\mathcal{O}(pq(p + q))$ time to compute this approximated value of the PCM.

2. In the second approach we compute an *exact minimum* over all rotation angles, but use a modified polar coordinate function. Namely, we introduce a linear approximation of the PCM, which also has all metric properties and is sufficient for our applications. This approximation is depicted in the right part of Fig. 6: the supporting points (that correspond to a polygon vertex or a local minima of the PCF) are connected by straight line segments to get a modified PCF. The minimum integral norm is then defined like in the non-approximated version of the PCM (see (1)). Following an idea of Arkin et al. [2], the actual computation of the minimum integral norm between the two *piecewise linear* functions can now be carried out much faster than the computation of the original PCM: Arkin et al. compute the minimum integral norm between two *piecewise constant* functions in time $\mathcal{O}(pq)$. This idea can be generalized to compute the approximated PCM in time $\mathcal{O}(pq \cdot (p+q))$. Of course, if we do not want to minimize over the rotations, the computation time is again in $\mathcal{O}(p + q)$ like for the non-approximated version.

Using the PCM as distance $D(V_1^*, V_2^*)$ Since the PCM has all metric properties, it particularly fulfills the triangle inequality. Therefore, we can use it not only for defining the scan-skeleton distance $d(\mathcal{S}, V^*)$, but also for defining a *compatible* skeleton-skeleton distance $D(V_1^*, V_2^*)$.

For this task, we again use for each pair of a skeleton V^* and its corresponding cell $\mathcal{C}_{V^*}$ the polygonal representative $\mathcal{V}_c$ as defined in the last section. Then, the triangle inequality

$$d(\mathcal{S}, V_2^*) \leq d(\mathcal{S}, V_1^*) + D(V_1^*, V_2^*)$$

follows immediately from the triangle inequality of the PCM,

$$\mathrm{pcm}(\mathcal{V}_S, \mathcal{V}_{c_2}) \leq \mathrm{pcm}(\mathcal{V}_S, \mathcal{V}_{c_1}) + \mathrm{pcm}(\mathcal{V}_{c_1}, \mathcal{V}_{c_2}) \ ,$$

and we can apply the Monotonous Bisector Tree to the set of skeletons for increasing the performance of the nearest-neighbor query.

The PCM in realistic scenarios Since the PCM is continuous and invariant under Euclidean transformations, we can hope that the first two problems mentioned in Sect. 3 (the noisy scans and the unknown robot orientation) are solved satisfactorily.

Also the third problem (unknown obstacles in the environment) does not strongly influnce the PCM as long as the obstacles take up only a small interval of the whole scanning angle. But the other case, where the obstacles occupy a large part of the robot's view, poses a problem to our localization method. Here, additional (heuristic) algorithms are needed to detect such cases.

In contrast to this, a possibly limited sensing range (the fourth problem addressed in Sect. 3) can be easily tackled in a straightforward manner: when we compute the distances $d(S, V^*)$ and $D(V_1^*, V_2^*)$, we simply cut off all distance values larger than the sensing range, that is, for all occuring polygons we use a modified PCF,

$$\mathrm{pcf}'_P(\varphi) := \min\{\mathrm{pcf}_P(\varphi), \text{sensing range}\} \ .$$

6   Our Implementation RoLoPro

We have implemented the two versions of the localization algorithm in C++ using the LEDA Library of Efficient Datatypes and Algorithms [8], namely the original method described in Sect. 2 for exact sensors as well as the modification for realistic scenarios introduced above. Here, the original algorithm was modified and simplified at some points, since we did not focus our efforts on handling sophisticated but rather complicated data structures and algorithmic ideas that were suggested by Guibas et al. Rather, we wanted to have an instrument to experiment with different inputs for the algorithm that is reasonably stable to be used in real-life environments in the future and that can serve as a basis for own modifications. A consequence is that the program does not keep to all theoretical time and space bounds proven in [5,6], as this would have required a tremendous programming effort. Nevertheless, it is reasonably efficient. Figure 7 shows some screen shots of our robot localization program RoLoPro processing localization queries in real as well as in simulated environments.

As distance function $d(S, V^*)$ we have implemented the Hausdorff distance and the polar coordinate metric described in Sect. 5. Only for the PCM the efficient skeleton management described in Sect. 4.2 was implemented, since for the Hausdorff distance we could not find a suitable distance $D(V_1^*, V_2^*)$. Therefore, in this case the scan has to be compared with *all skeletons*, which is

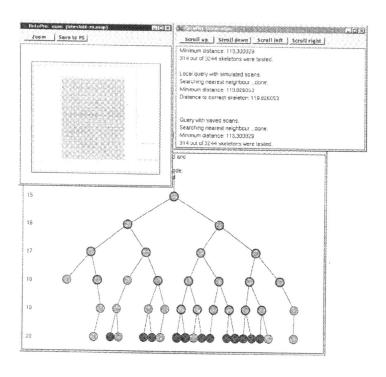

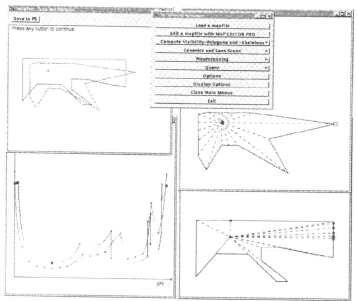

Fig. 7. Screen shots of RoLoPro

much more time consuming than the PCM approach, which uses the Monotonous Bisector Tree.

Furthermore, we have implemented some additional features into RoLoPro, described in the following.

Noisy compass We are able to model different kinds of compasses by using modified *minimization intervals* in (1), namely

- an exact compass: here the minimization interval consists only of a single value that represents the orientation of the robot with respect to the map,
- no compass at all: since we know nothing about the robot's orientation, we use $[0, 2\pi[$ as the minimization interval,
- a noisy compass: here we use a smaller interval $[o - \varepsilon, o + \varepsilon[\subset [0, 2\pi[$ that depends on the orientation of the robot and the uncertainty $\varepsilon < \pi$ of the compass.

In most cases we have to carefully adapt the distances $d(S, V^*)$ and $D(V_1^*, V_2^*)$ in different ways, such that the requirements from Sect. 4.1 and the triangle inequality remain fulfilled.

Partial range scans By modifying the *integration interval* in (1) we also can process partial range scans, where the scanning angle is less than 2π. As well as in the noisy-compass feature the distances have to be carefully modified, in particular if we want to use both features at the same time.

Additional coarsening step The complexity of the visibility cell decomposition, which is necessary for the geometrical approach described in Sect. 2 and which is the basis for our own approach, is quite high. But in practice many of the visibility cells need not be considered, because their skeletons differ only slightly from the skeletons of neighboring cells and the cells themselves are often very small.

Therefore, we have implemented an additional coarsening step, where neighboring cells with small distances $D(V_1^*, V_2^*)$ are combined into one bigger cell. This way, we need less space for storing the cells as well as less time for constructing the Monotonous Bisector Tree. Of course, this coarsening procedure reduces the quality of the localization, since we obtain both, a fewer number of distance values and larger visibility cells. Thus, the threshold value for the merging step must be carefully chosen by the user.

7   First Experimental Tests

First tests in small simulated scenes have shown a success rate of approximately 60 % for the directed Hausdorff distance, i.e., the scan origins of about 60

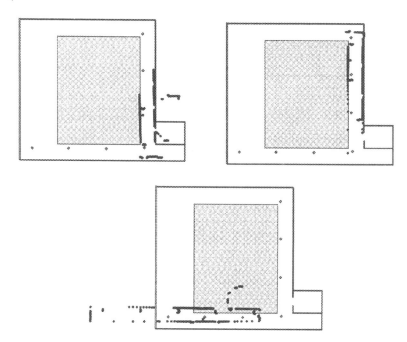

Fig. 8. Some localization examples using the PCM

out of 100 randomly generated scans were inside that cell with the smallest distance $d(\mathcal{S}, V^*)$ to the scan. In the same scenes the success rate of the polar coordinate metric was about 80-90 %.

We also have evaluated our approach in real environments (using the PCM as distance function), namely in our department and in a "less friendly" supermarket environment, where the scans were very noisy. In both cases the scans were generated by a SICK PLS/LMS 200 laser scanner and we always assumed that the robot has a compass.

Figure 8 shows some localization examples for the first environment: we used a partial map of our department, which consists of 76 vertices, 54 of them reflex. This led to a total number of about 4 300 visibility cells, which were finally reduced to about 3 300 cells by our coarsening step (using a suitable threshold). Each of the full range scans consists of 720 range measurements.

Almost all localization queries resulted in solutions like the first two examples in Fig. 8. Only about 5-10 % of the localization queries failed. For example, the reason for the bad result of the third query in Fig. 8 probably were some scan points at a distance of about 50 m in the right part of the scan (cut off in the figure), which were produced by scanning through a window. Since the PCM "tries" to minimize a kind of *average* quadratic error, the scan was shifted in the opposite direction to the left.

In the supermarket environment our current localization approach completely failed, probably because of the very noisy range scans, which produced effects like the third example of Fig. 8. Moreover, in this environment only half range scans were available such that the localization problem gets even harder.

8   Future Work

Our main goals for the future are to overcome some of the current limitations of our approach:

- We try to further reduce the large space requirements of our approach (caused by the space consuming visibility cell decomposition). For this task we want to use a more sophisticated approach than our current coarsening step, for example by considering also the scan resolution and the limited sensing range.
- Currently, the performance of our approach is not good enough, if we have no compass information, because the distance computation for this case takes $\mathcal{O}(pq(p+q))$ time, instead of linear time for the case of an exact compass. Therefore, we want to speed up the distance computation using a suitable scan preprocessing.
- For the case of extremely disordered scans we get very bad localization results (like in the supermarket environment described above). To overcome this problem, we try to modify the PCM appropriately and try to do an additional preprocessing for the scans.

Another goal is to implement a matching algorithm (see Sect. 4), which eventually determines the position of the robot from the scan and that skeleton with highest resemblance to the scan. Otherwise, we would only know the *visibility cell*, where the robot is located, that is, a possibly coarse approximation of its position.

Furthermore, we also want to integrate navigation algorithms into our approach, such that the robot can move around to eliminate ambiguous positions.

Acknowledgements

We gratefully acknowledge helpful comments given by Rolf Klein, Sven Schuierer, and Knut Verbarg.

This research is supported by the Deutsche Forschungsgemeinschaft (DFG) under project numbers No 88/14-1 and No 88/14-2.

References

1. H. Alt, B. Behrends, and J. Blömer. Approximate Matching of Polygonal Shapes. *Annals of Mathematics and Artificial Intelligence*, 13:251–265, 1995.
2. E. M. Arkin, L. P. Chew, D. P. Huttenlocher, K. Kedem, and J. S. B. Mitchell. An Efficiently Computable Metric for Comparing Polygonal Shapes. *IEEE Transactions on Pattern Analysis and Machine Intelligence*, 13:209–216, 1991.

3. I. J. Cox. Blanche — An Experiment in Guidance and Navigation of an Autonomous Robot Vehicle. *IEEE Transactions on Robotics and Automation*, 7(2):193–204, April 1991.
4. G. Dudek, K. Romanik, and S. Whitesides. Localizing a Robot with Minimum Travel. *SIAM Journal on Computing*, 27(2):583–604, 1998.
5. L. J. Guibas, R. Motwani, and P. Raghavan. The Robot Localization Problem. In K. Goldberg, D. Halperin, J.-C. Latombe, and R. Wilson, editors, *Algorithmic Foundations of Robotics*, pages 269–282. A K Peters, 1995. http://theory.stanford. edu/people/motwani/postscripts/localiz.ps.Z.
6. O. Karch. A Sharper Complexity Bound for the Robot Localization Problem. Technical Report No. 139, Department of Computer Science I, University of Würzburg, June 1996. http://www-info1.informatik.uni-wuerzburg.de/publications/karch/ tr139.ps.gz.
7. O. Karch and Th. Wahl. Relocalization — Theory and Practice. *Discrete Applied Mathematics (Special Issue on Computational Geometry), to appear*, 1999.
8. K. Mehlhorn and S. Näher. LEDA – A Platform for Combinatorial and Geometric Computing. *Communications of the ACM*, 38:96–102, 1995. http://www.mpi-sb. mpg.de/guide/staff/uhrig/ledapub/reports/leda.ps.Z.
9. H. Noltemeier, K. Verbarg, and C. Zirkelbach. A Data Structure for Representing and Efficient Querying Large Scenes of Geometric Objects: MB*–Trees. In G. Farin, H. Hagen, and H. Noltemeier, editors, *Geometric Modelling*, volume 8 of *Computing Supplement*, pages 211–226. Springer, 1993.
10. S. Schuierer. Efficient Robot Self-Localization in Simple Polygons. In R. C. Bolles, H. Bunke, and H. Noltemeier, editors, *Intelligent Robots – Sensing, Modelling, and Planning*, pages 129–146. World Scientific, 1997.
11. Th. Wahl. Distance Functions for Polygons and their Application to Robot Localization (in German). Master's thesis, University of Würzburg, June 1997.
12. C. M. Wang. Location estimation and uncertainty analysis for mobile robots. In I. J. Cox and G. T. Wilfong, editors, *Autonomous Robot Vehicles*. Springer, Berlin, 1990.

On-Line Searching in Simple Polygons[*]

Sven Schuierer

Institut für Informatik, Universität Freiburg
Am Flughafen 17, Geb. 051, D-79110 Freiburg, FRG
schuiere@informatik.uni-freiburg.de

Abstract. In this paper we study the problem of a robot searching for a visually recognizable target in an unknown simple polygon. We present two algorithms. Both work for arbitrarily oriented polygons. The search cost is proportional to the distance traveled by the robot. We use competitive analysis to judge the performance of our strategies. The first one is a simple modification of Dijkstra's shortest path algorithm and achieves a competitive ratio of $2n - 7$ where n is the number of vertices of the polygon. The second strategy achieves a competitive ratio of $1 + 2 \cdot (2k)^{2k}/(2k-1)^{2k-1}$ if the polygon is k-spiral. This can be shown to be optimal. It is based on an optimal algorithm to search in geometric trees which is also presented in this paper.

Consider a geometric tree with m leaves and a target hidden somewhere in the tree. The target can only be detected if the robot reaches the location of the target. We provide an algorithm that achieves a competitive ratio of $1 + 2m^m/(m-1)^{m-1}$ and show its optimality.

1 Introduction

An important problem in robotics is to plan a path from a starting point to a goal. Recently, this problem has been extensively investigated in the case that the robot does not have any prior knowledge of its environment and the position of the goal is unknown. In such a setting the robot gathers local information through its on-board sensors and tries to find a "short" path from the start point to the goal. The search of the robot can be viewed as an *on-line* problem since the robot has to make decisions about the search based only on the part of its environment that it has explored before. This invites the application of competitive analysis to judge the performance of an on-line search strategy [ST85,PY91]. In competitive analysis we compare the distance traveled by the robot to the length of the shortest path from its starting point s to the goal g. The ratio of the distance traveled by the robot to the length of a shortest path from s to g maximized over all possible locations s and g is called the *competitive ratio* of the search strategy.

One of the fundamental problems in this setting is searching on m concurrent rays [BYCR93,Gal80]. Here a point robot is imagined to stand at the origin of m rays and one of the rays contains the goal g whose distance to the origin

---

[*] This research is supported by the DFG-Project "Diskrete Probleme", No. Ot 64/8-2.

Christensen et al. (Eds.): Sensor Based Intelligent Robots, LNAI 1724, pp. 220–239, 1999.
© Springer-Verlag Berlin Heidelberg 1999

is unknown. The robot can only detect g if it stands on top of it. There is a simple optimal algorithm with a competitive ratio of $1 + 2m^m/(m-1)^{m-1}$. If the rays are allowed to contain branching vertices, then we obtain a geometric tree. Icking presents an algorithm to search in a geometric tree with m leaves that has a competitive ratio of $8m - 3$ [Ick94,Kle97]. This algorithm can be applied to search in a simple polygon and achieves a competitive ratio of $8n - 3$ if the polygon has n vertices.

In this paper we show how to adapt the optimal deterministic strategy to search on m concurrent rays to geometric trees, i.e., trees embedded in d-dimensional Euclidean space. We obtain an algorithm with a competitive ratio of $1 + 2m^m/(m-1)^{m-1} \leq 1 + 2em$ if the tree contains m leaves which can be shown to be optimal. The algorithm we obtain can be directly applied to searching in simple polygons which leads to a competitive ratio of $1 + 2 \cdot (2k)^{2k}/(2k-1)^{2k-1}$ if the polygon boundary can be decomposed into k convex and k reflex chains. It is not necessary for our algorithm to know the number k in advance. We also present a different strategy which is a modification of Dijkstra's shortest path algorithm with a competitive ratio of $2n - 7$.

Our algorithms work in any simple polygon. This should be contrasted with previous search algorithms that require restricted classes of polygons as their input [DHS95,DI94,Kle92,Kle94,LOS96,LOS97].

The paper is organized as follows. In the next section we introduce some definitions. In Section 3 we present the algorithm to search in a geometric tree and analyse its competitive ratio. Section 4 shows how the algorithm can be applied to searching in a simple polygon. It also introduces a second algorithm based on Dijkstra's algorithm for searching in a simple polygon. We conclude with some final remarks in Section 5.

2   Definitions

Let $\mathcal{C}$ be a closed curve without self-intersections consisting of n line segments such that no two consecutive segments are collinear. We define a *simple polygon* P to be the union of $\mathcal{C}$ and its interior. A vertex of P is the end point of a line segment on the boundary of P. A vertex is called *convex* if the subtending angle in the interior of P is less than $\pi/2$ and reflex otherwise.

Definition 1. *A geometric tree $T = (V, E)$ is a tree embedded into $\mathbb{E}^d$ such that each $v \in V$ is a point and each edge $e \in E$ is a polygonal path whose end points lie in V. The paths of E intersect only at points in V, and they do not induce any cycles.*

We only consider unbounded trees, that is, we assume that the leaves are rays that extend to infinity (see Figure 1).

If p and q are two points inside P, then we denote the line segment from p to q by $\overline{pq}$. The shortest path from p to q is denoted by $shp(p, q)$. The union of all shortest paths from a fixed point p to the vertices of P forms a geometric tree called the *shortest path tree* of p [Kle97]. It is denoted by T_p.

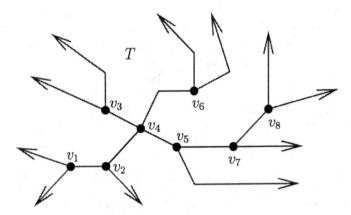

Fig. 1. A geometric tree T embedded in $I\!E^2$ with 11 leaf rays and 8 vertices.

If $\mathcal{P}$ is a path in P, then the *visibility polygon* $vis(\mathcal{P})$ of $\mathcal{P}$ is defined as set of all points in P that are seen by at least one point of $\mathcal{P}$. A maximal line segment of the boundary of $vis(\mathcal{P})$ that does not belong to the boundary of P is called a *window* of $vis(\mathcal{P})$. A window w splits P into two parts one of which contains $vis(\mathcal{P})$. The part that does not contain $vis(\mathcal{P})$ is called the *pocket of* w. In general, a line segment that touches the polygon boundary exactly in its end points is called a *chord*. A chord splits the polygon into two disconnected regions.

Let A be an algorithm to search for a target g in an environment class $\mathcal{E}$. We denote the length of the path traveled by the robot to find g using algorithm A starting at s in $E \in \mathcal{E}$ by $L_A(s,g,E)$ and the length of the shortest path from the robot location s to g by $L_{opt}(s,g,E)$. The algorithm A is called c-competitive if, for all environments $E \in \mathcal{E}$ and starting and goal positions $s, t \in E$,

$$L_A(s,g,E) \leq c \cdot L_{opt}(s,g,E).$$

The value c is called the *competitive ratio* of algorithm A.

3   Searching in Trees

In this section we consider the problem of searching in geometric trees. Before we describe our strategy to search in trees, we briefly discuss the optimal algorithm to search for a point in m concurrent rays.

3.1   Multi-way Ray Search

The model we consider is the following. A point robot is placed at the meeting point or origin of m concurrent rays and has to find a point g which is located in one of the rays. The distance of the point g is unknown to the robot but it is given a lower bound d_{min} on the distance to g. In the following we assume without loss of generality that $d_{min} = 1$. The robot can detect the point g only when it reaches g. A deterministic strategy which achieves the optimal competitive ratio works as follows [BYCR93,Gal80]. The robot visits the rays one by one in a round robin fashion until the point g is reached. On every ray, the robot goes a certain distance and turns back if the point g is not found and explores the next ray. The distance from the origin that the robot walks before the i-th turn is determined by the function

$$f(i) = \left(\frac{m}{m-1} \right)^i,$$

that is, each time the robot visits a new ray it increases the distance traveled by a factor of $m/(m-1)$.

It is easy to see that the worst case ratio of the distance traversed by the robot to the actual distance of g from the origin (i.e., the competitive ratio of this strategy) is

$$1 + 2 \frac{m^m}{(m-1)^{m-1}} \leq 1 + 2em.$$

For instance, if $m = 2$, then the competitive ratio is 9.

3.2 Geometric Trees

We assume that a point robot is located on an edge or at a vertex of the geometric tree T, and it has to find a target hidden somewhere in T. We assume that all the vertices of T have degree greater than 2 since a vertex of degree 2 can be removed by joining the polygonal paths of two edges into one.

At first it seems that the problem of searching in geometric trees can be solved trivially by using the optimal strategy to search on m rays if the tree has m leaves. However, there is one crucial difference. In the ray searching problem the number m of the rays to be searched is known in advance whereas in a geometric tree the number of branch vertices of the tree and their degree is unknown. Icking provides a simple way to avoid this problem by dividing the search into *phases* [Ick94]. In phase i all the branches are explored to a depth of 2^i. The phase ends when the last branch is visited. Then the search depth is doubled, and a new phase starts. It can be easily seen that the competitive ratio of this strategy is at most

$$\frac{2 \sum_{i=0}^{k} m2^i + 2(m-1)2^{k+1} + 2^k}{2^k} \leq 8m - 3,$$

as the robot explores at most $m - 1$ branches to a distance of 2^{k+1} before the target is detected on branch m at a distance of $2^k + \varepsilon$, for some $\varepsilon > 0$.

In the following we show that by adapting the strategy to search on m rays more carefully, one can achieve a competitive ratio of at most

$1 + 2m^m/(m-1)^{m-1}$; that is, searching in a tree with m leaves has the same competitive ratio as searching on m concurrent rays.

We assume again that we are given a lower bound d_{min} on the distance to the target as well as the distance to the closest branch vertex. In the applications that we present such a bound can be easily derived. As above we can set $d_{min} = 1$. Assume that in the beginning m_1 rays are incident to the root of the tree. We start off with the optimal search strategy to search m_1 rays. The initial step length is set to $d_1 = (m_1 - 1)^{m_1}/m_1^{m_1}$, that is, the robot explores only a fraction of the lower bound in the first m_1 steps. This guarantees that every ray has been visited at least once before a branch vertex is detected.

The strategy works as follows. Assume the currently explored tree has m leaves $l_1, \ldots, l_m$. The robot searches the m root-to-leaf paths by simulating the strategy to search on m concurrent rays. Let $\mathcal{P}_i$ be the path from the root to leaf l_i. The robot travels on each path $\mathcal{P}_i$ for a distance of d_i where d_i is the current step length and then returns to the origin. What happens if new edges branching off $\mathcal{P}_i$ are discovered is described below. The requirement that the robot should return to the origin is, of course, overly restrictive since the branching pattern of the tree often makes it unnecessary to return to the root. Instead, the robot only needs to return to the last vertex that belongs to both $\mathcal{P}_i$ and the new path $\mathcal{P}_{i+1}$ to be explored. While in practice a robot using our algorithm would of course make use of such shortcuts, it is conceptually more convenient to assume that the robot returns to the origin; that is, we treat the paths $\mathcal{P}_i$ as separate rays.

As indicated above, it may happen that the robot discovers new edges while traveling on $\mathcal{P}_i$. Once it discovers a new edge on $\mathcal{P}_i$, it reduces its step length d_i by a factor of $(1 - 1/m^2)^m$ and increases m by one where m is the number of leaves of the currently explored tree. This only happens the first time a new edge branching off $\mathcal{P}_i$ is discovered. All other new edges are just collected in a list without affecting d_i or m. After the robot has traversed a distance of d_i, it returns to the first vertex at which a new edge was detected. It chooses one of the newly discovered edges e, increases the step length by $m/(m-1)$ and follows e. If there are still other new edges that remain to be explored, then the step length d_i is reduced by a factor of $(1 - 1/m^2)^{m-1}$ and m is increased by one before e is explored. Once all unexplored branches are exhausted, the robot returns to the origin and chooses the next root-to-leaf path.

A description of our strategy in a more algorithmic notation can be found in Table 1. We use $\lambda_{\mathcal{P}}$ to denote the explored length of a path $\mathcal{P}$.

Algorithm *Tree-Search* makes use of a couple of procedure calls which we explain in the following. The procedures *enqueue* and *dequeue* are the standard queue operations and the procedures *push* and *pop* are the standard stack operations. The procedure *first* just returns the first element of list L. Finally, the procedure *shrink* is described below.

Note that the index i can be removed without affecting the computation of the algorithm. It is introduced in order to simplify the analysis of the algorithm. In particular, the assignments in Steps 35, 41, and 42 are unnecessary if the

Algorithm Tree-Search

Input: a geometric tree T with root r, a target g

Output: a path in T from r to g

1 let $e_1, \ldots, e_m$ be the edges incident to r

2 **for** $i = 1$ **to** m **do**

3   follow e_i for a distance of $\lambda_{e_i} = (1 - 1/m)^{m-i}$ and return

4 $i = m + 1$

6 $m_i = m$ $\triangleright$ m_i is the number of known leaves

7 $\alpha_i = m_i/(m_i - 1)$ $\triangleright$ the step length is multiplied by α_i in each new step

8 $d_i = \alpha_i \cdot 1$ $\triangleright$ d_i is the current step length

9 $d_i^* = (1/\alpha_i)^{m_i-1}$ $\triangleright$ d_i^* is a lower bound for $d(r, g)$ on the current path

10 $E_i = [\,]$ $\triangleright$ E_i is a stack containing the newly encountered edges and
                                           paths, resp.

11 $L_i = [e_1, \ldots, e_{m_i}]$ $\triangleright$ L_i is a queue containing the known paths

12 **loop**

13   **if** $E_i = [\,]$ **then** $\delta_i = 0$ $\triangleright$ δ_i is an indicator variable

14                          $\mathcal{P} = L_i.dequeue()$

15   **else** $\delta_i = 1$

16                          $\mathcal{P} = E_i.pop()$

17   **if** $E_i \neq [\,]$ **then** $d_i = (1 - 1/m_i^2)^{m_i-1}d_i$

18 $L_i.shrink(1 - 1/m_i^2, 1)$

19   follow $\mathcal{P}$ **until**

20   **Case 1:** g is detected: **goto** g; **exit loop**

21   **Case 2:** a new vertex v is reached:

22   **if** $E_i = [\,]$ **then** $d_i = (1 - 1/m_i^2)^{m_i}d_i$

23                                $L_i.shrink(1 - 1/m_i^2, \delta_i)$

24                                $v_i^* = v$

25   let $e_1, \ldots, e_k$ be the edges incident to v

26   **for each** edge e_j **do**

27   let $\mathcal{Q}_j$ be $\mathcal{P}$ concatenated with e_j

28            $E_i.push(\mathcal{Q}_2, \ldots, \mathcal{Q}_k)$

29            $\mathcal{P} = \mathcal{Q}_1$

30   **go to** 19

31   **Case 3:** the distance traveled on $\mathcal{P}$ is at least d_i:

32            $\lambda_{\mathcal{P}} = d_i$

33            $L_i.enqueue(\mathcal{P})$

34   **if** $E_i = [\,]$ **then** return to the root r

35                            $m_{i+1} = m_i$

36   **else** return to v_i^*

37                            $m_{i+1} = m_i + 1$

38            $\alpha_{i+1} = m_{i+1}/(m_{i+1} - 1)$

39            $d_{i+1}^* = \lambda_{L_i.first()}$

40            $d_{i+1} = \alpha_{i+1}d_i$

41            $E_{i+1} = E_i;\ v_{i+1}^* = v_i^*;\ L_{i+1} = L_i$

42            $i = i + 1$

43 **end loop;**

Table 1. Algorithm *Tree Search.*

index i is removed. Moreover, the value d_i^* introduced in the algorithm does not affect the computation; in particular, the assignments and calls in Steps 9, 18, 23, 32, and 39 could be omitted. But again, keeping track of the value d_i^* simplifies the analysis.

We also need to describe how the robot follows the path $\mathcal{P}$ in Step 19 since the length of $\mathcal{P}$ may be much smaller than the current step length. So first the robot follows $\mathcal{P}$ to its end point unless Case 3 occurs before. Let p be the end point of $\mathcal{P}$, and assume the robot reaches p. If p is not a vertex, then p belongs to a polygonal path of an edge e, and the robot continues to follow this polygonal path. If p is a vertex of T, then Step 26 ensures that there is a unique edge e that is incident to p and that belongs to $\mathcal{P}$. The robot follows the polygonal path of e until one of the three cases in Algorithm *Tree-Search* occurs.

The procedure *shrink* is defined as follows.

Algorithm Shrink
Input: a list $L = [\mathcal{P}_1 \ldots, \mathcal{P}_k]$ consisting of paths $\mathcal{P}_j$ and two numbers α and δ
Output: the list L where the explored length of $\mathcal{P}_j$ is reduced by a factor of $\alpha^{j-\delta}$

1 **for** $j = 1$ **to** k **do** $\lambda_{\mathcal{P}_j} = \alpha^{j-\delta} \lambda_{\mathcal{P}_j}$

It is easy to see that algorithm *Tree Search* is correct since clearly all the edges of T are visited. If no new edges are discovered anymore, then the step length of the root-to-leaf paths increases exponentially by a factor of $m/(m-1)$ after each step. This guarantees that the target is eventually found.

3.3   Analysis of Algorithm *Tree Search*

In order to analyse the strategy we show that a number of invariants are maintained. Recall that δ_i is an indicator variable that is defined as follows

$$\delta_i = \begin{cases} 0 \text{ if } E_i = [\,] \text{ in Step 12} \\ 1 \text{ otherwise.} \end{cases}$$

Invariant 1 (Algorithm *Tree Search*) *At Step 12 L_i contains $m_i - \delta_i$ paths $\mathcal{P}_j$, for $1 \leq j \leq m_i - \delta_i$, with*

$$\lambda_{\mathcal{P}_j} = \alpha_i^{j-1} d_i^*, \qquad \text{and, in addition,} \qquad d_i = \alpha_i^{m_i - \delta_i} d_i^*. \tag{1}$$

Proof: In order to show Invariant 1 we first note that the invariant clearly holds the first time the loop is entered. In this case $\lambda_{\mathcal{P}_j} = (1/\alpha_i)^{m_i - j}$, for $1 \leq j \leq m_i$, and $d_i^* = (1/\alpha_i)^{m_i - 1}$. So assume that the invariant holds before Step 12. We distinguish two cases.

Case 1 $E_i = [\,]$ *in Step 12.*
This implies that $\delta_i = 0$. In this case the first path $\mathcal{P}_1$ of L_i is removed in Step 14 and then explored. After Step 33 L_i consists of the paths $[\mathcal{P}_2, \ldots, \mathcal{P}_{m_i}, \mathcal{P}]$. We again distinguish two cases.

Case 1.1 *No new vertex v of T is encountered in Step 21 while the robot follows $\mathcal{P}$.*

The explored length of $\mathcal{P}$ is in Step 33

$$\lambda_{\mathcal{P}} = d_i = \alpha_i^{m_i} d_i^* = \alpha_i^{m_i-1} \lambda_{\mathcal{P}_2}$$

where the last two equalities follow from the invariant. Since $d_{i+1}^* = \alpha_i d_i^*$, it is easy to see that L_{i+1} again satisfies the invariant if the elements of L_{i+1} are renumbered appropriately. After Step 40 the value of d_{i+1} is

$$d_{i+1} = \alpha_{i+1} \alpha_i^{m_i} d_i^* = \alpha_{i+1}^{m_{i+1}} d_{i+1}^*$$

since $m_{i+1} = m_i$ as claimed.

Case 1.2 *A new branch vertex v is encountered the first time while the robot follows $\mathcal{P}$.*

The robot executes Step 21 and, when the end of the current path is reached, Step 36. Hence, $m_{i+1} = m_i + 1$. When the procedure *shrink* is called in Step 22, the path $\mathcal{P}_j$ is the $(j-1)$st path in L_i and, therefore, $\lambda_{\mathcal{P}_j}$ is reduced by a factor of $(1 - 1/m_i^2)^{j-1}$, for $2 \le j \le m_i$. The distance d_i is reduced by a factor of $(1 - 1/m_i^2)^{m_i}$ in Step 22. By the invariant we have after Step 22

$$\left(1 - \frac{1}{m_i^2}\right)^{j-1} \lambda_{\mathcal{P}_j} = \left(\frac{(m_i+1)(m_i-1)}{m_i^2}\right)^{j-1} \left(\frac{m_i}{m_i-1}\right)^{j-1} d_i^* = \alpha_{i+1}^{j-1} d_i^*, \quad (2)$$

for $2 \le j \le m_i$, and similarly $d_i = \alpha_{i+1}^{m_i} d_i^*$ after Step 22. After the path $\mathcal{P}$ is added to L_i in Step 33, the invariant for L_{i+1} holds again since now $E_{i+1} \ne [\,]$ and L_{i+1} contains $m_i = m_{i+1} - 1$ tuples which satisfy condition (1). After d_i is multiplied with α_{i+1}, we see as follows that d_{i+1} also satisfies Invariant 1 again. We have

$$d_{i+1}^* = \lambda_{\mathcal{P}_2} = \frac{m_i + 1}{m_i} d_i^* = \alpha_{i+1} d_i^*$$

and, hence,

$$d_{i+1} = \alpha_{i+1}\alpha_{i+1}^{m_i} d_i^* = \alpha_{i+1}^{m_{i+1}-\delta_{i+1}} d_{i+1}^*.$$

Moreover, L_{i+1} now contains the $m_{i+1} - \delta_{i+1}$ paths $\mathcal{P}_1', \dots, \mathcal{P}_{m_{i+1}-1}'$ where $\mathcal{P}_i' = \mathcal{P}_{i+1}$, for $1 \le i \le m_{i+1} - 2$, and $\mathcal{P}_{m_{i+1}-1}' = \mathcal{P}$. By equation (2) we have

$$\lambda_{\mathcal{P}_j'} = \lambda_{\mathcal{P}_{j+1}} = \alpha_{i+1}^j d_i^* = \alpha_{i+1}^{j-1} d_{i+1}^*$$

and the invariant holds.

Note that it may happen that d_i is reduced so much in Step 22 that the robot has actually already traveled farther than the new value of d_i. In this case the robot just remains at v and acts as if Case 3 occurs. If the new step length d_{i+1} is still less than $d(r, v)$, then there is again nothing to do, and the robot considers the next edge, and so on. Finally, it will encounter an edge for which d_i is larger than $d(r, v)$. To see this, note that if there have been k new edges in the beginning, then the new step length to explore the last edge is $(\frac{m_i+k}{m_i+k-1})^{m_i+k-1}d_{i+1}^* = (\frac{m_i+k}{m_i+k-1})^{m_i+k-1}(\frac{m_i+1}{m_i})d_i^*$ which is larger than $(\frac{m_i}{m_i-1})^{m_i}d_i^*$. Since $(\frac{m_i}{m_i-1})^{m_i}d_i^*$ is the step length of the iteration in which the robot encountered v, $d_i > d(v,r)$ in the last step.

Note also that the second time the robot encounters a new branch vertex, the new edges are only added to E_i and nothing else happens.

Case 2 $E_i \neq [\,]$ *in Step 12.*
In this case $\delta_i = 1$, and the first path $\mathcal{P}$ of E_i is removed in Step 16. Moreover, since $E_i \neq [\,]$, L_i contains the $m_i - 1$ paths $[\mathcal{P}_1, \ldots, \mathcal{P}_{m_i-1}]$ and $d_i = \alpha_i^{m_i-1}d_i^*$. First assume that after $\mathcal{P}$ is removed from E_i, $E_i = [\,]$ and no new vertex is encountered. Hence, after $\mathcal{P}$ is added to L_i in Step 33 and d_i is increased by a factor of $\alpha_{i+1} = \alpha_i$, the invariant holds again.

If after the removal of $\mathcal{P}$ from E_i in Step 16 still $E_i \neq [\,]$ or $E_i = [\,]$ and a new vertex is encountered by traveling on $\mathcal{P}$, then $\mathcal{P}_j$ is the jth path in L_i and, therefore, $\lambda_{\mathcal{P}_j}$ is reduced by a factor of $(1 - 1/m_i^2)^{j-1}$, for $1 \leq j \leq m_i - 1$ in the procedure *shrink*. We see as above that, after L_i is shrunk and d_i is reduced,

$$\lambda_{\mathcal{P}_j} = \alpha_{i+1}^{j-1}d_i^* \quad \text{and} \quad d_i = \alpha_{i+1}^{m_i-1}d_i^*,$$

for $1 \leq j \leq m_i - 1$, since $m_{i+1} = m_i$. Of course, now $\mathcal{P}_1$ is the first path of L_i. And as in the second case above we see that after the path $\mathcal{P}$ is added to L_i, the invariant for L_{i+1} holds again; and after d_i is multiplied with α_{i+1}, d_{i+1} also satisfies Invariant 1 again.

$\square$

Note that, for each path $\mathcal{P}$ in L_i, the robot has explored $\mathcal{P}$ at least up to $\lambda_{\mathcal{P}}$; this is true when $\mathcal{P}$ is added to L_i, and afterwards $\lambda_{\mathcal{P}}$ is only changed in the Steps 17 and 22 and in both steps $\lambda_{\mathcal{P}}$ is reduced.

Invariant 2 (Algorithm *Tree Search*)
If $E_i \neq [\,]$, then at Step 12 $d(v_i^*,r) \geq d_i^*/\alpha_i$.

Proof: To see this invariant we first note that as long as $E_i \neq [\,]$, $\mathcal{P}_1$ remains the first path in L_i, that is, d_i^* remains unchanged. Moreover, as can easily be seen the explored length of $\mathcal{P}_1$ is not reduced if the procedure *shrink* is invoked. Secondly, the only step where v_i^* is changed is Step 22. This happens if $E_i = [\,]$ and a new vertex is reached. If v_i^* was already defined before, then the search is started at v_i^* and the new vertex v_{i+1}^* has a distance to r that is greater than the distance of v_i^* to r.

Hence, $d(v_i^*,r)/d_i^*$ is only increased unless the robot returns to r. Therefore, it suffices to show that the inequality holds if $E_i = [\,]$ in Step 12 and a new vertex is encountered on a path starting at the root r. In this case the first new vertex that the robot encounters is on the part of $\mathcal{P}$ that comes after $\lambda_{\mathcal{P}}$. By Invariant 1 $\lambda_{\mathcal{P}}$ equals d_i^*. Since $\mathcal{P}$ is removed from L_i, the first element of L_{i+1} is now $\mathcal{P}_2$ and $d_{i+1}^* = \lambda_{\mathcal{P}_2}$. By Invariant 1 $\lambda_{\mathcal{P}_2}$ after the procedure *shrink* is given by $\alpha_{i+1}d_i^*$. Hence, before Step 42 $d(v_{i+1}^*,r) \geq d_i^* = d_{i+1}^*/\alpha_{i+1}$ which proves the invariant.

$\square$

Invariant 3 (Algorithm *Tree Search*) *If in Step 12 the robot is at a distance of d_{cur} to r, then the total distance traveled by the robot is at most*

$$2d_i^* \sum_{j=-\infty}^{m_i-1-\delta_i} \alpha_i^j + d_{cur}.$$

The above summation starts at $-\infty$ instead of 0 (as would be expected) in order to simplify the analysis; the index j can now be shifted up or down without affecting the starting point of the summation.

Proof: The invariant clearly holds when the loop is entered in Step 12. In the following we show that if the invariant holds at the beginning of the loop, then it also holds at the end. We distinguish two cases.

Case 1 $E_{i+1} = [\,]$.

In this case the robot returns to r, and no new vertex is discovered during the traversal of $\mathcal{P}$. We have $m_{i+1} = m_i$ and $\delta_{i+1} = 0$.. Once the robot reaches r, Invariant 3 implies that the total distance traveled by the robot is at most

$$2d_i^* \sum_{j=-\infty}^{m_i-1-\delta_i} \alpha_i^j + 2d_i, \tag{3}$$

where $d_i = \alpha_i^{m_i-\delta_i} d_i^*$ by Invariant 1. If $E_i \neq [\,]$, then $\mathcal{P}_1$ is still the first element of L_{i+1}, $d_{i+1}^* = d_i^*$, and by Invariant 1 expression (3) equals

$$2d_i^* \sum_{j=-\infty}^{m_i-2} \alpha_i^j + 2\alpha_i^{m_i-1} d_i^* = 2d_i^* \sum_{j=-\infty}^{m_i-1} \alpha_i^j = 2d_{i+1}^* \sum_{j=-\infty}^{m_{i+1}-1} \alpha_{i+1}^j.$$

Hence, the invariant holds.

If $E_i = [\,]$, then $\delta_i = 0$, and $\mathcal{P}_2$ is the first element of L_{i+1} which implies that $d_{i+1}^* = \lambda_{\mathcal{P}_2} = \alpha_i d_i^*$ by Invariant 1. expression (3) now equals

$$2d_i^* \sum_{j=-\infty}^{m_i-1} \alpha_i^j + 2\alpha_i^{m_i} d_i^* = 2d_{i+1}^* \sum_{j=-\infty}^{m_i-2} \alpha_i^j + 2\alpha_i^{m_i-1} d_{i+1}^* = 2d_{i+1}^* \sum_{j=-\infty}^{m_{i+1}-1} \alpha_j^j,$$

and the invariant holds again.

Case 2 $E_{i+1} \neq [\,]$.

In this case the robot returns to v_i^*, i.e., $d_{cur} = d(v_i^*, r)$, $m_{i+1} = m_i + 1$, and $\delta_{i+1} = 1$.

First assume that $E_i = [\,]$. Then, a new vertex is encountered while the robot follows $\mathcal{P}$. By Invariant 3 the total distance traveled by the robot is at most

$$2d_i^* \sum_{j=-\infty}^{m_i-1} \alpha_i^j + 2d_i - d(v_i^*, r), \tag{4}$$

since the robot starts at r, travels all the way to the end of $\mathcal{P}$ and then returns v_i^*. Here, $d_i = \alpha_{i+1}^{m_{i+1}-1} d_i^*$ by Invariant 1 and the fact that d_i has been shrunk in Step 22. As mentioned before, it may happen that the robot has traveled further than $\alpha_{i+1}^{m_{i+1}-1} d_i^*$ before the first new edge branching off $\mathcal{P}$ is discovered. In this case the robot only travels to v_i^* and stays there; now, the total distance traveled by the robot is at most

$$2d_i^* \sum_{j=-\infty}^{m_i-1} \alpha_i^j + d(v_i^*, r) \tag{5}$$

Since $E_i = []$, $\mathcal{P}_2$ is now the first element of L_i. In order to show that the invariant still holds at the end of the loop we need to prove that

$$2d_{i+1}^* \sum_{j=-\infty}^{m_{i+1}-1-\delta_{i+1}} \alpha_{i+1}^j + d_{cur} \tag{6}$$

is at least as large as expression (4) or (5). Here, $d_{cur} = d(v_i^*, r)$, $\delta_{i+1} = 1$, and $d_{i+1}^* = \lambda_{\mathcal{P}_2} = \alpha_{i+1} d_i^*$ by Invariant 1 and the fact that $\mathcal{P}_2$ has been shrunk in Step 22. If we subtract expression (4) from (6), then the term $2d_i = 2\alpha_{i+1}^{m_{i+1}-1} d_i^*$ cancels with $2\alpha_{i+1}^{m_{i+1}-2} d_{i+1}^*$, and we obtain

$$2d_i^* \left(\sum_{j=-\infty}^{m_i-1} \alpha_{i+1}^j - \sum_{j=-\infty}^{m_i-1} \alpha_i^j \right) + 2d(v_i^*, r) \geq 2d_i^* \left(\sum_{j=-\infty}^{m_i-1} \alpha_{i+1}^j - \sum_{j=-\infty}^{m_i-1} \alpha_i^j + 1 \right).$$

Similarly, if we subtract expression (5) from (6), then we obtain

$$2d_i^* \left(\sum_{j=-\infty}^{m_i} \alpha_{i+1}^j - \sum_{j=-\infty}^{m_i-1} \alpha_i^j \right) \geq 2d_i^* \left(\sum_{j=-\infty}^{m_i-1} \alpha_{i+1}^j - \sum_{j=-\infty}^{m_i-1} \alpha_i^j + 1 \right).$$

We want to show that the above expression is non-negative. Since $d(v_i^*, r) \geq d_i^*$ by Invariant 2 and $\alpha_{i+1}^j \geq \alpha_i^j$, for $j < 0$, both of the above expressions are bounded from below by

$$2d_i^* \left(\sum_{j=0}^{m_i-1} \alpha_{i+1}^j - \sum_{j=0}^{m_i-1} \alpha_i^j + 1 \right) = 2d_i^* \left(\frac{\alpha_{i+1}^{m_i} - 1}{\alpha_{i+1} - 1} - \frac{\alpha_i^{m_i} - 1}{\alpha_i - 1} + 1 \right) = 2d_i^* f(m_i),$$

where

$$f(m) = (m+1) \left(\frac{m+1}{m} \right)^{m-1} - m \left(\frac{m}{m-1} \right)^{m-1}.$$

We now show that the function f is non-negative. In order to do so let us define the function

$$g(m, k) = m \left(\frac{m}{m-1} \right)^k,$$

with $k \geq 1$, $m \geq 2$. If we consider the partial derivative of g w.r.t. m, then we obtain

$$\frac{\partial g}{\partial m}(m, k) = \left(\frac{m}{m-1}\right)^k \frac{m - 1 - k}{m - 1}$$

which has exactly one zero in the range $(1, \infty)$ at $m = k + 1$. Since $\lim_{m \to 1+} g(m, k) = \lim_{m \to \infty} g(m, k) = \infty$, g has a minimum at $g(k + 1, k)$. In particular, $g(m, m - 1) \leq g(x, m - 1)$, for all $x \in (1, \infty)$. Hence, $f(m) = g(m + 1, m - 1) - g(m, m - 1) \geq 0$ as claimed. This implies that the distance traveled by the robot is, indeed, bounded by expression (6).

Finally, we assume that $E_i \neq [\,]$. By Invariant 3 the total distance traveled by the robot is at most

$$2d_i^* \sum_{j=-\infty}^{m_i-2} \alpha_i^j + 2d_i - d(v_i^*, r). \tag{7}$$

We want to show that

$$2d_{i+1}^* \sum_{j=-\infty}^{m_{i+1}-2} \alpha_{i+1}^j + d(v_i^*, r) \tag{8}$$

is at least as large as expression (7). Since $E_i \neq [\,]$, $\mathcal{P}_1$ is still the first element of L_i and, therefore, $d_{i+1}^* = \lambda(\mathcal{P}_1) = d_i^*$. Moreover, since by Invariant 1 $\alpha_{i+1} d_i = d_{i+1} = \alpha_{i+1}^{m_{i+1}-1} d_{i+1}^*$, $d_i = \alpha_{i+1}^{m_{i+1}-2} d_i^*$. If we subtract expression (7) from (8), then we obtain again the sum

$$2d_i^* \left(\sum_{j=-\infty}^{m_i-1} \alpha_{i+1}^j - \alpha_i^j\right) + 2d(v_i^*, r)$$

which is non-negative as we have seen above which implies again that the distance traveled by the robot is, indeed, bounded by expression (8). $\square$

The Competitive Ratio Once we have established a bound on the distance traveled by the robot, it is easy to compute the competitive ratio of Algorithm *Tree Search*. So assume that the robot detects the target g. We again distinguish two cases. First assume that $E_i = [\,]$ at Step 12. Then, the current search path of the robot starts at the root r of T. By Invariant 1 the explored length of $\mathcal{P}$ is d_i^*. Hence, $d(r, g) > d_i^*$. On the other hand, by Invariant 3 the distance traveled by the robot is at most

$$2d_i^* \sum_{j=-\infty}^{m_i-1} \alpha_i^j + d(r, g).$$

Hence, the competitive ratio is given by

$$\frac{2d_i^* \sum_{j=-\infty}^{m_i-1} \alpha_i^j + d(r,g)}{d(r,g)} \leq 1 + 2\frac{d_i^*(m_i-1)\alpha_i^{m_i}}{d_i^*} = 1 + 2\frac{m_i^{m_i}}{(m_i-1)^{m_i-1}}$$

$$\leq 1 + 2\frac{m^m}{(m-1)^{m-1}}$$

where m is the number of leaves of T.

Now assume that $E_i \neq [\,]$ at Step 12. Then, the current search path of the robot starts at the vertex v_i^*. By Invariant 2 the distance of v_i^* to r is at least d_i^*/α_i. Hence, $d(r,g) > d_i^*/\alpha_i$. On the other hand, by Invariant 3 the distance traveled by the robot is at most

$$2d_i^* \sum_{j=-\infty}^{m_i-2} \alpha_i^j + d(r,g).$$

Hence, the competitive ratio is given by

$$\frac{2d_i^* \sum_{j=-\infty}^{m_i-2} \alpha_i^j + d(r,g)}{d(r,g)} \leq 1 + 2\frac{d_i^*(m_i-1)\alpha_i^{m_i-1}}{d_i^*/\alpha_i} = 1 + 2\frac{m_i^{m_i}}{(m_i-1)^{m_i-1}}$$

$$\leq 1 + 2\frac{m^m}{(m-1)^{m-1}}$$

as above.

The above algorithm is optimal since m concurrent rays also form a geometric tree. As we have seen before, $1 + 2m^m/(m-1)^{m-1}$ is a lower bound on the competitive ratio of searching on m concurrent rays [BYCR93,Gal80].

We have proven the following theorem.

Theorem 1. *There is a strategy to search in a geometric tree with m leaves that has a competitive ratio of $1 + 2m^m/(m-1)^{m-1}$. If we consider unbounded geometric trees, then there is no strategy with a lower competitive ratio.*

4 Searching in Simple Polygons

In this section we present an algorithm for a point robot to search for a target g inside a simple polygon P. We assume that the robot is equipped with a vision system that allows it to identify the target and a range finder which returns the local visibility polygon of the robot. Our algorithm is based on an approach by Icking [Ick94,Kle97]. However, we present a new analysis of the strategy. We assume that in the beginning the robot is located at a point s in a polygon P with n vertices.

At first we observe that there is a simple lower bound of $\Omega(n)$ for searching in a simple polygon [Kle97]. If we consider the polygon shown in Figure 2, then we obtain a lower bound on the competitive ratio of $n/2 - 1$. By placing the

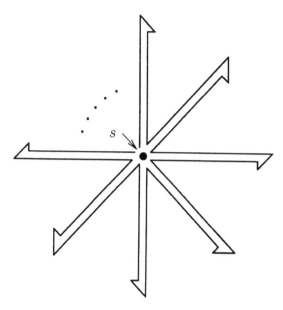

Fig. 2. A polygon for which every search strategy has a competitive ratio of $n/2 - 1$.

target g in the last spike that is explored by the strategy, an adversary can force any strategy to generate a path of length $(2(n/4) - 1)d$ whereas the shortest path from s to g has a length of $d + \varepsilon$. Here, d is the length of a spike in the polygon. This shows that no strategy can have a better competitive ratio than $n/2 - 1$ in an arbitrarily oriented simple polygons with n edges.

4.1   Searching on the Shortest Path Tree

Let T_s be the *shortest path tree* of s. Recall that the shortest path tree of s is defined as the union of all shortest paths from s to vertices of P. It consists of $n + 1$ nodes, which are s and the n vertices of P, of which $l \leq n + 1$ are leaves. We first show that it suffices to explore T_s in order to find g.

If g is not visible from s, then $shp(s, g)$ intersects a number of vertices. Let v be the last vertex that is intersected by $shp(s, g)$ (see Figure 3). Clearly, the line segment from v to g is contained in P. Once the robot has reached v, then it sees g and moves directly to g. Hence, if $L_A(s, v)$ denotes the length of the path that the robot travels in order to reach the vertex v using algorithm A, then the competitive ratio of A is given by

$$\frac{L_A(s, v) + d(v, g)}{d(s, v) + d(v, g)} \leq \frac{L_A(s, v)}{d(s, v)}.$$

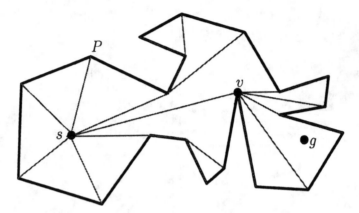

Fig. 3. There is a vertex on T_s from which g is visible.

This implies that we only need an algorithm to explore the vertices of P. This can be easily achieved by applying Algorithm *Tree Search* to T_s. That T_s is not known in the beginning is not a problem as the following lemma shows.

Lemma 1. *If $\mathcal{P}$ is a path in the simple polygon P, then the visibility polygon $vis(\mathcal{P})$ of $\mathcal{P}$ is geodesically convex, that is, the shortest path between any two points in $vis(\mathcal{P})$ is contained in $vis(\mathcal{P})$.*

Proof: The proof is by contradiction. So consider two points p and q that are contained in the visibility polygon $vis(\mathcal{P})$ of $\mathcal{P}$ and assume that the shortest path from p to q is not contained in $vis(\mathcal{P})$. Then, there is a window w of $vis(\mathcal{P})$ that is crossed by $shp(p,q)$.

It is easy to see that $vis(\mathcal{P})$ is a simply connected set. Since both p and q are contained in $vis(\mathcal{P})$, $shp(p,q)$ crosses w twice, once to enter the pocket of w and once to leave it again. But then the part of $shp(p,q)$ inside the pocket of w can be replaced by the part of w between the cross points. This yields a shorter path—a contradiction. □

Note that Lemma 1 is not true anymore if we allow the polygon to contain holes. But if the polygon is simple, then the robot can compute the distance between two points that it sees by Lemma 1.

Therefore, the robot can construct T_s on-the-fly as it executes Algorithm *Tree Search*. This immediately leads to a search algorithm with a competitive ratio of at most $1+2n^n/(n-1)^{n-1} \leq 1+2en$ for searching for a target in a simple polygon, which considerably improves the $(8n-3)$-competitive algorithm of Icking, which also uses a search algorithm for geometric trees to explore the shortest path tree [Ick94,Kle97].

However, the analysis of the above algorithm can be further improved by the following observation (see also [Kle97]).

Lemma 2. *If s and g are two points in P, then the last node of T_s that is intersected by $shp(s,g)$ and that is different from g is an internal node of T_s.*

Proof: If s sees g, then the claim is trivially true. So assume that s does not see g. Let v' be the vertex of $shp(s, g)$ before v. Since v' does not see g and v is contained in the shortest path from v' to g, it is easy to see that v is a reflex vertex. Let e be the edge of P incident to v that is not seen by v'. Clearly, the shortest path from v' to the other end point of e contains v as well. Hence, v is not a leaf of T_s. $\qquad\square$

Hence, in order to find the vertex v it is not necessary to consider the leaves of T_s. If we prune the m leaves of the tree, then the remaining tree T_s^* can have no more than m leaves as the pruning does not introduce new root-to-leaf paths.

Therefore, the number of leaves m^* of T_s^* is at most $n/2$. If we apply Algorithm *Tree Search* to T_s^*, then we obtain a strategy to search in polygons with a competitive ratio of $1 + en$. Note that since the robot can compute the shortest path between two points, it is easy to decide if a vertex is a leaf of T_s or not. Moreover, if we denote the number of reflex vertices of P by r, then the competitive ratio of our algorithm is $1 + 2r^r/(r-1)^{r-1} \leq 1 + 2er$ since the vertices of T_s^* obviously consist only of reflex vertices.

Thus, we have proven the following theorem.

Theorem 2. *If P is a polygon with r reflex vertices, then there is an algorithm to search in P with a competitive ratio of $1 + 2er$.*

Simulating Dijkstra's algorithm on T_s^* yields an even better algorithm with a competitive ratio of $1 + 2(r - 1)$. Here, the robot repeatedly chooses the closest unvisited vertex v, follows $shp(s, v)$, and returns if g is not visible from v.[1] Note again that even though P is not known to the robot, the closest unvisited vertex is visible to the robot. The competitive ratio is now just

$$\frac{2\sum_{v' \text{ visited}} d(s, v') + d(s, v)}{d(s, v)} \leq 1 + 2|\{v' \mid v' \text{ is visited}\}|$$

$$\leq 1 + 2(|T_s^*| - 1) \ \leq \ 1 + 2(r - 1)$$

since $d(s, v) \geq d(s, v')$, for all vertices v' that are visited during the search. However, the competitive ratio of the simulation of Dijkstra's algorithm depends on the *total number of vertices* of T_s^* whereas the competitive ratio of the algorithm based on Algorithm *Tree Search* only depends on the *number of leaves* m^* of T_s^*. In particular, if $|T_s^*| \geq m^*e + 1$, then Algorithm *Tree Search* yields a better competitive ratio. A simple example are spiral polygons where Algorithm *Tree Search* achieves a competitive ratio of 9 in the worst case whereas the simulation of Dijkstra's algorithm on T_s^* can have a competitive ratio of up to $2n - 7$. This is also an upper bound for the competitive ratio of the simulation of Dijkstra's algorithm since there are at least three convex vertices in a simple polygon.

4.2 k-spiral Polygons

We show in the following that in general the application of Algorithm *Tree Search* to T_s^* has a competitive ratio of $1 + 2(2k)^{2k}/(2k - 1)^{2k-1} \leq 1 + 4ek$ in polygons

---

[1] The same algorithm was independently suggested by I. Semrau [Sem97].

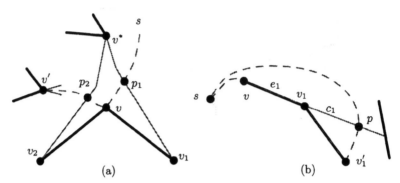

Fig. 4. (a) If v is not adjacent to v_1 and v_2, then v is a leaf of T_s. (b) The point p is on the extension of the last link of $shp(s, v_1)$.

whose boundary can be decomposed into k convex and k reflex chains. As we already observed above, if v is a convex vertex of a polygon, then v is a leaf of T_s and, therefore, does not belong to T_s^*. Hence, we only need to consider reflex vertices.

Lemma 3. *Let v be a reflex vertex of P and v_1 and v_2 the vertices that are adjacent to v in P. If v is an internal node of T_s, then either $shp(s, v_1)$ or $shp(s, v_2)$ contains v.*

Proof: Let v be a reflex vertex that is an internal node of T_s. We show that for one of the vertices v_i, $i = 1, 2$, the shortest path from s to v_i contains v by contradiction. Let e_i be the edge from v to v_i, for $i = 1, 2$, and v^* be the lowest common ancestor of v_1 and v_2 in T_s. We assume that neither $shp(v^*, v_1)$ nor $shp(v^*, v_2)$ contains v. The concatenation of $shp(v^*, v_1)$ with e_1, e_2, and $shp(v_2, v^*)$ forms a simple closed polygonal curve C. Since v is an internal node of T_s, there is one vertex v' of P such that $shp(s, v')$ contains v. Hence, $shp(s, v')$ intersects C once in a point p_1 to enter C and to reach v and once in a point p_2 to leave C and to reach the vertex v' (see Figure 4a).   Assume that p_2 belongs to $shp(s, v_2)$. Since subpaths of shortest paths are shortest paths themselves, there are two distinct shortest paths from s to p_2, one path that is contained in $shp(s, v')$ and that intersects v and one path that is contained in $shp(s, v_2)$ and that does not intersect v—a contradiction since shortest paths are unique in a simple polygon.   □

With Lemma 3 we can prove the following result.

Lemma 4. *Let v be a vertex of P and v_1 and v_2 the vertices that are adjacent to v in P. If v, v_1, and v_2 are reflex vertices, then either v is a leaf of T_s or v is not a leaf T_s^*.*

Proof: Let v be a reflex vertex of P such that the adjacent vertices v_1 and v_2 of v are also reflex. Assume that v is not a leaf of T_s. We show that v is also not a leaf of T_s^*.

By Lemma 3 there is one vertex of v_1 and v_2, say v_1, such that $shp(s, v_1)$ contains v. We show that v_1 is not a leaf of T_s. This implies that v_1 also belongs to T_s^* and is a child of v, which proves the claim.

Let e_1 be the edge of P that connects v and v_1 and v_1' the vertex that is adjacent to v_1 and different from v. Let c_1 be the chord that is collinear with e_1 and incident to v_1 (see Figure 4b). Since v_1 is a reflex vertex, c_1 exists and v is contained in the region of c_1 that contains s whereas v_1' is contained in the other region of c_1. Hence, the shortest path from v_1' to s crosses c_1 in a point p. Since p is on the extension of the last edge $\overline{vv_1}$ of $shp(s, v_1)$, $shp(s, p)$ contains v_1 and v. Therefore, p equals v_1 and v_1 is not a leaf of T_s.   □

Lemma 4 implies that if v is a leaf of T_s^*, then v is a reflex vertex such that at least one of its adjacent vertices is a convex vertex. Hence, v is the end vertex of a reflex chain. If the boundary of P can be decomposed into k convex and k reflex chains, then there are $2k$ end vertices of reflex chains. Therefore, if we apply Algorithm *Tree Search* to T_s^*, then the competitive ratio is at most $1 + 2 \cdot (2k)^{2k}/(2k - 1)^{2k-1}$ as claimed. The polygon shown in Figure 5 can be easily generalized to a polygon P_k whose boundary consists of k convex and k reflex chains such that the competitive ratio of any search strategy for P_k is at least $1 + 2 \cdot (2k)^{2k}/(2k - 1)^{2k-1}$. In this sense our algorithm to search in simple polygons is optimal.

We have shown the following theorem.

Theorem 3. *There is an algorithm to search in simple polygons that has a competitive ratio of $1 + 2 \cdot (2k)^{2k}/(2k - 1)^{2k-1}$ on the class of polygons whose boundary can be decomposed into k convex and k reflex chains.*

5   Conclusions

We present the Algorithm *Tree Search* to search for a target in an unknown geometric tree T. Algorithm *Tree Search* simulates the optimal search strategy to search in m concurrent rays and achieves an optimal competitive ratio of $1 + 2m^m/(m - 1)^{m-1}$ if m is the number of leaves of T.

In the second part we show how to apply the algorithm to the problem of searching for a target in unknown simple polygons. We present two algorithms, one based on Algorithm *Tree Search* and the other based on a simulation of Dijkstra's algorithm. The first algorithm achieves a competitive ratio of $1 + en$ whereas the second achieves a competitive ratio of $2n - 7$. However, if the polygon boundary can be decomposed into k convex and reflex chains, then the search algorithm based on Algorithm *Tree Search* achieves a competitive ratio of $1 + 4ek$ whereas the competitive ratio of the algorithm based on the simulation of Dijkstra's algorithm may be arbitrarily high.

The problem of searching in polygons is far from solved. There is a lower bound of $n/2$ for the competitive ratio which still leaves a significant gap compared to algorithms presented above. Also most environments in practice cannot be modeled by simple polygons. Hence, future research should also address the problem of searching in polygons with holes.

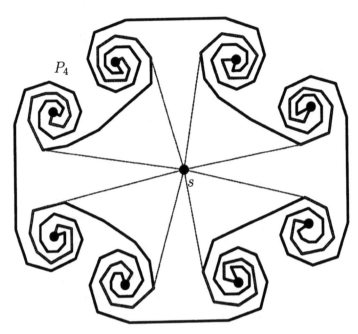

Fig. 5. The boundary of the polygon P_4 can be decomposed into four convex and four reflex chains. A search strategy in P_4 has a competitive ratio of at least $1 + 2 \cdot (8^8/7^7) \sim 41.7$.

References

BYCR93. R. Baeza-Yates, J. Culberson, and G. Rawlins. Searching in the plane. *Information and Computation*, 106:234–252, 1993.

DHS95. A. Datta, Ch. Hipke, and S. Schuierer. Competitive searching in polygons—beyond generalized streets. In *Proc. Sixth Annual International Symposium on Algorithms and Computation*, pages 32–41. LNCS 1004, 1995.

DI94. A. Datta and Ch. Icking. Competitive searching in a generalized street. In *Proc. 10th Annu. ACM Sympos. Comput. Geom.*, pages 175–182, 1994.

Gal80. S. Gal. *Search Games*. Academic Press, 1980.

Ick94. Ch. Icking. *Motion and Visibility in Simple Polygons*. PhD thesis, Fernuniversität Hagen, 1994.

Kle92. R. Klein. Walking an unknown street with bounded detour. *Comput. Geom. Theory Appl.*, 1:325–351, 1992.

Kle94. J. M. Kleinberg. On-line search in a simple polygon. In *Proc. of 5th ACM-SIAM Symp. on Discrete Algorithms*, pages 8–15, 1994.

Kle97. R. Klein. *Algorithmische Geometrie*. Addison-Wesley, 1997.

LOS96. A. López-Ortiz and S. Schuierer. Generalized streets revisited. In J. Diaz and M. Serna, editors, *Proc. 4th European Symposium on Algorithms*, pages 546–558. LNCS 1136, 1996.

LOS97. A. López-Ortiz and S. Schuierer. Position-independent near optimal searching and on-line recognition in star polygons. In F. Dehne, A. Rau-Chaplin,

J.-R. Sack, and R. Tamassia, editors, *Proc. 5th Workshop on Algorithms and Data Structures*, pages 284–296. LNCS 1272, 1997.

PY91.   C. H. Papadimitriou and M. Yannakakis. Shortest paths without a map. *Theoretical Computer Science*, 84(1):127–150, 1991.

Sem97.   I. Semrau. Personal communications, 1997.

ST85.   D. D. Sleator and R. E. Tarjan. Amortized efficiency of list update and paging rules. *Communications of the ACM*, 28:202–208, 1985.

Distributed Algorithms for Carrying a Ladder by Omnidirectional Robots in Near Optimal Time[*]

Yuichi Asahiro[1], Hajime Asama[2], Satoshi Fujita[3], Ichiro Suzuki[4], and Masafumi Yamashita[1]

[1] Kyushu University, Fukuoka, Fukuoka 812-8581, Japan
{asahiro,mak}@csce.kyushu-u.ac.jp
[2] The Institute of Physical and Chemical Research (RIKEN)
Wako, Saitama 351-01, Japan
asama@cel.riken.go.jp
[3] Hiroshima University, Kagamiyama, Higashi-Hiroshima 739-8511, Japan
fujita@se.hiroshima-u.ac.jp
[4] University of Wisconsin – Milwaukee, Milwaukee, WI 53201, USA
suzuki@cs.uwm.edu

Abstract. Consider two omnidirectional robots carrying a ladder, one at each end, in the plane without obstacles. Given start and goal positions of the ladder, what is a time-optimal motion of the robots subject to given constraints on their kinematics such as maximum acceleration and velocity? Using optimal control theory, Chen, Suzuki and Yamashita solved this problem under a kinematic constraint that the speed of each robot must be either 0 or a given constant v at any moment during the motion. Their solution, which requires complicated calculation, is centralized and off-line. The objective of this paper is to demonstrate that even without the complicated calculation, a motion that is sufficiently close to time-optimal can be obtained using a simple distributed algorithm in which each robot decides its motion individually based on the current and goal positions of the ladder.

1 Introduction

Recently distributed autonomous robot systems have attracted the attention of many researchers as a new approach for designing intelligent and robust systems [6]. A distributed autonomous robot system is a group of robots that individually and autonomously decide their motion. In particular, in such a system no robot is allowed to act as a leader that controls the other robots in the group.

[*] This paper has been prepared as a summary of a presentation given at the 1998 Dagstuhl Workshop on the authors' on-going research project. This work was supported in part by a Scientific Research Grant-in-Aid from the Ministry of Education, Science and Culture of Japan under Grants, the National Science Foundation under Grant IRI-9307506, the Office of Naval Research under Grant N00014-94-1-0284, and an endowed chair supported by Hitachi Ltd. at Faculty of Engineering Science, Osaka University.

A crucial issue in designing such a system is to have the robots autonomously coordinate their motion so that the system achieves a certain goal, since in many applications,

1. the interest of an individual robot can conflict with that of the entire system, and
2. each robot may be unaware of the global system state.

Typically there are two kinds of coordination problems; conflict resolution and cooperation. Roughly speaking, conflict resolution attempts to avoid certain (undesirable) situations such as collision of robots. Cooperation on the other hand aims to eventually generate a certain (favorable) situation.

The problem we discuss in this paper is a motion coordination problem for producing a time optimal behavior of the system.

Centralized and distributed approaches are two major paradigms for designing motion coordination in multi-robot systems [9], each with its own merits and demerits. It is not our objective here to discuss them in detail, and we only mention that roughly speaking a centralized approach tends to yield efficient solutions, while solutions obtained by a distributed approach can be more robust against failures. Most of the work in the literature, however, takes either a centralized or a leader-follower approach [2,8,12,13]. The work on distributed approachs often assumes the existence of some navigation devices [15,7], and only a few take a fully distributed approach [1,11,14,16,17]. Among these, Miyata, et al. [11] and Ahmadabadi and Nakano [1] discuss how a group of robots may carry and handle an object, which is the subject of this paper. They investigate how the robots can coordinate their motion to carry an object, while in this paper we investigate how quickly the robots can carry an object.

Recently a time optimal motion of robots has been investigated by some researchers. In [4], Chen, et al. discuss the problem of computing a time-optimal motion for two omnidirectional robots carrying a ladder from an initial position to a final position in a plane without obstacles, and calculate an optimal path using a method based on variational calculus, a branch of functional analysis [5], under the assumption that the speed of each robot must be either 0 or a given constant v at any moment during the motion. In [10] Mediavilla, et al. describe a path planning method for three robots for obtaining collision free time-optimal trajectories using a mathematical programming method. Both of these papers have adopted an off-line (and therefore centralized) setting: a motion is computed and given to each robot in advance.

The objective of this paper is to demonstrate that even without the complicated calculation of [4], a motion that is sufficiently close to time-optimal can be obtained for two omnidirectional robots carrying a ladder, using a simple *distributed algorithm* in which each robot decides its motion individually based on the current and goal positions of the ladder. The algorithms we propose are based on the following simple idea:

Basically we let each robot pursue its individual interest (e.g., mainly moving toward the goal) when deciding its motion. Since the robots are carrying a ladder, however, their intended motions may be "incompatible," making it impossible for the robots to move as intended. Now, suppose that there is a (virtual) "coordinator" that resolves the conflict in a "fair" manner for both, allowing the robots to continue to move. Then the resulting motion may be sufficiently close to time-optimal.

We will present two such algorithms. The first one is obtained by a naive application of this idea and works well for many, but not all, instances. The second one requires a slightly more complicated calculation, and works well for many of the instances for which the first algorithm does not.

The paper is organized as follows: Section 2 gives an outline of the method of [4] for computing an exact optimal motion under some (rather strong) assumptions. Section 3 presents the two distributed algorithms we propose. Section 4 describes the physical robots we plan to use and reports the results of some preliminary experiments using them. Section 5 presents the computer simulation results. We then conclude the paper by giving some remarks in Section 6.

2   Time-Optimal Motion for Carrying a Ladder

Chen, et al. discussed the problem of computing a time-optimal motion for two omnidirectional robots carrying a ladder from an initial position to a final position in a plane without obstacles [4]. In order to explain a formidable nature of the problem, suppose that robots A and B must move to A' and B', respectively, as is shown in Fig. 1, carrying a ladder. (We use "A" and "B" to refer to the robots as well as their initial positions.) A time optimal motion is shown in Fig. 2. You can observe that the path of robots B is not a straight line segment, indicating that B yields to give enough time for A to complete the necessary rotation. Thus this time optimal motion is not attainable if one of them insists on its individual interest of reaching its goal as quick as possible.

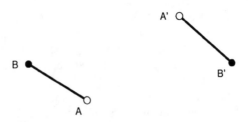

Fig. 1. Robots A (empty circle) and B (filled circle) move to A' and B', respectively, carrying a ladder.

Let v be the maximum speed of a robot. In [4] it is assumed that the speed of a robot is either 0 or v at any moment. (Computing a time-optimal motion in the

Fig. 2. Time-optimal motion for the instance shown in Fig. 1. The figure consists of a series of snapshots. In the first few shots the segment increases in grayscale as time increases.

general case is still open.) Without loss of generality assume that $|BB'| \geq |AA'|$. For convenience, we take AB to be of unit length ($|AB| = 1$), and let $L = |BB'|$ be the distance between B and B'. We set up a Cartesian coordinate system as is shown in Fig. 3, where B and B' are at the origin $(0,0)$ and $(L,0)$, respectively. Let α and β, respectively, be the angles that AB and $A'B'$ make with the x-axis.

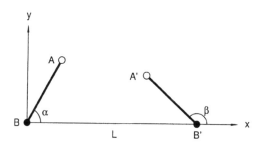

Fig. 3. The setup of the Cartesian coordinate system with reference to the initial and final positions of the robots.

Since the robots can move only at speed v or 0 at any moment, a lower bound on the time it takes to move AB to $A'B'$ is L/v. Intuitively, this lower bound is achievable if A can complete the necessary rotation around B within time L/v while B moves straight to B' at speed v. The following theorem characterizes the case.

Theorem 1. *AB can be moved to $A'B'$ in optimal time L/v if and only if either*

1. *$\alpha = \beta$, or*
2. *$0° < \alpha < \beta < 180°$ and $\tan(\beta/2) \leq \tan(\alpha/2)e^{2L}$,*

where L, α and β are as defined above.

Now we consider the case in which the lower bound L/v is not attainable. Let (x, y) be the coordinates of B during a motion, and ϕ the angle between the x-axis and the direction of the velocity of B. See Fig. 4.

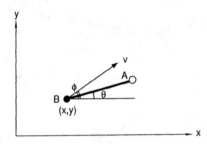

Fig. 4. The definitions of θ and ϕ. The arrow represents the velocity of robot B.

Since $|AB| = 1$, robot A rotates around robot B with angular speed $\omega = \dot{\theta}$, where the angle between the velocity of B and the velocity of A due to rotation around B is $90° + \theta - \phi$. Therefore V, the resultant speed of A, is given by

$$V^2 = v^2 + \omega^2 - 2\omega v \sin(\theta - \phi). \tag{1}$$

Since $V = v$ and $\omega = \dot{\theta}$, Eq. 1 can be rewritten as

$$\dot{\theta} - 2\dot{x}\sin\theta + 2\dot{y}\cos\theta = 0, \tag{2}$$

which describes the constraint on the optimal motion.

Now our task is to minimize the integral

$$F = \int_0^t \sqrt{\dot{x}^2 + \dot{y}^2}\, dt \tag{3}$$

subject to the constraint

$$\dot{\theta} - 2\dot{x}\sin\theta + 2\dot{y}\cos\theta = 0 \tag{4}$$

and the boundary conditions

$$\begin{cases} x(0) = 0 \\ y(0) = 0 \end{cases} \tag{5}$$

$$\begin{cases} x(t) = L \\ y(t) = 0 \end{cases} \tag{6}$$

By applying calculus of variations we can find the differential equations that an optimal trajectory obeys. In the next theorem $F(\phi, k)$ and $E(\phi, k)$ denote the Legendre elliptic integrals of the first and second kind, respectively, defined by

$$F(\phi, k) = \int_0^\phi \frac{d\theta}{\sqrt{1 - k^2 \sin^2\theta}} \tag{7}$$

$$E(\phi, k) = \int_0^\phi \sqrt{1 - k^2 \sin^2 \theta} d\theta. \tag{8}$$

Theorem 2. *Suppose that the lower bound L/v is not attainable. An optimal motion in which segment AB rotates about B counterclockwise can be obtained from*

$$\dot{x} = \frac{\dot\theta}{2} \sin\theta + cv \cos\theta \sin(\theta + \delta) \tag{9}$$

$$\dot{y} = -\frac{\dot\theta}{2} \cos\theta + cv \sin\theta \sin(\theta + \delta) \tag{10}$$

$$\dot\theta = 2v\sqrt{1 - c^2 \sin^2(\theta + \delta)}, \tag{11}$$

were constants c and δ are numerically calculated from

$$\frac{1}{2}(\cos\alpha - \cos\beta) + \frac{\cos\delta}{2c}\left[\sqrt{1 - c^2 \sin^2(\alpha + \delta)} - \sqrt{1 - c^2 \sin^2(\beta + \delta)}\right]$$
$$+ \frac{\sin\delta}{2c}\left[F(\beta + \delta, c) - F(\alpha + \delta, c) - E(\beta + \delta, c) + E(\alpha + \delta, c)\right] = L \tag{12}$$

and

$$\frac{1}{2}(\sin\alpha - \sin\beta) - \frac{\sin\delta}{2c}\left[\sqrt{1 - c^2 \sin^2(\alpha + \delta)} - \sqrt{1 - c^2 \sin^2(\beta + \delta)}\right]$$
$$+ \frac{\cos\delta}{2c}\left[F(\beta + \delta, c) - F(\alpha + \delta, c) - E(\beta + \delta, c) + E(\alpha + \delta, c)\right] = 0. \tag{13}$$

The time needed to execute the motion is obtained from

$$t = \frac{1}{2v}\left[F(\beta + \delta, c) - F(\alpha + \delta, c)\right]. \tag{14}$$

Figs. 2 and 2 show two optimal motions computed by the method.

Fig. 5. Optimal motion in which the lower bound L/v is attained.

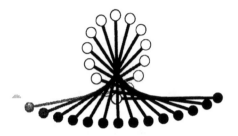

Fig. 6. Optimal motion in which the ladder turns around by nearly 180°.

3   Distributed Algorithms for Carrying a Ladder

In this section we describe two distributed algorithms for two omnidirectional robots carrying a ladder. They are designed based on the following idea, which is a restatement of the idea mentioned in Section 1.

> At any moment both robots intend to move toward their respective goal positions, and in case of a conflict they both yield equally by adjusting the directions of motion so that the distance between them is always equal to the length of the ladder.

Consider two robots A and B carring a ladder of length ℓ (see Fig. 3).[1] They both execute the same algorithm. Let A and B, and A' and B' be the start and goal positions, respectively, where $|\overline{AB}| = |\overline{A'B'}| = \ell$. Let $|\overline{AA'}| = L'$ and $|\overline{BB'}| = L$, where we assume without loss of generality that $L' \leq L$. Angles α and β are as defined in Fig. 3.

We view an algorithm for robot R as a mechanism that takes as input the current and goal positions of the ladder and produces as output a force vector $\mathbf{f}_R$ by which we attempt to drive R. Since R is connected to another robot by a rigid ladder, however, most likely it cannot move in the direction given by $\mathbf{f}_R$. Specifically, if R is about to move closer to (or away from) the other robot, then the ladder pushes (or pulls) R by some force $\mathbf{h}_R$. So in this paper we assume that R's actual motion is determined by force $\mathbf{f}_R + \mathbf{h}_R$. Note that the term $\mathbf{h}_R$ in $\mathbf{f}_R + \mathbf{h}_R$ effectively forces the robots to "yield equally" in case of a conflict. The assumption that an algorithm's output $\mathbf{f}_R$ does not depend on $\mathbf{h}_R$ somewhat simplifies the task of designing algorithms. (In contrast, physical robots are likely to have a force sensor for detecting the motion of the ladder, and the sensor output will be used explicitly in a feedback control scheme that drives the robot.)

Both algorithms ALG1 and ALG2 we present below are memoryless in the sense that their output is a function of the current input (and is independent of the motions in the past). It is therefore sufficient to view A and B of Fig. 3 as

---

[1] In Section 2, we assumed that $\ell = 1$.

the robots' current positions and specify the output for input A, B, A' and B'. As we will see shortly ALG2 is an extension of ALG1.

As we mentioned earlier, an algorithm's output is a force vector $\mathbf{f}_R$. In the following, however, we describe ALG1 and ALG2 in terms of *the target velocity vector* that the force they compute should allow the robot to achieve, if force $\mathbf{h}_R$ were not present.

Algorithm ALG1: Both A and B move toward their respective goals A' and B' at speeds v_A and v_B, respectively, where $v_A = v(L'/L)^s$ and $v_B = v$. Note that v is the robots' maximum speed and $s \geq 0$ is a parameter of ALG1.

Algorithm ALG2: Let $\mathbf{v}_A$ and $\mathbf{v}_B$ be the velocity vectors for robots A and B that ALG1 computes. ALG2 uses auxiliary velocity vectors $\mathbf{r}_A$ and $\mathbf{r}_B$ for robots A and B, respectively, that rotate the ladder couterclockwise. The direction of $\mathbf{r}_A$ (or $\mathbf{r}_B$) is $\alpha + \pi/2$ (or $\alpha - \pi/2$), i.e., perpendicular to the ladder $\overline{AB}$, and $||\mathbf{r}_A|| = ||\mathbf{r}_B|| = c(\beta - \alpha)$, where constant $c \geq 0$ is a parameter of ALG2. Then the velocity vectors that ALG2 computes for A and B are respectively $v(\mathbf{v}_A + \mathbf{r}_A)/a$ and $v(\mathbf{v}_B + \mathbf{r}_B)/a$, where $a = max\{||\mathbf{v}_A + \mathbf{r}_A||, ||\mathbf{v}_B + \mathbf{r}_B||\}$. Observe that ALG2 coincides with ALG1 when $c = 0$.

4   The Physical Robot System

For experiments we use a physical robot system involving two omnidirectional robots developed at RIKEN [3]. Fig. 7 shows the configuration of used in a preliminary experiment where the two robots are connected to a ladder via a force sensor placed on top through which each can sense the motion of the other. The connection via the force sensor is rather rigid, but it at least allows a human operator to hold one end of the ladder attached to a robot and gently push and pull the ladder so that the robot smoothly follows the motion of the ladder.

The results of the experiment using two robots in this configuration, however, has indicated that a more flexible link between the robots and the ladder are required. The relatively rigid connection between the robots and the ladder requires the robots to react to the motion of the other much more quickly than they actually can, resulting in an undesirable motion that is not smooth. A promising solution for this problem is to place a flexible multi-link mechanism that acts as a buffer between the ladder and the force sensors. We have designed such a mechanism consisting of springs and dumpers, and are now installing them.

5   Performance Evaluation by Simulation

5.1   Simulation Model

We conducted computer simulation to analyze the behavior of the robot system introduced in the last section executing the algorithms proposed in Section 3.

Fig. 7. Two RIKEN robots connected by a ladder. The ladder is fixed to each robot via a force sensor placed at the top.

We model the robots by a disk of radius 10 with a maximum speed of 1 per unit time. The length of ladder is 100. The multi-link mechanism is modeled by an ideal spring.

We use discrete time $0, 1, \dots$ and denote by $\mathbf{p}_R(t)$ and $\mathbf{v}_R(t)$ the position of R and the target velocity vector of R that an algorithm specifies at time t, respectively. Also, we let $\mathbf{x}_R(t)$ denote the vector from the center of disk R to the position of the corresponding endpoint of the ladder at t. To simplify the simulation we assume:

1. Force $\mathbf{h}_R(t)$ that robot R receives from the ladder at time t is given by $k\mathbf{x}_R(t)$, where k is a spring constant.
2. The velocity of R at time t is $\mathbf{w}_R(t) = \mathbf{v}_R(t) + k\mathbf{x}_R(t)$. (We ignore the kinematic constraints of R and assume that R can attain any target velocity instantaneously.) If the length of $\mathbf{w}_A(t)$ or $\mathbf{w}_B(t)$ exceeds 1 (the robots' maximum speed), then we normalize them by dividing both by $\max\{\|\mathbf{w}_A(t)\|, \|\mathbf{w}_B(t)\|\}$.
3. At time $t + 1$, R is at position $\mathbf{p}_R(t + 1) = \mathbf{p}_R(t) + \mathbf{w}_R(t)$.

The constant k is a parameter that controls the stiffness of the link between a robot and the ladder and therefore the amount of freedom of a robot's motion. When $k = 0$, for example, the robots can move freely irrespective of the position of the ladder. In this paper we used $k = 10.0$ which we found to be sufficiently large to keep the endpoints of the ladder within the disks of radius 10 representing the two robots (i.e., $\|\mathbf{x}_R(t)\| \leq 10$).

5.2 Algorithm ALG1

To investigate the effect of parameter s on the performance of Algorithm ALG1, we examine the number of steps N necessary for the robots to reach their goals using the setup of Fig. 3 for $L = 200, 400, \ldots, 1600$, $0° < \alpha \le 90°$ and $s = 0.0, 0.5, \ldots, 4.0$. To reduce the number of instances to examine, however, we consider only those cases in which $\alpha = 180° - \beta$ and (by symmetry) $\alpha \le 90°$. Note that $N \ge L$ holds since the robots' maximum speed is 1.

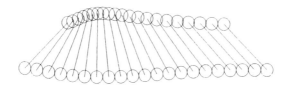

Fig. 8. The motion of the robots obeying ALG1 for $s = 3.0, L = 400$ and $\alpha = 45°$.

Fig. 8 shows a motion obtained by ALG1 for $s = 3.0, L = 400$ and $\alpha = 45°$.

Note that the robot further from its destination moves nearly straight, as in an optimal motion shown in Fig. 2 for the same instance. In fact, ALG1 is expected to perform well when L and α are large since the robots need not rotate the ladder too quickly, and our simulation results confirm this. We present only the results for $L = 200$ in Fig. 9, where we observe the following:

1. The performance of ALG1 is not very sensitive to the value of s. Setting $s = 3$ seems to work particularly well for most cases.
2. ALG1 shows sufficiently good performance for $\alpha \ge 20°$. For $\alpha \ge 65°$ it always achieves the lower bound of $N = L = 200$.
3. The performance of ALG1 is noticeably worse when $\alpha \le 10°$ and rapidly degrades as α approaches 0.

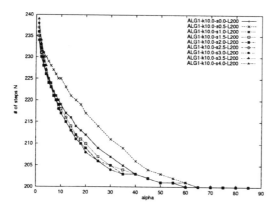

Fig. 9. The number N of steps needed by ALG1 for $L = 200$ and various values of α.

Comparing the motion generated by ALG1 for $s = 3.0, L = 200$ and $\alpha = 1°$ shown in Fig. 10 with an optimal motion shown in Fig. 2 computed by the method of Section 2, we notice that in the former the ladder starts to rotate only toward the end of the motion, while in the latter rotation starts as soon as the ladder starts moving. This observation has motivated us to introduce in ALG2 auxiliary velocity vectors r_A and r_B that help rotate the ladder during motion.

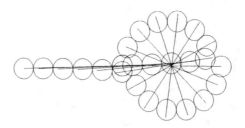

Fig. 10. The motion generated by ALG1 for $s = 3.0, L = 200$ and $\alpha = 1°$.

5.3   Algorithm ALG2

Recall that ALG2 has a parameter c and reduces to ALG1 when $c = 0$. Using the same set of instances as in Subsection 5.2 and for various values of c, we again measure the number N of steps necessary for the robots to reach the goal. Throughout this section we use $s = 3.0$ which was found to work well in ALG1.

Fig. 11 shows the results for $L = 200$ and $0° < \alpha \leq 90°$, for several values of c. We observe the following:

1. ALG1, i.e., ALG2 with $c = 0$, performs better than ALG2 with $c > 0$ for all $\alpha \geq 20°$.
2. When $\alpha < 20°$ ALG2 with $c = 0.5$ shows the best performance.

Figs. 12 and 13 show the motions generated by ALG2 with $c = 0.5$ for the instances examined for ALG1 in Figs. 8 and 10, respectively.

Again, in Fig. 12 the robot further from its destination moves nearly straight, as in an optimal motion shown in Fig. 2 for the same instance. The motion shown in Fig. 13 is fairly similar to an optimal motion shown in Fig. 2 and takes much less time than the motion in Fig. 10 generated by ALG1.

Based on the results shown in Fig. 11 we conclude that small effort to rotate the ladder is sufficient to avoid undesired motion such as the one shown in Fig. 10. The results also indicate that larger values of c tend to increase N. We verified this through additional simulation, and present only the results for $L \geq 400$ in

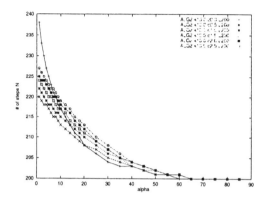

Fig. 11. The number N of steps needed by ALG2 for $L = 200$ and various values of α.

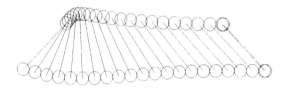

Fig. 12. The motion generated by ALG2 with $c = 0.5$, for $L = 400$ and $\alpha = 45°$.

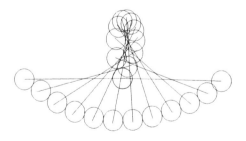

Fig. 13. The motion generated by ALG2 with $c = 0.5$, for $L = 200$ and $\alpha = 1°$.

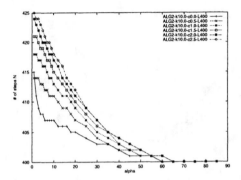

Fig. 14. The number N of steps needed by ALG2 for $L = 400$ and various values of α.

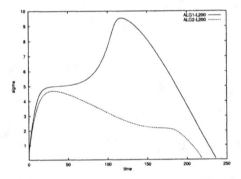

Fig. 15. The change of σ during the motions of Figs. 10 and 13 for $L = 200$ and $\alpha = 1°$.

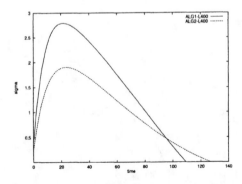

Fig. 16. The change of σ during the motions of Figs. 8 and 12 for $L = 400$ and $\alpha = 45°$.

Fig. 14, which shows that ALG1 performs in general better than ALG2 with $c > 0$.

Finally, we compare ALG1 and ALG2 with $c = 0.5$ from a different point of view. The measure is $\|\mathbf{x}_R(t)\|$, i.e., the distance between the center of robot R and the corresponding endpoint of the ladder at time t. If an algorithm produces a motion with small $\|\mathbf{x}_R(t)\|$, then it might be said that the actions of the two robots are coordinated well, and consequently, physical robots using the algorithm can be expected to move smoothly. Figs. 15 and 16 show the ratio $\sigma = 100 \times (\|\mathbf{x}_R(t)\|/\ell)(\%)$ taking time as the parameter, where $\ell = 100$ is the length of ladder. Fig. 15 shows the case $L = 200$ and $\alpha = 1°$ (motions in Figs. 10 and 13), and Fig. 16 the case $L = 400$ and $\alpha = 45°$ (motions in Figs. 8 and 12). Clearly ALG2 performs better than ALG1 in both cases.

6   Conclusions

Time-optimal motion of two omnidirectional robots carrying a ladder can be computed using optimal control theory in a off-line and centralized manner. This paper has demonstrated that even without the complicated calculation of such an approach, a motion that is sufficiently close to time-optimal can be obtained using a simple distributed algorithm in which each robot decides its motion individually based on the current and goal positions of the ladder.

We presented two such algorithms, ALG1 and ALG2, and using computer simulation demonstrated that the former performs sufficiently well for large L and α (two of the parameters describing a given instance), while for small L and α the latter algorithm performs better. We also observed that the smoothness of motion is another advantage of ALG2.

This paper has reported only the preliminary results of the authors' ongoing project on multi-robot coordination. Issues to be investigated in the future include the following:

1. Implementing ALG1 and ALG2 on the RIKEN robots.
2. Coordination of robots having different capabilities.
3. Obstacle avoidance while carrying a ladder.
4. Extension to the problem of carrying an n-gon.

References

1. M. N. Ahmadabadi and E. Nakano, "Constrain and move: a new concept to develop distributed transferring protocols," *Proc. IEEE Int. Conf. on Robotics and Automation,* 2318–2325 (1997).
2. R. Alami, S. Fleury, M. Herrb, F. Ingrand, and S.Qutub, "Operating a large fleet of mobile robots using the plan-merging paradigm," *Proc. IEEE Int. Conf. on Robotics and Automation,* 2312–2317 (1997).
3. H. Asama, M. Sato, L. Bogoni, H. Kaetsu, A. Matsumoto, and I. Endo, "Development of an Omni-directional mobile robot with 3 DOF decoupling drive mechanism," *Proc. IEEE Int. Conf. on Robotics and Automation,* 1925–1930 (1995).

4. Z, Chen, I. Suzuki, and M. Yamashita, "Time optimal motion of two robots carrying a ladder under a velocity constraint," *IEEE Trans. Robotics and Automation* *13*, 5, 721–729 (1997).
5. R. Courant and D. Hilbert, *Methods of Mathematical Physics*, Interscience Publishers, New York, 1953.
6. Proceedings of the International Symposium on Distributed Autonomous Robot Systems (DARS) in 1992, 1994, 1996, and 1998.
7. B.R. Donald, "Information Invariants in Robotics: Part I – State, Communication, and Side-Effects," *Proc. IEEE Int. Conf. on Robotics and Automation*, 276–283 (1993).
8. K. Kosuge and T. Oosumi, "Decentralized control of multiple robots handling and objects," *Proc. IEEE/RSJ Int. Conf. on Intelligent Robots and Systems*, 556–561 (1996).
9. J-C. Latombe, "Robot Motion Planning," Kluwer Academic Publishers, Boston, 1991.
10. M. Mediavilla, J. C. Fraile, J. R. Perán, and G. I. Dodds, "Optimization of collision free trajectories in multi-robot systems," *Proc. IEEE Int. Conf. on Robotics and Automation*, 2910–2915 (1998).
11. N. Miyata, J. Ota, Y. Aiyama, J. Sasaki, and T. Arai, "Cooperative transport system with regrasping car-like mobile robots," *Proc. IEEE/RSJ Int. Conf. on Intelligent Robots and Systems*, 1754–1761 (1997).
12. Y. Nakamura, K. Nagai, and T. Yoshikawa, "Dynamics and stability in coordination of multiple robotics mechanisms," *The Int. J. of Robotics Research*, 8,2, 44–60 (1989).
13. D. J. Stilwell and J. S. Bay, "Toward the development of a material transport system using swarms of ant-like robots," *Proc. IEEE Int. Conf. on Robotics and Automation*, 766–771 (1995).
14. I. Suzuki and M. Yamashita: "A Theory of distributed anonymous mobile robots –formation and agreement problems," *SIAM J. Computing* (to appear).
15. L.L. Whitcomb, D.E. Koditschek, and J.B.D. Cabrera, "Toward the Automatic Control of Robot Assembly Tasks via Potential Functions: The Case of 2-D Sphere Assemblies," *Proc. IEEE Int. Conf. on Robotics and Automation*, 2186–2191 (1992).
16. H. Yamaguchi, "Adaptive formation control for distributed autonomous mobile robot groups," *Proc. IEEE Int. Conf. on Robotics and Automation*, 2300–2305 (1997).
17. H. Yamaguchi, "A cooperative hunting behavior by mobile robot troops," *Proc. IEEE Int. Conf. on Robotics and Automation*, 3204–3209 (1998).

Modeling and Stability Analysis of the Dynamic Behavior in Load Sharing Systems

Paul Levi, Michael Schanz, Viktor Avrutin, and Robert Lammert

University of Stuttgart, Institute of Parallel and Distributed High-Performance Systems, Breitwiesenstr. 20-22, D-70565 Stuttgart, Germany

1 Introduction

The main goal of this work is the development of suitable multi-agent based models of real technical systems. As an example we use a simple load sharing system consisting of cooperating robotic units. In order to get a deeper insight and better understanding of the dynamics occurring in such systems, we have to simulate and analyze the dynamical behavior of this models. This requires an adequate mathematical representation of the multi-agent systems. For this representation we use in our work dynamical systems with non-continuous step-functions. However, this kind of representation as well as the applied analysis methods are not specific for multi-agent systems, but also suitable for a much broader class of technical systems.

In this paper, we are interested in analyzing the dynamical properties of our models, using concepts and adapted analysis methods from the theory of non-linear dynamical systems. Our research and analysis work is mainly concerned with the stability of the system at given parameter settings. Additionally we perform also extended parameter studies in order to find appropriate parameter settings leading to an optimally working multi-agent system. It is expected, that the results of our investigations, especially the parameter studies, are to some extent transferable to the real technical systems.

2 Motivation

The continuously increasing competition on the global market and a larger growing spectrum of products are the reasons for introducing more intelligent production methods and technical systems in industry and manufacturing [1]. The main idea is to allow a rapid switch (transition) between different production methods and products because the requirements of the customers and their needs are frequently changing. Under these circumstances, one is interested in having an adaptive production in order to accomplish the adaption to the new conditions as fast and as well as possible. A solution to the problem described above consists of using multi-functional and programmable agents (robotic units), which have to deal with the production issues. In this case it is possible to change the process of production by only reprogramming the involved agents, i.e., without replacing the existent hardware. We would like to call this concept a multi-agent

Christensen et al. (Eds.): Sensor Based Intelligent Robots, LNAI 1724, pp. 255–271, 1999.
© Springer-Verlag Berlin Heidelberg 1999

system and assume, that any agent owns a functionality which implies a cyclic (iterative) behavior [2].

As one can describe the dynamical behavior by evolution equations, we suppose that the theory of non-linear time-discrete dynamical systems could be a useful tool for simulating and analyzing the complex behavior of a particular multi-agent system. Our work is basically concerned with modeling, simulation and analysis of a simple multi-agent system, called $\mathcal{R}n$, in order to determine the applicability of this concept in practice and to detect the domains of interest, which should be the subject of further investigations. From our (mathematical) point of view, one can change the program of a single agent by adjusting some of its parameters. This is the main reason, which indicates that we need the exploration of the parameter space in order to analyze the behavior of the system and hence to find the parameter settings which are appropriate to our purposes.

3   System Description

Rather from a technical point of view, our system consists of n single agents, or robotic units R_i, $i = 1, \ldots, n$, and therefore we would like to refer to our system as a $\mathcal{R}n$-system, in order to always indicate the number of involved robots by only replacing n with the desired natural number. The robotic units we have mentioned, have all the same skills, but the efficiencies of the involved robotic units may be variable and can be regarded as parameters in our $\mathcal{R}n$-system. Our simple model allows only one kind of robot activity, which we abstractly called *production*. In this sense, the ability to produce is the only robot skill in our system and we would like to designate the related efficiency as the *productivity* of the respective robotic unit. So we can consider each productivity α_i of the robotic unit R_i as belonging to the parameter space of the $\mathcal{R}n$-system. This parameter space contains only one more parameter, namely the job size C. In our model, we assume that the robotic units have to execute these jobs of constant amount C, which are assigned to our multi-agent system. Furthermore one can regard the incoming jobs as periodical events. This is the main reason for dividing the time axis of the $\mathcal{R}n$-system in discrete time units and designing our mathematical model of the system by time-discrete evolution equations.

The jobs in our system are defined in a very abstract way, because we left their physical properties unspecified. The only important aspect here is the job size C, which is highly correlated with the amount of time, that a robotic unit needs to perform the job. In the $\mathcal{R}n$-system, the incoming jobs must be *immediately* directed to a particular robotic unit, according to the distribution strategy. To avoid the loss of jobs, we have equipped our robotic units with buffers, which allow a job buffering in FIFO-order. For the case of n robotic units we obtain the respective, potentially unbounded n buffers, as our state space.

During an execution time interval, the robotic unit R_i always reduces the content of its buffer B_i by the respective productivity α_i, as long as $B_i \geq \alpha_i$. In the special case $B_i < \alpha_i$, the buffer B_i will get empty ($B_i = 0$) during execution and the robotic unit will be idle for some time interval.

In order to enhance the productivity of the whole system, one will intend to avoid situations like described above by seeking for an optimally working system under the given preconditions. Subsequently we would like to refer to this aim as our optimality criterion.

Until now, we have described only the processes occurring in the execution time interval, which is situated between two consecutive arriving jobs. To pursue, we would like to focus our attention to the distribution of the incoming jobs, a process that takes place instantly for any job assigned to the system. The distribution strategy provided for the $\mathcal{R}n$-system can be described by the rules presented below:

▶ The jobs are indivisible at the distribution level and must be handled as entities. That means, that the entire job has to be assigned to exactly one robotic unit.

▶ An incoming job will be assigned to that robotic unit, which terminates first with the execution of this job. The termination time of each robotic unit is the time it needs to empty its buffer, assumed it gets the job and there are no further incoming jobs.

▶ A conflicting situation arises, if the computation delivers identical termination times for two or more different robotic units. In this case, the distribution strategy selects the robotic unit with the lowest index number, i.e. the one who appears on the first position in the ordered list of conflicting robotic units.

4   The Evolution Equations of the $\mathcal{R}n$-System

The rules for the execution and the distribution of jobs are both integrated in the evolution equations of our $\mathcal{R}n$-system, which consist of difference equations for each robotic unit. The evolution equation of a single agent i can be described by:

$$B_{k+1}^i = \begin{cases} B_k^i - \alpha_i + C & \text{if agent } i \text{ gets the job} \\ B_k^i - \alpha_i & \text{otherwise} \end{cases} \tag{1}$$

The lower index k indicates the discrete time in the dynamic evolution of the system. One can see, that the buffer content at the subsequent time point $(k+1)$ depends only on the former buffer content B_k^i, the parameters of the system and a so-called *decision function* (the if-statement above). Decision functions are non-continuous, with a logical expression as argument. They also play the role of semantic functions, because they have to firstly evaluate their argument expression, in order to generate the corresponding numerical value (0 or 1). A decision function χ will be defined by:

$$\chi[b] = \begin{cases} 1, & \text{if } (b \Leftrightarrow \text{true}) \\ 0, & \text{if } (b \Leftrightarrow \text{false}) \end{cases}$$

Additionally we need the simple operator $(x)^+$ to indicate, that each valid buffer content must be represented by a non-negative, real number. This operator is therefore defined by

$$(x)^+ = \begin{cases} x \text{ if } x \geq 0 \\ 0 \text{ if } x < 0 \end{cases}$$

or equivalently

$$(x)^+ = x \cdot \chi[x \geq 0]$$

Now, we are finally able to summarize all the informations presented above in a mathematical formulas (our evolution equations):

$$B_{k+1}^1 = \left(B_k^1 - \alpha_1 + C \cdot \prod_{j=2}^n \chi \left[(\tau_1(B_k^1 + C) \leq \tau_j(B_k^j + C)) \right] \right)^+$$

$$\vdots$$

$$B_{k+1}^i = \left(B_k^i - \alpha_i + C \cdot \prod_{j=1}^{i-1} \chi \left[(\tau_i(B_k^i + C) < \tau_j(B_k^j + C)) \right] \right.$$
$$\left. \cdot \prod_{j=i+1}^n \chi \left[(\tau_i(B_k^i + C) \leq \tau_j(B_k^j + C)) \right] \right)^+ \quad (2)$$

$$\vdots$$

$$B_{k+1}^n = \left(B_k^n - \alpha_n + C \cdot \prod_{j=1}^{n-1} \chi \left[(\tau_n(B_k^n + C) < \tau_l(B_k^l + C)) \right] \right)^+$$

The expression $\tau_i(B_k^i + C)$ computes the termination time of the robotic unit R_i, according to the descriptions given above, where τ_i is the reciprocal value of the parameter α_i (i.e. $\tau_i = \frac{1}{\alpha_i}$). The introduction of τ_i makes sense, because some analytical investigations are easier to accomplish by using the reciprocal parameter values τ_i.

5   Confinement of the Parameter Space

Without loss of generality we assume that the robotic units are ordered according to their productivities. Let R_1 be the fastest robotic unit and R_n the slowest one, i.e.:

$$\alpha_1 \geq \alpha_2 \geq \ldots \geq \alpha_n. \quad (3)$$

We denote the system suitably loaded if all robotic units together are able to perform all incoming jobs and any subset of robotic units is not able to do this. It makes sense to investigate the system only in this case, because otherwise there occurs either an overflow of the buffers or some of the robotic units are permanently idle. We can express this mathematically by

$$\sum_{i=1}^{n-1} \alpha_i < C \leq \sum_{i=1}^n \alpha_i \quad (4)$$

and the constraint is assumed to be fulfilled.

6   Simulation Results

In order to guarantee a stable and suitable mode of operation of our system we have to investigate its dynamical behavior depending on the parameters. The standard methods of stability analysis in the field of nonlinear dynamics are not applicable here due to the presence of the non-continuous decision functions in our system. However, we are able to find regions with qualitatively different properties in parameter space using the method of symbolic dynamics [3], [4], [5]. The main property we can investigate with this method is the stability of the system with respect to small deviations of the parameters. To obtain a deeper insight in the underlying principles of the dynamics, we started our investigations with the simulation of the most simple system, consisting of two robotic units and hence denoted as the $\mathcal{R}2$-system. Although this system is very simple, it is not possible to analyze its dynamic behavior completely analytically. Therefore one has to investigate the dynamics using numerical simulations. In this article we present mainly the results of this simulations. However we have obtained some interesting analytical results (bifurcation points and scaling constants of the period-increment scenario), which we will present in a forthcoming paper [6].

6.1 $\mathcal{R}2$-System

The evolution equations of the $\mathcal{R}2$-system are given by Eq. 2, which read in the special case $n = 2$:

$$
\begin{aligned}
B_{k+1}^1 &= \left(B_k^1 - \alpha_1 + C \cdot \chi \left[(\tau_1(B_k^1 + C) \leq \tau_2(B_k^2 + C)]\right)\right)^+ \\
B_{k+1}^2 &= \left(B_k^2 - \alpha_2 + C \cdot \chi \left[(\tau_2(B_k^2 + C) < \tau_1(B_k^1 + C)]\right)\right)^+
\end{aligned}
\tag{5}
$$

Hence this system has two state variables B_k^1, B_k^2, and three parameters α_1, α_2 and C. However it can be shown by an appropriate scaling transformation, that one parameter is in fact unnecessary, so that we have only two relevant control parameters in the $\mathcal{R}2$-system. For reasons of simplicity, we use in the following τ_1, τ_2 (the reciprocal values of the parameters α_1, α_2) as the control parameters of the $\mathcal{R}2$-system and choose for the parameter C the value 1.

Time series and assignment sequences:
Firstly we have simulated the $\mathcal{R}2$-system with specific settings of the parameters. Fig. 1 shows typical time series of the buffer contents B_k^1, B_k^2. As one can see from Fig. 1.a, where the parameters τ_1, τ_2 are equal, the jobs will be assigned to the both robotic units alternately. That means if an incoming job is assigned to one robotic unit, the next incoming job will be assigned to the other robotic unit and so forth. For other parameter settings, some jobs will be assigned to the faster unit R_1 and after that one job will be assigned to the slower unit R_2, and so forth.

To pursue further we use the following symbolic coding scheme for this dynamics. We write the symbol 1 at the time points, when jobs will be assigned to the unit R_1, and the symbol 2 at the time points, when jobs will be assigned to the unit R_2. With this symbolic coding we obtained the following two results:

▶ After a job is assigned for the first time to the faster robotic unit R_1, there will never be an assignment of two jobs, which follow each other, to the slower robotic unit R_2. It follows, that in the corresponding symbolic sequence two or more symbols 2 which follow each other can occur only in the beginning part of the sequence. A proof of this will be presented in [6].

▶ The dynamic behavior and hence the symbolic sequence can be divided in two parts, a transient and an asymptotic one. If the parameters τ_1, τ_2 are rational than the asymptotic dynamics of the system is periodic with a finite length of the period and the corresponding trajectory in the state space represents a limit-cycle. We call the symbolic sequence belonging to this limit-cycle a periodic sequence.

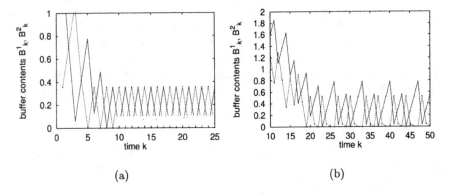

(a)   (b)

Fig. 1. *Time series of the buffer contents* $B_k^{1,2}$ *of the* $\mathcal{R}2$-*system*

Parameter settings: $C = 1$, (a) $\tau_1 = 1.55, \tau_2 = 1.55$; (b)$\tau_1 = 1.35, \tau_2 = 2.1$
In both cases a periodic mode of operation of the robotic units occur, the corresponding assignment sequences are (12) *for sub-figure (a) and* (1121112) *for sub-figure (b). One can clearly recognize the transient and asymptotic part of the dynamic.*

The time series of the buffer contents B_k^1, B_k^2, which correspond to the periodic sequence (12), is presented in Fig. 1.a. This is the most simple case of periodic behavior in the $\mathcal{R}2$-system. A more complex example of a periodic behavior with a larger period is shown in Fig. 1.b. The trajectory leads in this case to the symbolic sequence:

$$\underbrace{22212121121211212112}_{\text{transient dynamics}}\underbrace{(1121112)(1121112)(1121112)}_{\text{asymptotic dynamics}}\ldots$$

The periodic sequence is given by (1112112) and therefore the period is obviously 7.

One dimensional bifurcation diagrams:
The two examples in section 6.1 show that the asymptotic dynamic of the system depends on the parameters τ_1, τ_2. It is interesting to investigate this dependency

in detail. Therefore, we firstly keep one of the parameters, namely τ_2, fixed and simulate the system varying the other parameter τ_1. Depending on the quantity we observe during this variation of the parameter τ_1, we get different kinds of bifurcation diagrams. In Figs. 2 and 3 the dependencies of the following observed quantities on the parameter τ_1 at two different values of parameter τ_2 are presented:

▶ buffer content B_k^1 (Fig. 2.a, Fig. 3.a)
▶ length $N(P)$ of the periodic sequence P (Fig. 2.b, Fig. 3.b)
▶ ratios $\rho_i(P) = \frac{N_i(P)}{N(P)}$, where $N_i(P)$ denotes the number of symbols i in the periodic sequence P (Fig. 2.c, Fig. 3.c)

Case 1: $(\tau_2 = 1.5)$

As one can see from the Figs. 2.a and 2.b, there occurs with τ_1 decreasing an interesting phenomenon. The system undergoes a sequence of bifurcations, where at each bifurcation point the length of the periodic sequence increases by 1. In analogy to the well-known period-doubling scenario, occurring in many nonlinear dynamical systems, we denote this scenario a period-increment scenario. (In the literature however this scenario will be also denoted as period-adding scenario.) Like in the period-doubling scenario, the range of the parameter, for which certain period occurs, decreases as the period increases (see Fig. 2.b).

In addition one can observe in Fig. 2.c that with τ_1 decreasing the ratio ρ_1 increases monotonically whereas the ratio ρ_2 decreases monotonically. Thereby this increase respectively decrease takes place in steps. If τ_1 finally reaches the value 1, that means that the robotic unit R_1, according to (4), is able to perform all incoming jobs by itself, then obviously the ratio ρ_1 will reach the value 1 and the ratio ρ_2 will reach the value 0.

Case 2: $(\tau_2 = 2.1)$

Fig. 3.a shows for this value of τ_2 a similar structure as in the case $\tau_2 = 1.5$, but with interesting sub-structures. Like in Fig. 2.a, there exists a period-increment scenario in the interval $1 \leq \tau_1 \leq 1.5$, which we call main scenario. Furthermore there exists an infinite sequence of parameter ranges where in each one a separate period-increment scenario occurs. In Fig. 3.b the corresponding lengths of the periodic sequences of each scenario are presented.

Looking at the Fig. 3.c one can observe, that like in the case $\tau_2 = 1.5$ the ratio ρ_1 increases monotonically and the ratio ρ_2 decreases monotonically with the parameter τ_1 decreasing. The increase of ρ_1 and decrease of ρ_2 takes place in steps again, but due to the infinite sequence of period-increment scenarios the steps are smaller. In the parameter range between two steps of the main scenario there exists a period-increment scenario which leads to an infinite sequence of steps that joins the steps of the main scenario.

The characteristic difference between both cases presented so far can be understood only if one looks at the bifurcation diagram in the two-dimensional parameter space.

Bifurcation diagram in the two-dimensional parameter space:

In Fig. 4 the bifurcation diagram of our $\mathcal{R}2$-system in the two-dimensional pa-

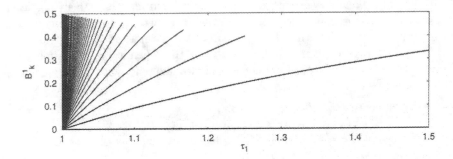

(a) Buffer content B_k^1 after the transient behavior

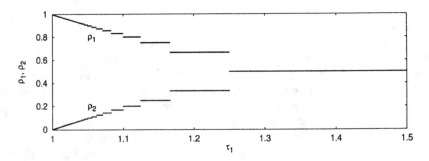

(b) Length N of the corresponding periodic sequences

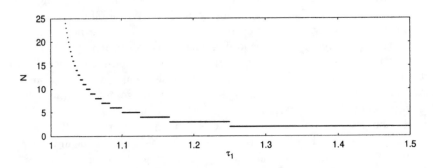

(c) Ratios $\rho_{1,2}$ of the symbols 1 and 2 in the corresponding periodic sequences

Fig. 2. *Bifurcation diagrams of the $\mathcal{R}2$-system (case 1, $\tau_2 = 1.5$). The parameter τ_1 is varied.*

rameter space $(\tau_1 \times \tau_2)$ is shown. One can see regions in the two-dimensional parameter space, where the parameter setting leads to the same periodic sequences. Because we have assumed that the robotic units are ordered according

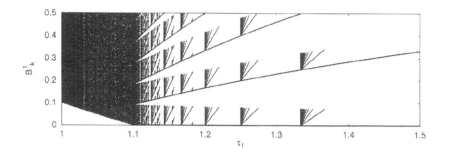

(a) Buffer content B_k^1 after the transient behavior

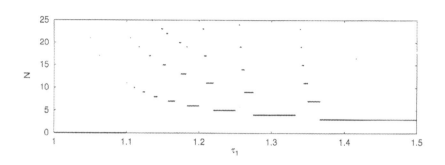

(b) Length N of the corresponding periodic sequences

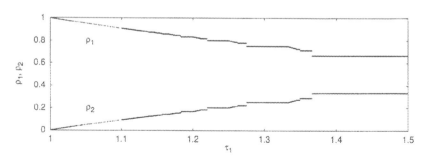

(c) Ratios $\rho_{1,2}$ of the symbols 1 and 2 in the corresponding periodic sequences

Fig. 3. *Bifurcation diagrams of the $\mathcal{R}2$-system (case 2, $\tau_2 = 2.1$). The parameter τ_1 is varied.*

to their productivities (3), it follows for the $\mathcal{R}2$-system that $\tau_1 \leq \tau_2$. This corresponds to the part of the two-dimensional bifurcation diagram located above the angle's bisector. As already mentioned, the ordering of the robotic units does not lead to a loss of generality. Therefore the bifurcation diagram in the two-dimensional parameter space has to be symmetric.

Apart from the angle's bisector there exists another confinement, namely the inequality (4). For the $\mathcal{R}2$-system, this inequality leads to the constraint $\frac{1}{\tau_1} + \frac{1}{\tau_2} \geq C$, which corresponds to the region below the hyperbola $\tau_2 = \frac{\tau_1}{C\tau_1 - 1}$. Our investigations within this region show three remarkable results:

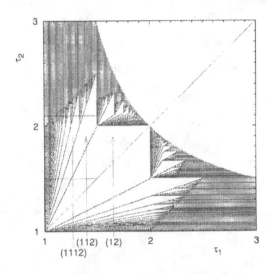

Fig. 4. *Bifurcation diagram of the $\mathcal{R}2$-system in the two-dimensional parameter space $\tau_1 \times \tau_2$.*

The region in which we investigate the behavior of the $\mathcal{R}2$ system is confined by the τ_2-axis, the hyperbola $\tau_2 = \frac{\tau_1}{C\tau_1 - 1}$ and the angle's bisector $\tau_1 = \tau_2$ (see also Fig. 5). The assignment sequences occurring in some triangular areas are shown, as well as two cuts corresponding to the one-dimensional bifurcation diagrams presented in Figs. 2 and 3.

▶ In the middle part of the bifurcation diagram some large fan-shaped set of triangular regions can be observed. Each triangle rooted at the point $\tau_1 = 1$, $\tau_2 = 1$ represents those values of the parameters which lead to the same period, whereby the periodic sequences are given by: $1^n 2$ with $n = 1, 2, \ldots$ (1^n denotes hereby the concatenation of n symbols of 1). The Fig. 2 corresponds to a cross section of this fan-shaped set of triangular regions at the value $\tau_2 = 1.5$. This set represents a period-increment scenario in the two-dimensional parameter space and is therefore the reason for the occurrence of period-increment scenarios in the one-dimensional parameter space. Like in the one-dimensional case, the ranges of the parameters, for which certain period occurs, (i.e., the areas of the triangles) decrease with the length of the period increasing.

▶ One see also in Fig. 4 that the parts of the bifurcation diagram located above the large fan-shaped set of triangular regions possess a fractal self-similar structure. Between two adjacent triangles of the large fan-shaped triangular regions a smaller fan-shaped set of triangular regions is located, where also a period-increment scenario takes place. However in this scenarios consecutive periods will be increased by a value larger than 1.

▶ All triangular regions together build up an area, which we call the P^2-region (see Fig. 5). Within each triangle in this area deviations of the parameters cause no qualitative change of the behavior as far as the corresponding point in parameter space remains in the same triangle. Especially the periodic sequence remains the same. Within the region above the P^2-region and below the hyperbola the behavior of system is different from that in the P^2-region. The system is here sensitive with respect to deviations of the parameter τ_1. That means that any infinitesimal deviation of this parameter leads to a change in the period. A change of the other parameter τ_2 in the range from the hyperbola to the border of the P^2-region cause no change of the period. We call the area which is built up from all this one-dimensional regions, i.e., lines between the hyperbola and the border of the P^2-region, the P^1-region (see Fig. 5). The remaining part of the parameter space we investigate, namely the hyperbola, we finally denote as the P^0-region (see Fig. 5). Here any infinitesimal deviation of both parameters, i.e., τ_1 as well as τ_2, changes the period.

Interpretation of the investigation results of the $\mathcal{R}2$-system:

Concerning the two robotic units R_1 and R_2, the mathematical results presented so far have to be interpreted. First we have to keep in mind that the parameters τ_1 and τ_2 are the reciprocal values of the productivities α_1 and α_2 of the robotic units. Second we have to realize, that in any technical system it is impossible to keep parameters exactly fixed. Consequently some regions in the parameter space of such systems, where the behavior or mode of operation is sensitive on even infinitesimal deviations of the parameters, are not usable for practical purposes. As already mentioned, this sensitivity occurs in the P^1- and P^0-region

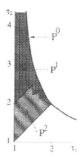

Fig. 5. P^2, P^1 and P^0-regions in the parameter space of the $\mathcal{R}2$-system. Of practical relevance is only the P^2-region. For a detailed description see text.

of the $\mathcal{R}2$-system, hence the parameter settings must be located in the P^2-region. Furthermore any technical system has to obey some additional constraints or optimization criteria. For instance the load of the system has to be maximized. We have already mentioned, that the $\mathcal{R}2$-system reaches its maximal load, if both robotic units are not idle. This is only the case, if the parameters of the $\mathcal{R}2$-system are located on the hyperbola $\tau_2 = \frac{\tau_1}{C\tau_1-1}$. However it turned out, that setting the parameter values exactly on the hyperbola makes no sense, because in this case the system shows a high sensitivity with respect to deviations of the parameters. Setting other parameter values leads to idle times of one or both robotic units which are proportional to the distance between the point in parameter space defined by the parameter values and the hyperbola itself. Consequently one has to make a compromise between the systems load on the one hand and the sensitivity of the system with respect to deviations of the parameters on the other hand.

In addition we are also interested in determining the maximal extent of the vicinity related to a particular parameter setting, for which the parameters can be varied without causing qualitative changes of the behavior. Figure 6.a shows the system load depending on the position of the parameters in the two-dimensional parameter space, whereas in Fig. 6.b the admissible deviation depending on the position of the parameters in the two-dimensional parameter space is presented. As one can see, the largest admissible deviations occur, if the parameters are equal ($\tau_1 = \tau_2$) and hence lead to the periodic sequence (12). For practical purposes one has to set the parameters on the angle's bisector $\tau_1 = \tau_2$ as close to the hyperbola as allowed by the parameter deviations, caused by the real robotic units.

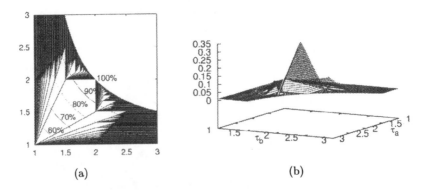

(a)   (b)

Fig. 6. *System load (a) and admissible deviations of the parameters causing no qualitative changes of the behavior (b), depending on the position of the parameters in the two-dimensional parameter space*

6.2 $\mathcal{R}n$-System

For the general $\mathcal{R}n$-model with $n > 2$ the methods presented so far are also applicable. One is mainly interested in analyzing the model and interpreting the results. However, one have to deal with two inherent problems, occurring in all higher-dimensional dynamical systems. The first one arises in dynamical systems with a high-dimensional state space, where the analytical investigations become more complex and possibly lead to unsolvable tasks. The second problem occurs in dynamical systems with a high-dimensional parameter space. During numerical simulations, one have to regard the entire parameter space, but the representation of the corresponding results is quite difficult, due to the fact that one is not able to produce complete diagrams for n-dimensional spaces, if $n > 3$ holds. Concerning the $\mathcal{R}n$-system, we obviously encounter both problems, if regarding systems with more than three robotic units. Therefore, we would like to present here some interesting results of our work, which refer to the $\mathcal{R}3$-system only. In this case we are not only able to present those results in a comprehensible way but also to compare them with the already obtained results of the $\mathcal{R}2$-system.

Fig. 7 shows a sequence of two-dimensional cuts in the three-dimensional parameter space of the $\mathcal{R}3$-system at several values of the parameter τ_3. One can see, that the underlying structure in this case is also fractal and self-similar, but more complex as the one of the $\mathcal{R}2$-system. According to the inequality (4) and in analogy to the $\mathcal{R}2$-system the region we investigate is confined by the hyperbolic surface given by $\frac{1}{\tau_1} + \frac{1}{\tau_2} + \frac{1}{\tau_3} = C$. Cuts of this surface, that means the respective hyperbolas, are clearly visible in each part of Fig. 7. With τ_3 increasing the productivity α_3 tends towards 0 and hence the corresponding two-dimensional cut of the three-dimensional parameter space of the $\mathcal{R}3$-system looks more and more like the two-dimensional bifurcation diagram of the $\mathcal{R}2$-system (see Fig. 7.f and Fig. 4).

Like in the two-dimensional case, that means $\mathcal{R}2$-system, we also have a decomposition of the parameter space in regions with different parameter sensitivity. In analogy to the regions P^2, P^1 and P^0, which we introduced for the $\mathcal{R}2$-system, we assume that the parameter space of the $\mathcal{R}n$-system consists of a sequence $P^n, \ldots P^0$ of regions with analogous characteristics. In the region P^n, the $\mathcal{R}n$-system behaves like the $\mathcal{R}2$-system within P^2, i.e., there exist no sensitivity on deviations of any parameter. Within P^{n-1}, the $\mathcal{R}n$-system has a certain similarity to the region P^1 of the $\mathcal{R}2$-system, i.e., the system is sensitive on deviations of exactly one parameter, and so forth. We would like to mention, that P^n is the only region with practical meaning, because in all the other regions one cannot guarantee a stable periodic behavior.

Real bifurcation diagrams of the $\mathcal{R}3$-system are naturally three-dimensional, but their complete representation is not possible, due to the complexity of intersecting surfaces. Therefore in Fig. 8 we have only represented some regions of the parameter space, where the parameter settings lead to periodic sequences with the same length.

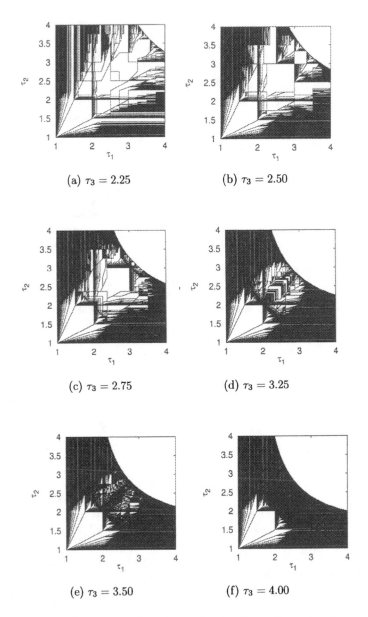

Fig. 7. *Sequence of two-dimensional cuts in the three-dimensional parameter space of the $\mathcal{R}3$-system. For a detailed description see text.*

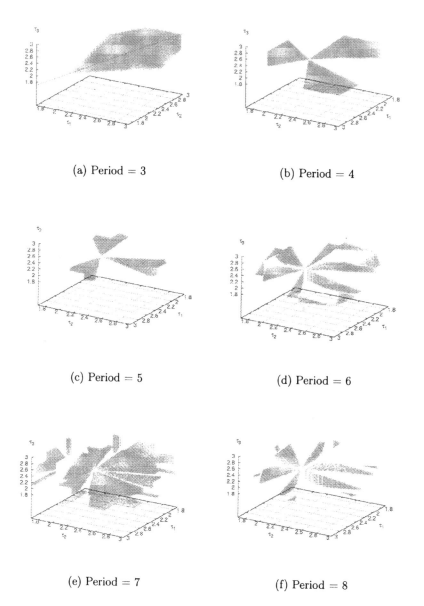

(a) Period = 3

(b) Period = 4

(c) Period = 5

(d) Period = 6

(e) Period = 7

(f) Period = 8

Fig. 8. *Regions with the same period in the three-dimensional parameter space of the $\mathcal{R}3$-system*

7   Summary

The results of our investigations show, that the behavior of our model depends qualitatively on the position of the involved parameters in parameter space. For the simplest model consisting of 2 robotic units ($\mathcal{R}$2-system), we have found areas in the parameter space with different periodical behavior. We have classified these areas depending on the stability of solution with respect to the variation of the parameters. In this sense we have found extended two-dimensional areas, where the system is not sensitive with respect to deviations of the parameters (P^2-region). We have also found large areas in parameter space, where the deviation of one parameter cause no change of period (P^1-region). In this case the system is sensitive with respect to the other parameter. Finally, we detected a region in parameter space, where each variation of the parameters implies a change of the period (P^0-region).

Our results from the analysis for systems consisting of more than two robotic units show that in this case we have a classification, which is analogue to the one in the $\mathcal{R}$2-system.

It is interesting in this context, that even such simple systems we have investigated as models of real cooperating robotic systems, show a complex dynamical behavior. It is therefore expected, that in cooperating robotic systems an analogue complex dynamical behavior occurs. Remarkable is the high sensitivity of the modeled system with respect to the parameters in some regions in parameter space. Because we assume that such kind of sensitivity occurs also in real robotic systems, our parameter studies are useful in order to detect also the critical regions in the parameter space, i.e., the critical parameter settings.

8   Outlook

In this work we have presented the results of the investigations of the $\mathcal{R}n$-model. This model is a simple one, but it has some advantages, so that we could efficiently perform extended parameter studies using numerical simulation techniques. Furthermore we also obtained some analytical results, basically due to the simplicity of the model.

Real technical systems are usually more complex than the one modeled in this work, but it is our assumption that we can simplify them so that we end up with models which we can regard as dynamical systems. Furthermore we assume that most of these dynamical systems need decision functions in their evolution equations, in order to properly simulate the behavior of the modeled technical devices.

The numerical simulation of these modeled systems and the analysis of their dynamical behavior is necessary, for instance if one has to verify its specified functionality. In addition the analysis allows the detection of appropriate parameter settings which ensure a suitable mode of operation.

References

1. P. Levi and S. Hahndel. Modeling distributed manufacturing systems. In *Task oriented Agent Robot Systems, Proc. of Int. Conf. on Intelligent Autonomous Systems (IAS-4)*, 1995.
2. P. Levi. Architectures of individual and distributed autonomous agents. In *Proceedings of the 2nd International Conference on Intelligent Autonomous Systems IAS-2*, Amsterdam, 1989.
3. B.-L. Hao. Symbolic dynamics and characterization of complexity. *Physica D*, 51, 1991.
4. H.P. Fang. Universal bifurcation property of two- or higher-dimensional dissipative systems in parameter space: why does 1D symbolic dynamics work so well? *J.Phys.A: Math.Gen.*, 28:3901–3910, 1995.
5. B.-L. Hao, J.-X. Liu, and W.-M. Zheng. Symbolic dynamics analysis of the Lorenz equations. *Physical Review E*, 57(5):5378–5396, May 1998.
6. R. Lammert, V. Avrutin, M. Schanz, and P. Levi. Symbolic dynamics of periodic orbits of a simple self-organized multi-agent system. to be published.

Recent Advances in Range Image Segmentation

Xiaoyi Jiang

Department of Computer Science
University of Bern, CH-3012 Bern, Switzerland
jiang@iam.unibe.ch

Abstract. In this paper we discuss recent advances in range image segmentation concerning two important issues: experimental comparison of segmentation algorithms and the potential of edge-based segmentation approaches.

The development of a rigorous framework for experimental comparison of range image segmentation algorithms is of great practical importance. This allows us to assess the state of the art by empirically evaluating range image segmentation algorithms. On the other hand, it help us figure out collective weaknesses of current algorithms, so as to identify requirements of further research.

By means of a simple adaptive edge grouping algorithm we show the potential of edge-based segmentation techniques to outperform region-based approaches in terms of both segmentation quality and computation time. The aspect of edge-based complete range image segmentation has not been fully explored so far in the literature and deserves more attention in the future.

1 Introduction

Vision tasks based on range image analysis have been relying in most cases upon scene representations in terms of surface patches. Examples are object recognition [3,8], model construction [26], configuration analysis [23], motion analysis [16,28], automated visual inspection [25], and robotic grasping operations [1,2]. Accordingly, the range image segmentation problem has generally been defined as one of dividing range images into closed regions with application domain specific surface properties. Formally, segmentation [9] is a process that partitions the entire image region R into n subregions, $R_1, R_2, \ldots, R_n$, such that

1. $\cup_{i=1}^{n} R_i = R$,
2. R_i is connected region, $i = 1, 2, \cdots, n$,
3. $R_i \cap R_j = \emptyset$, for all i and j, $i \neq j$,
4. $P(R_i) =$TRUE for $i = 1, 2, \cdots, n$, and
5. $P(R_i \cup R_j) =$FALSE, R_i and R_j are disjoint,

where $P(R_i)$ is a logical predicate over the points in set R_i and $\emptyset$ is the empty set. One exception is introduced to handle *nonsurface* pixels which correspond to noisy outliers, no range measurement (due to occlusion), etc. It is convenient to

Christensen et al. (Eds.): Sensor Based Intelligent Robots, LNAI 1724, pp. 272–286, 1999.

use the same region label for all nonsurface pixels in the range image, regardless of whether they are spatially connected. The terms 2 and 4 of the above definition are violated by nonsurface regions.

Range image segmentation turns out to be a difficult problem. Numerous algorithms have been proposed in the literature. In this paper we will discuss some recent advances in this area. Our discussion considers two important issues: experimental comparison of segmentation algorithms and the potential of edge-based segmentation approaches, that have only received their due attention recently. Another important development concerns the application of robust estimation techniques from statistics to range image segmentation [5,18,20,27,29] and is not the topic of this paper.

2   Experimental Comparison of Segmentation Algorithms

In general computer vision suffers from a lack of tradition of sound experimental work. As pointed out by R. Jain and T. Binford [10]: "The importance of theory cannot be overemphasized. But at the same time, a discipline without experimentation is not scientific. Without adequate experimental methods, there is no way to rigorously substantiate new ideas and to evaluate different approaches." The dialogue on "Ignorance, Myopia, and Naivité in Computer Vision Systems" initiated by R. Jain and T. Binford [10] and the responses documented the necessity of evaluating theoretical findings, vision algorithm, etc. by using empirical data.

For range image segmentation many algorithms have been proposed in the literature. However, these methods have been typically verified on a small number of range images that were either synthetic or acquired by quite different range sensors, and are not publicly available in general. Of fundamental importance is a common comparison framework that contains a large number of real range images and a set of evaluation criterion. As a consequence, it will allow an assessment of the relative performance of the segmentation algorithms on a (more) objective basis.

Recently, such an experimental performance comparison framework has been proposed for range image segmentation [9,22]. In the following we give a brief review of this work which will be used in Section 3 to explore the potential of edge-based range image segmentation.

2.1   Imagery Design

The range image segmentation problem is "high-dimensional", i.e., there are a large number of factors to be considered in the test imagery design:

- type of range scanners, the associated number of bits per pixel (quantization level), type and amount of noise (besides quantization);
- type of surfaces, number of surfaces, size of surfaces;
- jump edges: amount of depth discontinuity between two adjacent surfaces;
- crease edges: angle between two adjacent surfaces, incident angle of edge to viewpoint.

	ABW	Perceptron	K2T
#images	40	40	60
resolution	512×512	512×512	480×640
surfaces	planar	planar	planar/curved
imaging volume	table-top size	room-size	table-top size

Table 1. Characteristics of the three image subsets.

Ideally, a segmentation algorithm should be tested on an image set spanning the full ranges of all these design parameters. However, acquiring, ground-truthing, processing and analyzing the necessary image data, and comparing the results would require a prohibitive amount of effort. Since triangulation using some kind of structured light and time-of-flight are by far the most popular range acquisition techniques in computer vision, a compromise of three image subsets has been made:

- 40 images of planar surfaces acquired by an ABW structured light scanner,
- 40 images of planar surfaces acquired by a Perceptron time-of-flight laser scanner, and
- 60 images of planar and curved surfaces acquired by a K2T model GRF-2 structured light scanner.

Table 1 summarizes the characteristics of the three test image subsets. Ground truth was created for each image by a human operator. Figure 1 shows one example image from each image subset together with the segmentation ground truth.

2.2   Performance Measures

Assume a machine segmentation (MS) and the corresponding ground truth (GT). Based on a parameter T, a total of five performance measures have been defined in [9,22]:

- Number of correctly detected regions: $R_m \in$ MS and $R_g \in$ GT are classified as an instance of correct detection if at least T percent of the pixels in R_m are marked as pixels in R_g in GT and at least T percent of the pixels in R_g are marked as pixels in R_m in MS.
- Number of over-segmented regions: a region $R_g \in$ GT is classified as over-segmented if there exist regions $R_{m1}, R_{m2}, \cdots, R_{mk} \in$ MS such that at least T percent of the pixels in each region R_{mi} in MS are marked as pixels in R_g in GT and at least T percent of the pixels in R_g are marked as pixels in the union of R_{mi} in MS.
- Number of under-segmented regions: a region $R_m \in$ MS is classified as under-segmented if there exist regions $R_{g1}, R_{g2}, \cdots, R_{gk} \in$ GT such that at least T percent of the pixels in each region R_{gi} in GT are marked as pixels in R_m in MS and at least T percent of the pixels in R_m are marked as pixels in the union of R_{gi} in GT.

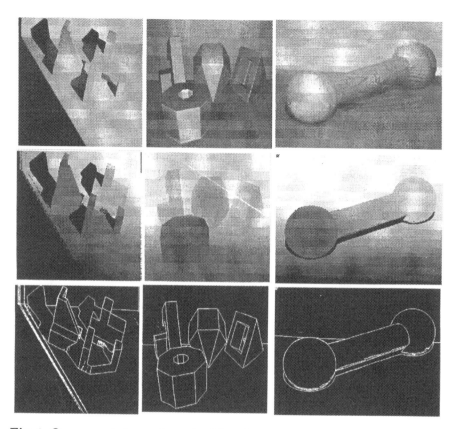

Fig. 1. One example image from the ABW (left), Perceptron (middle), and K2T (right) image subset. Shown are the respective intensity image (top), range image (middle), and ground truth (bottom).

- Number of missed regions: $R_g \in$ GT is classified as missed if R_g does not participate in any of instance of correct detection, over-segmentation and under-segmentation.
- Number of noise regions: $R_m \in$ MS is classified as noise if R_m does not participate in any of instance of correct detection, over-segmentation and under-segmentation.

In essence the parameter T defines the degree of the overlap between a GT region and a MS region to be regarded as a corresponding region pair. Note that the classification for a region is not unique in general. In [9] a simple optimization procedure has been introduced to generate a unique final classification for all regions.

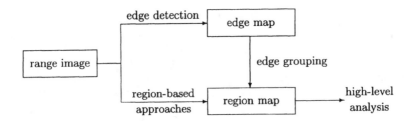

Fig. 2. Two paradigms of range image segmentation.

2.3   Comparison Procedure

Each image subset is disjointly divided into a set of training images and a set of test images. The number of training images is 10 (ABW), 10 (Perceptron), and 20 (K2T), respectively. The training images are used to fix the parameters of a segmenter. Using the fixed parameter values, the segmenter is run on all the test images. The performance measures are then computed based upon the segmentation results of the test images and serve as the basis for a comparison of different segmentation algorithms.

3   Edge-Based Complete Range Image Segmentation

In principle, a complete image segmentation into regions can be achieved by both region- and edge-based approaches, see Figure 2. In the region-based range image segmentation methods, pixels of similar properties are grouped together. On the other hand, discontinuities are extracted in the edge-based approaches and the segmentation is guided by the obtained contours. In general edge detection methods cannot guarantee closure of boundaries for surface extraction, resulting in the need of a subsequent grouping and completion process.

Compared to region-based techniques, edge detection actually has some very appealing properties. Usually, the algorithms possess simple control structure and regular operators like convolution, and thus lend themselves to an efficient implementation on special-purpose image processors and parallel computers. In addition, edge detection techniques are able to localize surface boundaries more precisely in general. We believe that, despite of their incomplete nature, edge detection methods are of significant potential for a successful image segmentation.

Although a number of edge detection methods have been proposed for range images, the edge map completion problem has not received the due attention in the literature. In this section we discuss a simple adaptive technique which has been suggested recently for grouping edges into regions [12] and demonstrated superior performance to region-based algorithms with respect to both segmentation quality and computation time.

3.1   Brief Literature Review

Several methods have been proposed in the literature to perform edge linking, or edge aggregation [7,21,30]. The task is to form more contiguous contours by filling in edge gaps and connecting nearby contour segments. This is different from the edge completion problem considered here that intends to achieve a complete segmentation by grouping edges into closed regions.

The edge map completion problem has been ignored in most of the papers describing edge detection techniques. If considered, *ad hoc* rules have been typically applied [2,15]. In [2], for instance, a linking operation is performed once to close one-pixel gaps. The authors reported the success of this simple strategy in dealing with their (very few) test images. Obviously, such approaches don't contribute to solving the edge grouping problem in general.

Systematic grouping methods have been proposed in [13,19,26]. The contour closure algorithm proposed in [13] is embedded in the hypotheses generation and verification paradigm. Plausible contour closure hypotheses are generated by a search algorithm and then verified by a region test. The main disadvantage of this approach lies in the complex search procedure and the inherent high computational burden of such a search.

Parvin and Medioni [26] formulate the boundary closure problem as a dynamic network of long and short term variables. Long term variables represent the connected contours from an initial binary edge map, whereas short term variables correspond to the competing closure hypotheses that cooperate with the long term variables. Based on smoothness and geometric cohesion constraints, an energy function is defined and the dynamic network is iteratively updated to convergence. The behavior of this system crucially depends on the control parameters. Its performance has only been demonstrated on range images of good quality.

The work [19] uses an edge strength map and makes the assumption that a region can always be closed by choosing an appropriate threshold. Since the closure of the regions of a range image generally requires different thresholds, region hypotheses are generated by binarizing the edge strength map using a series of thresholds, and verified by means of a lack-of-fit test against a polynomial surface function. Finally, the surviving region hypotheses resulting from different thresholds are combined to obtain an overall region map. This approach is not applicable to those edge detection methods that don't provide a quantitative characterization of edge strengths [6,17]. In addition, the basic assumption of region closure by means of an appropriate threshold may be violated by noise and other anomalies of the range finder, in which case it may happen that some boundary pixels receive very low edge responses from an edge detector, necessitating the probe of very low thresholds. This tends to generate an over-segmentation.

3.2   Adaptive Grouping of Edges into Regions

A binary edge map is assumed as input to the adaptive grouping technique proposed in [12]. We observe that any boundary gap can be closed by dilating

```
perform component labeling on input edge map;
RegionList = { connected regions of size > T_size };
while (RegionList != ∅) {
    delete arbitrary region R from RegionList;
    verify R;
    if (successful)
        record region R;
    else
        perform dilation within R;
        perform component labeling within R;
        RegionList += { connected regions of size > T_size };
    }
}
postprocessing;
```

Fig. 3. Outline of the adaptive grouping algorithm.

the edge map. If the largest gap of a region has a length L, then $L/2$ dilations will successfully complete the region. However, we have no idea about the actual value of L before the grouping process is finished. In order not to miss any region, we potentially have to select a high value for L as maximum allowable gap length, resulting in a consistently large number of dilations applied to all regions. This is not only a unnecessary overhead in dealing with regions that are (almost) closed. But also relatively small-sized or thin regions will disappear. In the following an adaptive approach is described that carries out the minimum number of dilations necessary for each particular region. This method is embedded in a hypothesis-and-verification framework. It increases the number of dilations only for those regions that cannot be successfully verified. An outline of the adaptive grouping technique in C-style pseudo-code is given in Figure 3.

From the input edge map, region hypotheses can be found by a component labeling. Usually, this initial region map contains many instances of under-segmentation, i.e., multiple true regions are covered by a single region hypothesis. To recognize the correctly segmented and under-segmented regions, we perform a region test for each region R of the initial segmentation. If the region test is successful, the region R is registered. Otherwise, there still exist open boundaries within R. In this case we perform one dilation operation within R, potentially completing the boundaries. Again, a component labeling is done for R to find new region hypotheses, and these are verified in the same manner as for the initial regions. This process of hypotheses generation (component labeling) and verification (region test) is recursively repeated until the generated regions have been successfully verified or they are not further considered because of a region size smaller than a preset threshold T_{size}.

The region test starts with a plane test. The principal component analysis technique [24] is used to compute a plane function by minimizing the the sum of squared Euclidean (orthogonal) distances. The region is regarded a plane if the RMS orthogonal distance of region pixels to the plane is smaller than a preset threshold. If this test fails, we compute a second surface approximation by means of a biquartic polynomial function

$$f(x,y) = \sum_{i+j \leq 4} a_{ij} x^i y^j.$$

This is done by a least-square method that minimizes the sum of squared fit errors

$$\sum_{k=1}^{n} (f(x_k, y_k) - z_k)^2$$

where the points $(x_k, y_k, z_k), k = 1, \ldots, n$, comprise the region under consideration. Again, the region acceptance is based on the RMS fit error.

Conceptually, a region test based on the biquartic surface function alone suffices for the verification purpose. Using regression theory it can be shown that such a function is an unbiased estimate of any underlying region models of order less than four including planes [19]. For two practical reasons, however, we have added the plane test. First, a quartic function approximation is computationally much more expensive than the plane test. Therefore, an initial plane test enables us to exclude the planar regions of a range image from the expensive second test. Moreover, in the experiments the quartic surface function encountered difficulties to reasonably approximate highly sloped planes, mainly due to the use of fit error in the minimization instead of the Euclidean distance. Since the desired Euclidean distance is minimized by the principal component analysis method, the plane test turns out to be more suitable for handling planar surface patches in general.

Three postprocessing steps are introduced to complete the adaptive grouping algorithm. Until now, the edge pixels are not considered to be part of regions. Also, the dilations necessary for the boundary closure discard pixels near region boundaries. These unlabeled pixels should be added to their corresponding regions. For this purpose we merge each unlabeled pixel to an adjacent region if the orthogonal distance (for a plane) or the fit error (for a biquartic surface patch) is tolerable.

Occasionally, it happens that the dilations link some noise edge pixels within a true region to a connected contour, and produces an over-segmentation of the region. Instances of over-segmentation can be easily corrected by merging adjacent regions. The region test described above is performed for the union of two adjacent regions, and they are merged in case of success. This operation is repeated until no more merge is possible. In the current implementation, this postprocessing step is only done for planar regions. In this case the region test is very fast. More importantly, we have a simple criterion to exclude most of the pairs of adjacent regions from an actual region test. A pair should undergo the region test only if the angle of their normals takes a small value.

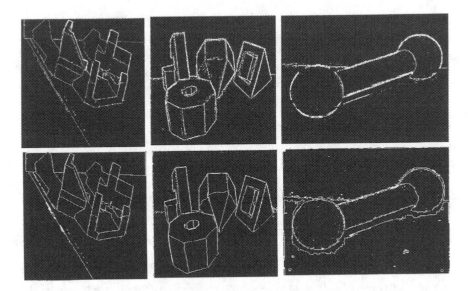

Fig. 4. Results of edge detection (top) and grouping (bottom).

Another potential problem with dilations is that small-sized or thin regions may disappear or become smaller than the minimum region size threshold T_{size}. To recover these regions, we consider all the pixels that don't belong to any region found so far. Again, region hypotheses are generated by a component labeling and verified by the region test. The successful regions complete the overall region segmentation.

3.3   Comparison with Region-Based Approaches

The adaptive grouping algorithm has been implemented in C on a Sun SparcStation. In this implementation the edge detection method reported in [14] has been used to generate binary edge maps. Tests have been carried out in the framework described in [9,22] in order to compare the performance of this edge-based approach with region-based methods.

Segmentation into planar surface patches

So far, four region-based algorithms for segmentation into planar regions have been tested within the experimental comparison framework [9], using the ABW and Perceptron image subsets. In the following we call them USF, WSU, UB, and UE; see [9] for a description of these algorithms. The average performance metrics per image for the four segmenters are graphed in Figure 5 against the parameter T.

For the ABW and Perceptron image in Figure 1, the binary edge map and the result of the adaptive grouping are shown in Figure 4. There are noise edge pixels, especially in the Perceptron images, which are noisier than the structured light

ABW 30 test images

algorithm	GT regions	correct detection	over-segmentation	under-segmentation	missed	noise
USF	15.2	12.7	0.2	0.1	2.1	1.2
WSU	15.2	9.7	0.5	0.2	4.5	2.2
UB	15.2	12.8	0.5	0.1	1.7	2.1
UE	15.2	13.4	0.4	0.2	1.1	0.8
EG	15.2	13.5	0.2	0.0	1.5	0.8

Perceptron 30 test images

algorithm	GT regions	correct detection	over-segmentation	under-segmentation	missed	noise
USF	14.6	8.9	0.4	0.0	5.3	3.6
WSU	14.6	5.9	0.5	0.6	6.7	4.8
UB	14.6	9.6	0.6	0.1	4.2	2.8
UE	14.6	10.0	0.2	0.3	3.8	2.1
EG	14.6	10.6	0.1	0.2	3.4	1.9

Table 2. Average results of all five segmenters on ABW and Perceptron image subsets for $T = 80\%$.

data in general. Many open boundaries can be observed. Partly, very long gaps exist; see, for instance, the U-shaped object in the ABW image. The adaptive grouping technique has been successful in dealing with both problems.

The performance metrics for the adaptive grouping method, referred to as EG, over the test sets are drawn in Figure 5. On the ABW image set, it reached the same performance as the UE algorithm, which is the best among the four region-based approaches tested so far on this set. In terms of over-segmentation, it shows even an improvement. On the Perceptron image set, the adaptive grouping method beats all four region-based algorithms with respect to four performance metrics; the only exception is the under-segmentation metric which is located in the middle field. Table 2 presents the average results on all performance metrics for all five algorithms on both test sets for $T = 80\%$, demonstrating the superior performance of the adaptive grouping algorithm.

The average processing time for the four region-based segmentation algorithms on the ABW and Perceptron test sets, per image, is tabulated in Table 3. Also included is the computation time of the adaptive edge grouping method, which includes all processing steps from edge detection to postprocessing.

Segmentation into curved surface patches

For the K2T image in Figure 1, the binary edge map and the result of edge grouping are shown in Figure 4. On the K2T image subsets, the performance of the adaptive edge grouping method has been compared with the classical

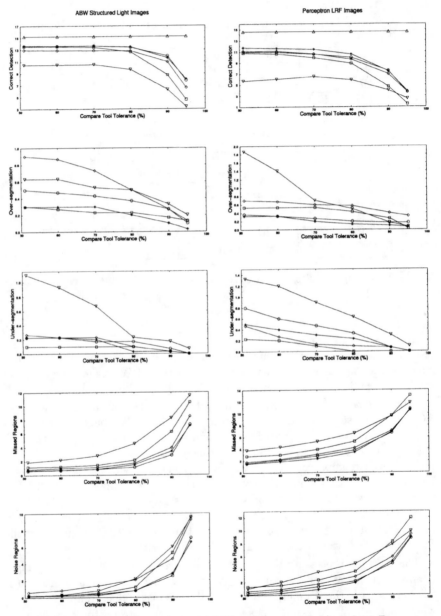

Fig. 5. Performance metrics for 30 ABW test images (left) and 30 Perceptron test images (right). Symbols: △ = average number of regions per image (maximum number of correct detections), □ = USF, ▽ = WSU, ◇ = UB, ○ = UE, * = EG.

Alg.	Computer	ABW	Perceptron	K2T
USF	Sun SparcStation 20	78 min.	117 min.	
UE	Sun SparcStation 5	6.3 min.	9.1 min.	
WSU	HP 9000/730	4.4 min.	7.7 min.	
UB	Sun SparcStation 20	7 sec.	10 sec.	
EG	*Sun SparcStation 5*	*15 sec.*	*15 sec.*	*23 sec.*
BJ	Sun SparcStation 20			hours

Table 3. Comparison of computation time.

comparison tolerance T	algorithm	correct detection	over-segmentation	under-segmentation	missed	noise
51%	BJ	15.3%	62.9%	5.5%	10.1%	18.8%
	EG	68.6%	2.3%	12.6%	2.0%	1.4%
60%	BJ	16.1%	61.9%	4.3%	12.8%	20.8%
	EG	68.6%	2.3%	12.6%	2.6%	1.4%
70%	BJ	16.7%	58.6%	4.3%	15.9%	24.3%
	EG	68.6%	2.3%	12.6%	2.7%	1.4%
80%	BJ	16.1%	54.5%	3.6%	21.8%	30.8%
	EG	68.2%	2.2%	12.3%	4.5%	2.6%
90%	BJ	16.1%	43.3%	1.7%	37.2%	44.9%
	EG	64.0%	0.9%	10.1%	14.9%	14.0%
95%	BJ	14.2%	35.1%	1.0%	48.7%	53.3%
	EG	56.6%	0.0%	9.1%	25.2%	24.8%

Table 4. Average results of the algorithms BJ and EG on K2T image subset.

work by Besl and Jain [4], referred to as BJ. The performance metrics for both algorithms over the entire K2T subset are recorded in Table 4, where each metric represents the average percentage per image. The region growing algorithm BJ has surprisingly much more difficulties on this image set than the simple grouping technique. The only exception is that EG demonstrates a higher percentage of under-segmentation. The reason lies in the nature of such an edge grouping approach. The edge detection method [14] used in the current implementation is able to detect jump and crease edges but not smooth edges (discontinuities only in curvature). Therefore, two surfaces meeting at a smooth boundary will not be separated and an under-segmentation occurs. This happened in some of the test images. Since it seems that no edge detection method known from the literature can deal with smooth edges, a potential under-segmentation is not a particular weakness of the adaptive grouping technique, but an inherent problem of any edge grouping method. The computation time of the two segmentation algorithms is recorded in Table 3. The difference in processing time is remarkable.

Dilations	0	1	2	3	4	5	6	7	8	9	10
Regions	42	208	75	16	12	0	2	2	0	2	0

Table 5. Histogram of number of dilations.

	0	1	2	3	4	5	6	7	8	9	10
CD	0.7	6.4	8.7	9.0	*9.2*	8.4	8.1	7.9	7.7	7.3	6.7

Table 6. Performance metric CD: fixed number of dilations, T=80%.

Discussions

Dilation of an edge map is able to close the inherent gaps in general. The main motivation behind the adaptive edge grouping algorithm is that the actual number of dilations depends on the length of a contour gap. A fixed number of dilations is wasteful and tends to cause problems in small or thin regions. In the following the Perceptron image subset is used to demonstrate the effectiveness of the adaptive selection of the number of dilations.

For the Perceptron image subset, the number of dilations necessary for the successful grouping of regions is histogrammed in Table 5. Among all regions, for instance, 208 regions need a single dilation, while two regions could be completed after nine dilations. This histogram obviously show the necessity of an adaptive selection of the number of dilations. As a comparison, the performance metric CD using a fixed number of dilations is listed in Table 6. The optimal result CD=9.2 has been achieved for four dilations, which is substantially lower than the value 10.6 from the adaptive edge grouping algorithm.

4   Conclusions

In this paper we have discussed recent advances in range image segmentation concerning two important issues: experimental comparison of segmentation algorithms and the potential of edge-based segmentation approaches, that have only received their due attention recently.

The development of a rigorous framework for experimental comparison of range image segmentation algorithms is of great practical importance. This allows us to assess the state of the art by empirically evaluating range image segmentation algorithms on a large number of real range images of various surface types acquired by different range scanners. On the other hand, it help us figure out collective weaknesses of current algorithms, so as to identify requirements of further research. As pointed out in [9,22], an actual comparison of six segmentation algorithms has provided some interesting insights in their relative performance.

We have also discussed a simple adaptive grouping algorithm to solve the boundary closure problem. Its effectiveness has been extensively evaluated on three range image sets. It turned out that this edge-based approach almost consistently outperforms all the region-based segmentation algorithms tested on the image sets with respect to both segmentation quality and computation time. Considering the algorithmic and implementation ease in general, we believe that edge-based techniques have a large potential of complete segmentation into regions. This aspect has not been fully explored so far in the literature and deserves more attention in the future.

Acknowledgments

We thank K. Bowyer and D. Goldgof, Univ. of South Florida, Tampa, for providing us the Perceptron and GRF-2 images.

References

1. F. Ade *et al.*, Grasping unknown objects, in *Modelling and Planning for Sensor Based Intelligent Robot Systems* (H. Bunke *et al.*, Eds.), 445–459, World Scientific, 1995.
2. E. Al-Hujazi and A. Sood, Range image segmentation with applications to robot bin-picking using vacuum gripper, IEEE. Trans. on SMC, 20(6): 1313–1325, 1990.
3. F. Arman and J.K. Aggarwal, Model-based object recognition in dense-range images – A review, ACM Computing Surveys, 25(1): 5–43, 1993.
4. P.J. Besl and R.C. Jain, Segmentation through variable-order surface fitting, IEEE Trans. on PAMI, 10(2): 167–192, 1988.
5. K.L. Boyer *et al.*, The robust sequential estimator: A general approach and its application to surface organization in range data, IEEE Trans. on PAMI, 16(10): 987–1001, 1994.
6. J.-C. Cheng and H.-S.Don, Roof edge detection: A morphological skeleton approach, in *Advances in Machine Vision: Strategies and Applications* (C. Archibald *et al.*, Eds.), 171–191, World Scientific, 1992.
7. I.J. Cox *et al.*, A Bayesian multiple-hypothesis approach to edge grouping and contour segmentation, Int. Journal of Computer Vision, 11(1): 5–24, 1993.
8. P.J. Flynn and A.K. Jain, Three-dimensional object recognition, in *Handbook of Pattern Recognition and Image Processing: Computer Vision* (T.Y. Young, Ed.), 497–541, Academic Press, 1994.
9. A. Hoover *et al.*, An experimental comparison of range image segmentation algorithms, IEEE Trans. on PAMI, 18(7): 673–689, 1996.
10. R. Jain and T. Binford, Ignorance, myopia, and naivité in computer vision systems, CVGIP: Image Understanding, 53(1): 112–117, 1991.
11. X. Jiang and H. Bunke, Fast segmentation of range images into planar regions by scan line grouping, Machine Vision and Applications, 7(2): 115–122, 1994.
12. X. Jiang and H. Bunke, Range image segmentation: Adaptive grouping of edges into regions, in *Computer Vision – ACCV'98* (R. Chin, T.-C. Pong, Eds.), 299–306, Springer-Verlag, 1998.
13. X. Jiang and P. Kühni, Search-based contour closure in range images, Proc. of 14th Int. Conf. on Pattern Recognition, 16–18, Brisbane, Australia, 1998.

14. X. Jiang and H. Bunke, Edge detection in range images based on scan line approximation, Computer Vision and Image Understanding, 73(2): 183–199, 1999.
15. S. Kaveti *et al.*, Second-order implicit polynomials for segmentation of range images, Pattern Recognition, 29(6): 937–949, 1996.
16. N. Kehtarnavaz and S. Mohan, A framework of estimation of motion parameters from range images, Computer Vision, Graphics, and Image Processing, 45: 88-105, 1989.
17. R. Krishnapuram and S. Gupta, Morphological methods for detection and classification for edges in range images, Journal of Mathematical Imaging and Vision, 2: 351–375, 1992.
18. K.-M. Lee *et al.*, Robust adaptive segmentation of range images, IEEE Trans. on PAMI, 20(2): 200–205, 1998.
19. S.-P. Liou *et al.*, A parallel technique for signal-level perceptual organization, IEEE Trans. on PAMI, 13(4): 317–325, 1991.
20. P. Meer *et al.*, Robust regression methods for computer vision: A review, Int. Journal of Computer Vision, 6(1): 59–70, 1991.
21. V.S. Nalwa and E. Pauchon, Edgel aggregation and edge description, Computer Graphics and Image Processing, 40: 79–94, 1987.
22. M. Powell *et al.*, Comparing curved-surface range image segmenters, submitted for publication.
23. P.G. Mulgaonkar *et al.*, Understanding object configurations using range images, IEEE Trans. on PAMI, 14(2): 303–307, 1992.
24. T.S. Newman *et al.*, Model-based classification of quadric surfaces, CVGIP: Image Understanding, 58(2): 235–249, 1993.
25. T.S. Newman and A.K. Jain, A system for 3D CAD-based inspection using range images, Pattern Recognition, 28(10): 1555–1574, 1995.
26. B. Parvin and G. Medioni, B-rep object description from multiple range views, Int. Journal of Computer Vision, 20(1/2): 81–11, 1996.
27. G. Roth and M.D. Levine, Extracting geometric primitives, CVGIP: Image Understanding, 58(1): 1273–1283, 1993.
28. B. Sabata and J.K. Aggarwal, Surface correspondence and motion computation from a pair of range images, Computer Vision and Image Understanding, 63(2): 232–250, 1996.
29. X. Yu *et al.*, Robust estimation for range image segmentation and reconstruction, IEEE Trans. on PAMI, 16(5): 530–538, 1994.
30. Q. Zhu *et al.*, Edge linking by a directional potential function (DPF), Image and Vision Computing, 14: 59–70, 1996.

ISR: An Intelligent Service Robot

Magnus Andersson, Anders Orebäck, Matthias Lindström, and
Henrik I. Christensen

Centre for Autonomous Systems,
Royal Institute of Technology
S-100 44 Stockholm, Sweden
http://www.cas.kth.se

Abstract. A major challenge in mobile robotics is integration of methods into operational autonomous systems. Construction of such systems requires use of methods from perception, control engineering, software engineering, mathematical modelling, and artificial intelligence. In this paper it is described how such a variety of methods have been integrated to provide an autonomous service robot system that can carry out fetch-and-carry type missions. The system integrates sensory information from sonars, laser scanner, and computer vision to allow navigation and human-computer interaction through the use of a hybrid deliberative architecture. The paper presents the underlying objectives, the underlying architecture, and the needed behaviours. Throughout the paper, examples from real-world evaluation of the system are presented.

1 Introduction

Service robotics is an area of research that is rapidly expanding. We strongly believe that we will have small robots roaming around in our houses in the near future. An excellent example of such a device is the autonomous Electrolux TriLobote vacuum-cleaner, that was revealed to the public during spring 1998. The application potential of robots is enormous, ranging from boring tasks like vacuuming, to advanced household tasks such as cleaning up after a dinner party. Recent progress, particularly in sensor-based intelligent robotics, has paved the way for such domestic robots. It is however characteristic that relatively few robots are in daily use anywhere, and very few mobile robot systems are being mass produced. Examples of mass produced mobile systems include the Help-Mate Robotics platform for delivery of food and x-ray plates at hospitals, and the RoboKent floor sweeper produced by the Kent Corporation. Both have only been produced in relatively small series (in the order of hundreds).

The primary obstacles to the deployment of robots in domestic and commercial settings, are flexibility and robustness. The robots must be flexible so that they are relatively easy to deploy under different conditions, and so that they can be used by non-experts. This requires a rich set of control functions, combined with an intuitive user interface and automatic task acquisition functions (like automatic learning). Robustness, in terms of sensory perception, is needed

Christensen et al. (Eds.): Sensor Based Intelligent Robots, LNAI 1724, pp. 287–310, 1999.

to allow for operation 365 days a year under different environmental conditions. Typically, research robots have been demonstrated in a single environment under ideal working conditions, like an in-door environment with artificial light (and no windows). Such systems provide a proof of concept, but there is a long way from such systems to commercially viable systems. Some might claim that the gap is purely a development process with no or very little research involved. We claim that there is a need for fundamentally new methods to empower deployment of robots in everyday settings, and there is thus a need for fundamentally new research to enable use in the above mentioned market segments.

To pursue research on robust, flexible, and easy-to-use robot systems for everyday environments, an intelligent service robot project has been initiated at the Centre for Autonomous Systems at KTH. The long-term goal of the project is deployment of an intelligent robotic assistant in a regular home. The system must be able to perform fetch-and-carry operations for the human operator. To accommodate such tasks, it must be able to understand commands from a non-expert in robotics. This requires an intelligent dialogue with the user. Having received an instruction, the robot must be able to plan a sequence of actions and subsequently execute these actions to carry out the task. In a realistic scenario, the robot will encounter unexpected events such as closed doors and obstacles. To be perceived as a useful appliance, the robot must cope with such ambiguities in an 'intelligent' manner. To perform fetch and carry missions, which include opening of doors, picking up and delivery of objects, etc, the robot must be equipped with actuators that allow for manipulation. A basic functionality for such a robotic system is the ability to perform robust navigation in a realistic in-door environment. The first phase of the *Intelligent Service Robot* project has thus been devoted to the development of a flexible and scalable navigation system, that can carry out simple tasks like delivery of mail in an office environment, that includes rooms similar to a regular living room.

In this paper we describe the results of this initial phase of the project. The robot is equipped with a speech and gesture interface for human computer interaction. For navigation, a combination of ultra-sonic ranging and laser-based ranging is used. The different navigational functions are implemented as a set of behaviours that provide direct coupling between sensory input and actuator control. The output from different behaviours are integrated using simple superposition. To control the execution of a mission, a state manager, and a planner are used. For the integration of the overall system, a hybrid deliberative architecture has been developed, which allows for easy integration of the different system components.

Initially, related research on service robotics and navigation in an in-door environment is reviewed in Section 2. We then outline the overall architecture of the system in Section 4. Each of the components in the system are described in the following sections to give an impression of the complexity and diversity of the overall system. A number of issues related to the implementation of the system are reviewed in Section 8. The system has been used in a large number of experiments in laboratory and living room settings. Some of the results from these experiments are provided in Section 9. Finally a summary and issues of future research are provided in Section 10.

2   Related Work

The area of service robotics has recently received significant attention. The area is, however, closely related to general in-door navigation, where one of the first efforts was the Stanford cart by Moravec [1,2]. The area gained significantly in popularity with the change from sense-plan-act type systems to reactive behaviour-based systems, as motivated by Brooks subsumption [3] and Arkin's AuRA system [4].

In more recent time, some of the most dominating efforts have been those of Simmons in terms of his Xavier system [5], Reactive Action Packages by Firby [6,7], the RHINO system from Bonn [8], the ROMAN system from Munich [9], and the Minerva [10] and Sage [11] systems from CMU.

The Xavier system was specifically designed for autonomous navigation in an in-door setting. The system has been designed to carry out point-to-point missions, navigating between known places in an in-door environment. It is well-known that sensory perception is non-robust, and a principal issue has thus been to provide the needed functionality to cope with such uncertainty. Xavier is built around a hybrid deliberative architecture. A topological map of the environment is used for initial planning of missions, while obstacle handling is based on potential field methods. The system includes Partially Observable Markov Processes (POMPS) for managing uncertainty. The system has been reported to have a success rate of more than 90% for missions, that together cumulates to a total travel length of several kilometres.

Based on the successful use of Markov models in Xavier, the team at University of Bonn has built a museum tour guiding system [8]. The system has a probabilistic map of the environment that allows for automatic localisation and error recovery in structured environments. The system is also based on a hybrid deliberative architecture that allows for automatic handling of obstacles, like humans, in its immediate proximity. The RHINO system has later been updated and recoded for the Minerva system [10] that gave tours at the American Museum of Natural History in New York. Overall, the probabilistic framework has turned out to be very successful. All of the above systems rely on ultra-sonic sensing as the primary modality for localisation and navigation.

The RAP (Reactive Action Packages) framework developed by Firby [6,7] has been used in a number of different robot systems. The RAP system does situation specific action selection. Based on sensory interpretation of the context, an index is generated that enables action selection. The method relies heavily on explicit internal models of the environment, but in such situations it is a very powerful and intuitive framework for robot control. The methodology is accompanied by a framework for programming of reactive systems. The system has been used extensively at University of Chicago for robots like the Animate Agent [12]. This system includes colour based vision for obstacle detection, gesture interpretation and object recognition. Robust performance is reported for a laboratory setting without windows.

In Munich the ROMAN (RObot MANipulator) system has been developed over the last ten years [9]. This is a system that performs fetch and carry missions in a laboratory setting. The system includes a manipulator for interaction with the environment. The system uses a laser scanner for localisation, and sonar for detection of obstacles. The system relies on vision for recognition of objects (and obstacles), servoing on natural landmarks like doors, and manipulation of objects. The system is based on a sense-plan-act type of architecture. Planning is a key component to facilitate interaction with the environment for tasks like opening doors or picking up objects in a drawer. For interaction with objects, the system relies heavily on a priori specified geometric models. The system has been demonstrated in a laboratory setting, where is it able to navigate and perform error recovery in the presence of dynamic obstacles like humans. Error handling is here performed using a plan library that explicitly encodes all of the situations to be handled.

The Sage system developed by Nourbakhsh et al. [11] at CMU is derived from ideas originating in the Dervish system [13]. The original Dervish system uses the layout of an office building to detect symmetries, that allow relative localisation in combination with odometry. The system is controlled explicitly by a planner, based on local map information. In this sense, the system is highly deliberative and only obstacle avoidance, using potential fields, is performed reactively. The system uses map abstraction, rather than behavioural abstraction, for control, which is a departure from the present trend towards behavioural systems. The Sage system has replaced the planner by pure sequential/imperative programming of missions in combination with artificial landmarks, for navigation in the Dinosaur Hall at the Carnegie Museum of Natural History. A unique characteristic of this system is that it is fully autonomous, in the sense that it does automatic docking for recharging, and it can thus be used for extended periods of time without human intervention.

All of the above systems illustrate how robotics gradually has reached a stage where in-door navigation in natural environments is within reach. Many of the systems rely on prior defined maps, but when such maps are available, the systems can perform autonomous navigation. A few of them also include manipulation of objects. The trend has been towards hybrid deliberative architectures, where low level behaviours are used for control, while the actual invocation and configuration is carried out by a supervisor, that in turn receives missions specifications from a planner. This is also the approach that has been adopted in the system described in this paper.

3   Goals

The intelligent service robot demonstrator is intended to show that it is possible to build a useful robot for a house or office environment. The system therefore has to meet realistic demands concerning task complexity, human-robot interaction and the capabilities of acting in a realistic environment.

The specification of the operating environment for the robot has been performed in a very pragmatic way. By looking at homes and our own laboratory environment (mainly offices), specifications of room sizes, floor types, lighting conditions, etc have been made. To be able to test the robot in a realistic environment, our main laboratory has been turned into a living room, furnished with sofas, tables, book shelves, etc., see Figure 1.

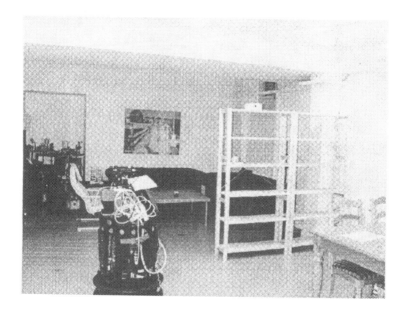

Fig. 1. The living room. Our main lab has been turned into a living room in which there are sofas, tables, book-shelves, and so on, to make it possible to test the robot in a realistic setting.

We believe that in a typical home or office, changing the environment for the sake of the robot will either be too costly or undesirable for other reasons. Therefore, our intentions are for the robot to work in an unmodified environment without artificial landmarks. The goal is to have the robot navigate safely, combining data from several sensors, and using a minimum of a priori information about its environment.

To be useful, the robot must also be able to manipulate different objects. The work is focused on recognition of everyday objects like cups, books, soda cans, etc., but it is equally important to be able to open doors, drawers etc. as this will be necessary in a home or an office. The challenge is to be able to do this in an unstructured and partly changing environment where objects positions are unknown or only partially known. Manipulation work is not described in this paper as it has only recently been started.

An important skill of home or office robots is the ability to communicate with its operator or coworker. Therefore a study of the human-robot interface is performed within the project. The interface includes speech and gesture recognition. Both of these communication modalities are natural for humans and therefore easy to learn for a non-specialist. A more detailed description is given in Section 5.2.

There are a variety of possible tasks for a domestic or office robot. The tasks we are currently considering are mainly fetch-and-carry tasks like

- Go to the refrigerator and bring back the milk.
- Deliver mail to a person.
- Setting and cleaning a table.
- Riding an elevator.

4   System Architecture

Most software architectures for mobile robots are layered. By layers, we mean distinct parts that have different levels of authority. The distinction is never totally clear in a modern robot system; a resulting motion of a robot is normally the consequence of a complex command sequence. It is seldom a direct path from the top (highest authority) layers to the motors in the lowest layer. However, a system description can be easier to explain if we divide it into layers.

We have chosen to divide our explanation of the system into three layers; the *deliberate layer*, the *task execution layer*, and the *reactive layer*. The deliberate layer makes deliberation decisions according to the state of the system. These can be derived from robot objectives or orders from a human user. The task execution layer makes sure that the plans of the deliberate layer are carried out. The last layer consists of sensor, behaviour, and actuator modules that can be configured and connected/fused together in a flexible network. The task execution layer will decide on the configuration of the network, that will solve the task at hand. This describes a hybrid architecture design of the selection paradigm. In other words, planning is viewed as configuration, according to the definition by Agre and Chapman [14]. An overview of the layers in the architecture can be seen in Figure 2, where each part is described in more detail below.

4.1   Deliberate Layer

The deliberate layer consists of a *planner* and a *human robot interface* (HRI). The HRI interprets human commands and intentions using speech, gestures, and keyboard as input. The commands are relayed to the planner that makes a plan for their execution.

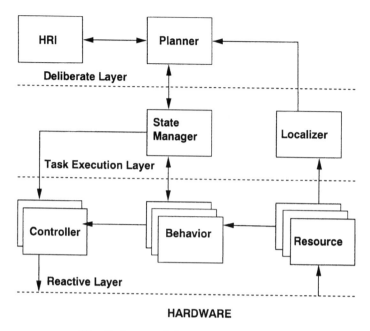

Fig. 2. Layers of the system architecture

4.2   Task Execution Layer

The task execution layer consists of a *state manager* and a *localiser*. The state manager configures the modules in the reactive layer to solve the task. The configuration is at the moment done using a lookup table for construction of the appropriate network. The localiser keeps track of the current position and provides metric and topological map data. These data are also used by the planner for path planning.

4.3   Reactive Layer

The reactive layer is closest to the hardware. It consists of a large set of modules that can be configured into a network connecting the sensors and actuators of the robot. The network defines tight sensorimotor connections, which results in fast reflexive behaviours without any involvement of deliberation.

The modules of the reactive layer are of three types: *resources, behaviours,* and *controllers*. The resources read and preprocess sensory data to extract essential information, which is forwarded to the behaviours. The behaviours act as mappings between sensing and actuation. They present propositions for the control of the actuators, depending on the data from one or more resources. Finally, the controllers, the modules closest to the actuators, fuse the propositions of the behaviours to a unified control of one or more actuators.

5   Deliberation

5.1   Planning

The planner belongs to the deliberate layer of the robot system architecture. It receives orders from the human-robot interface or from keyboard input. The planner will, depending on the current state of the robot, accomplish the given orders. During execution, the planner will give information about the state of the robot to the user. This consists of current action, success or failure to execute it, the position, and the immediate route.

The planner has access to a topological map (provided by a map server) with information about important nodes in the environment and metric distances between nodes. The topological map is augmented by an estimate of the travelling time between nodes. These time intervals are not necessary proportional to the actual metric distance, since the area in between the nodes can be cluttered with obstacles that the robot have to spend time to swirl. When a destination is given to the planner it will plan the route that has the shortest time estimate to completion. The plan is created using Dijkstra's algorithm for directed graph search. The route will normally contain a number of intermediate nodes. The robot traverses the nodes in the given order, while avoiding obstacles. The estimate in travelling time is updated when the path between two nodes has been traversed. Initial estimates of the travelling time are presently set to a time proportional to the metric distance.

5.2   Human-Robot Communication

A service robot working in a house or office environment will need a user-friendly, safe, simple-to-use and simple-to-learn human-robot interface (HRI). The most important reason is that the robot is going to work together with non-experts as opposed to most industrial robots of today. Possible ways of communicating with the robot include keyboards, touch-screens, joy-sticks, voice, and gesture commands. The preferred modes depend upon a combination of environment, user skill, task, and cost. In a noisy environment keyboard input is probably preferred over voice input, while in a situation where the operator needs to use his/her hands for other tasks, voice input is a better choice. However, for the environments and applications discussed within the intelligent service robot project, we are focusing our research efforts on a user interface combining both speech and gesture recognition. These modes of communication are natural for humans and complement each other well.

An overview of the current system is shown in Figure 3. The input is provided through a camera and a wireless microphone. The camera is connected to the gesture recognition module, which will search the images for gestures. The base unit of the microphone is connected to the speech recognition program, that will analyse and convert the spoken words into text. Speech and gestures are then combined into fully working commands in the command constructor. If a user says "Go there", a pointing gesture can be used to recognise in which direction to go and the result of this is, e.g., the command "Go left", which is then forwarded to the planner. The robot is also equipped with a speech synthesiser, which today is primarily used for acknowledgement by repeating the input command. In the future, the speech capabilities will be used in a more dialogue based communication with the end-user, similar to [15].

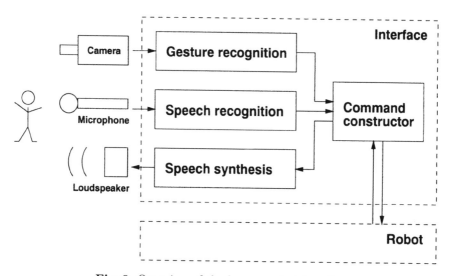

Fig. 3. Overview of the human-robot interface.

The language interpretor of the robot is built around a finite state automata framework. The valid voice commands are defined by the grammar listed below. Capital words are the actual spoken words (terminals). Lower case words are groups of possible words or sequences (non-terminals). Words in curly brackets are optional. All voice commands given in a session are described by a voice sequence.

```
voice:
      ATTENTION {command-sequence} IDLE {voice}
command-sequence:
      ROBOT command {command-sequence}
command:
      STOP
      EXPLORE
      HOW ARE YOU DOING TODAY
      FIND THE BOX
      FOLLOW ME
      DELIVER MAIL IN {THE} absolute
      LOCATE IN {THE} absolute
```

```
        GO relative
        GO TO {THE} absolute
relative:
        BACKWARD
        FORWARD
        LEFT
        RIGHT
        THAT WAY
absolute:
        OLLES ROOM
        DINNER TABLE
        LAB
        LIVING ROOM
        MAILBOX
        SCIENTIFIC AMERICAN
```

The current gesture recognition module is able to find pointing gestures, that allows it to identify if a person is pointing left, right, or not at all. The gesture is recognised through identification of the relation between the head and the hands. Skin colour segmentation is used for identification and tracking of the head and the hands. Recognition of skin colour is not stable enough, there are many objects that have similar colours. Therefore other cues need to be integrated into the system and recently *depth from stereo*, has been incorporated in order to increase performance. We have so far not used the gesture interpretation in real tasks, but a probable scenario could be the following. A person is standing in front of the camera, telling the robot to *Pick up that*, and pointing at an object in the scene. The robot then identifies the spoken command, finds out that it also needs gesture input and uses the image information to identify which object the operator is pointing at. It is important that the gesture and speech interpretation is done in parallel as the operator needs to be pointing and talking concurrently.

6   Mission Execution

The Mission Execution layer consists of two parts, a *state manager* and a *localiser*. These two components are described in below.

6.1   The State Manager

The state manager is a finite state automaton that controls the reactive layer. At startup, the state manager reads a file that associates states and behaviours. This file can be reread during execution, to allow for online reconfiguration. Behaviours that are not connected to a controller, are controlled directly by the state manager.

A new task is initiated by a message from the planner to the state manager. This message contains a *state* and *data*. The state constitutes a certain set of behaviours that should be run in ensemble. An example of a state is *GoPoint* which translates into the behaviours *Gopoint* and *Obstacle Avoidance*. The data is usually a goal point, but for some states, such as *Explore*, the data field is irrelevant. Each behaviour is associated with a controller. Upon a change of state, the state manager informs the controllers which behaviours it should integrate. A task-list may also be supplied by the user through a file when the system is debugged without a planner.

The state manager awaits messages from the behaviours that tell whether the task was successfully carried out or if it failed. In the case of success, the planner is informed and a new command may be issued. Otherwise the state manager goes into a Stop-state, which is also sent to the controllers. A backup-state may defined for each state. If a failure occurs, this backup-state is initiated if it exists, otherwise the planner is notified. At any time a new command from the planner may be received and execution is then pre-emptied or modified. Especially the command to stop is given high priority. The state manager will also forward any goal point updates originating from the localiser.

6.2   Localisation

One of the most important capabilities of a robot is to know where it is. The knowledge of the current location can be expressed in many different ways. Here, we will say that the robot is localised if it knows where in a room it is (metric information). We have used two different approaches to localisation, one using sonar sensors and one using a laser range finder.

In the sonar approach the robot initially has a topological map of the environment. To be able to get metric information it uses its sonar sensors. The basic steps in the localisation process are the following:

- Let the robot explore a room to collect sonar data. The robot can either automatically perform the exploration or an operator can use the joystick to guide the robot. The data is used to automatically create a sonar landmark map. During this process it is also possible for an operator to assign names to different places in the room (goal points).
- When a room is revisited, the robot collects new sonar data, again by exploring the room, and these are matched to landmarks in the original map. From this mapping, the position and the orientation of the robot is estimated.
- Following the previous step, the robot iteratively updates its current position estimate.

The process is described below, for more details see [16].

Sonar data using triangulation Our Nomad 200 robot, used for most of our experiments, has a ring of 16 ultrasonic sensors. Sonars are known to provide noisy data. The sound sent out by the sonar is spread in a cone emerging from the source (actually the process is even more complicated). The emitted sound is then reflected and the time difference between emission and reception can be used to calculate the distance to the object that reflected the sound. Due to the spreading of the emitted sound, the reflecting point(s) can be anywhere on a surface. To remedy the problem, a triangulation technique has been developed, which fuses sonar readings from multiple positions. The basic idea is that if two sonar readings come from the same object, the position of that object must be in the intersection of the two cones. By accumulating such intersection points, it is possible to use simple voting to identify the most stable landmarks.

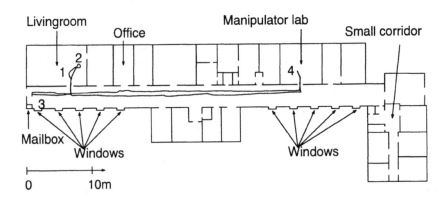

Fig. 4. A layout of the CAS lab floor showing the path of the robot executing some commands. At point 1 the robot is told to localise itself in the living-room. It then wanders around randomly, collecting sonar data and matching these to a previously stored map. At point 2 it knows where it is and stops. It is then told to deliver mail in the manipulator lab, which means that it will go to the mail slots, point 3, pick up the mail, and then go down the hallway to the lab, point 4. Finally it is told to go back to the living room.

Map acquisition The robot is started at an arbitrary position in a room. This position will become its reference point in the room. It then moves around, collecting sonar data using the triangulation scheme outlined above. Some of the collected sonar data will be stable over an extended time interval. By selecting the most stable data combined with readings from the odometry, it is possible to create a landmark map of the room. The landmarks are selected automatically by the robot itself. However, when studying the data a posteriori, it is evident that corners on bookshelves, tables, chairs and door posts are likely to become landmarks. At present, the 20 most stable landmarks are stored for each room. During this process, it is also possible for an operator to name different positions, called goal points, in the room. The goal points are symbolic references that can be used by the user when giving commands to the robot.

This map making process is repeated in each room. In larger rooms, for example a long corridor, the limited number of landmarks is inadequate. Such rooms are divided into a number of smaller regions that each have a separate set of landmarks.

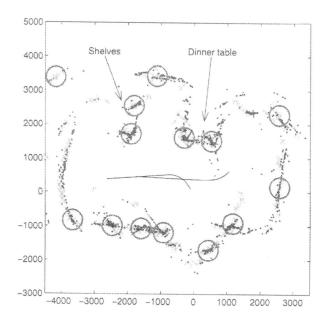

Fig. 5. A sonar map of the living room. The circles indicate where a map land-mark has been matched to a point from the current run. In this case 13 landmarks have been matched.

Map matching When the robot revisits a room, it will use the previously stored map to localise itself. The topology map, used by the planner, is used to determine switches between rooms. Upon entering a room the robot has the option to carry out a relocalisation process. If the robot encounters an error or looses track of its position it can perform an absolute localisation in a room. The robot carries out exploration in the room and collects a set of at least 10 landmarks. The set of collected landmarks are then matched against the stored map, and the associated rigid transformation (translation and rotation) is estimated based on the best possible match between the two maps.

6.3   Localisation Using a Laser Range Finder

The robot also has an on-board laser scanner. The sensor is a SICK PLS-200 scanner, that provides a 180° scan of its environment, represented as 361 measurements in the range 0.7 m – 100 m and with an accuracy of 5 cm. In this particular work, each room is represented as a rectangle (represented by width and length). The laser scan is then matched against the room model and an estimate of the position is feed into a Kalman filter that maintains an estimate of the robot position. Given an initial estimate of the position, the following steps are taken, see also Figure 6:

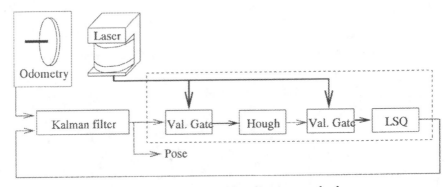

Fig. 6. The laser based localisation method

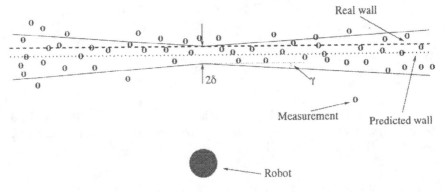

Fig. 7. Selection of reasonable measurements. The concept of a *validation gate* is in this case an area, points outside this area will be rejected.

1. Using odometry, the last position estimate is updated. This is the time update phase for the Kalman filter.
2. The updated position estimate is used in combination with the estimated uncertainty, to generate a set of validation gates that allows filtering of data, to reject outliers, as shown in Figure 7.
3. The filtered data are feed into a range weighted Hough transformation. Using an estimate of the position of walls it is now possible to perform model based segmentation.
4. The estimated Hough lines are used for definition of more accurate validation gates, that allow segmentation of data points into segments.
5. A least square line fitting is performed on the segmented data points.
6. A rigid transformation between fitted line segments and the model is used for updating of the Kalman filter. This is the measurement update step of the Kalman filter.

For a more detailed description of the procedure see [17] and [18].

7 Reactive Control

The reactive control layer consists of resources, behaviours, and controllers.

A resource is a server that distributes data to client behaviours. The data typically originates from a sensor, such as sonars, cameras, etc. A resource can also be a client to another resource, and compute higher level information.

A behaviour is a transformation between perception and action. Each behaviour in the system delivers proposals for the control of actuators. Depending on the function of the behaviour, it can use a number of resources to obtain the necessary information for the decision making. In view of the client/server concept, a behaviour is a server with controllers as clients. It is also a client with respect to resource-servers.

In this system, there are basically two types of behaviours. One type, e.g., the behaviour *GoPoint*, receives a goal-value (typically a goal point) from the state manager and reports back success or failure. The other type requires no goal-value. Consequently, it never reports to the state manager. One example of this type of behaviour is *AvoidObstacle*.

The controllers are responsible for sending direct commands to the robot's actuators. They will fuse the output from a set of behaviours defined by the state manager. This behaviour fusion mechanism has to be specified in the implementation of the controller. It could for example be an arbitration or a weighted vector summation of the proposals from the behaviours. Other schemes use fuzzy logic or voting.

Upon a change of state, the controller receives new directives from the state-manager that specifies which behaviours the controller should listen to.

The set of behaviours implemented in the system is outlined below.

7.1 GoPoint

The *GoPoint* behaviour steers towards a certain (x,y)-position in space. The odometry is used to determine the steering angle needed to reach the position.

7.2 Avoid

The *Avoid* behaviour detects obstacles and provides control to avoid them. The sonars are used for obstacle detection. Avoid uses a protection shield. It will suggest the robot to move away from any obstacle entering this shield. One parameter, the forward protection radius, specifies how close obstacles are allowed to come in front of the robot. A second parameter, the side protection radius, specifies how close obstacles are allowed to come to the side of the robot. The fusion of *GoPoint* and *Avoid* will drive the robot smoothly around small obstacles.

7.3   Explore

The *Explore* behaviour moves the robot away from positions it has visited before. If the robot for instance starts within a room, this behaviour will try to make the robot visit every open space in the room. When the behaviour believes there are no places left to visit, it will clear it's memory and start over again. This behaviour uses the sonar resource. Explore together with Avoid, is used during the localisation phase. When the localiser reports that enough landmarks have been found, the explore behaviour is terminated.

7.4   MailDocking

The *MailDocking* behaviour is used to pick up mail from a mail slot. It assumes that the robot is positioned exactly at the mailbox, where the forklift is used to pick up a mail-tray at a specified height. Currently, pure dead reckoning is used, but will later be done using visual servoing.

7.5   DoorTraverser

The *DoorTraverser* behaviour takes the robot through a doorway. The position of the doorway is fed into the behaviour by the state manager. The behaviour itself is incorporating avoidance facilities that have been trimmed for the purpose of narrow doorway traversal. DoorTraverser can tolerate corrupted doorway position information. In the current state, the behaviour can find the doorway if the doorway position error is less than 1.5 meters. Two versions of DoorTraverser have been implemented. One using sonars and another using laser range data.

7.6   FollowPerson

The *FollowPerson* behaviour will locomote the robot towards a human. At the moment, the human must face the robot while it follows. Monocular vision is normally used to detect and track the skin-colour of the head of the person. The *Avoid* behaviour is normally used simultaneously, so that the robot does not run into the person. The behaviour is described in more detail in [19].

7.7   CornerDocking

The *CornerDocking* behaviour uses the laser range finder to position itself relative to a corner. The method deployed is to extract the two walls defining the corner. Based on an estimate of the corner position from the laser data, a proportional control scheme is used to servo the robot to the desired position. The behaviour requires that the corner is in the "field of view" of the laser scanner when it is initiated.

7.8   Behaviour Fusion

The data received from the behaviours is fused or arbitrated by the controller to give a control signal to the actuators. Upon receiving the command to stop from the state manager, the controller stops the robot, disconnects the behaviours and enters an idle mode.

The controller implemented so far, is a vehicle controller. Future work includes integration of an arm controller, when the system is ported to our Nomad XR4000. The vehicle controller controls the steering and drive motors of the robot. Two formats of input data from the behaviours have been implemented.

In the first method, each behaviour provides a histogram containing 72 cells. The index of each cell represents an absolute direction from the robot centre, thus resulting in an angular resolution of five degrees (this is similar to a vector field histogram). The value of each cell indicates how much the behaviour wants to travel in that direction. The histograms are fused by component-wise summation and smoothed by convolution with a normalised, truncated, and discretised Gaussian function g. The resulting fused histogram is given by

$$f = \left(\sum h_j\right) * g,$$

where the discrete convolution with the Gaussian is performed on the circular buffer of cells. The robot will set course in the direction of the maximum value of f.

The second implemented data format is a polar representation (magnitude and direction). The vector outputs from the behaviours are fused by vector summation according to

$$v = \sum v_j,$$

where v_j is the output from behaviour j. The robot will head in the direction of the sum vector v. Both formats also include a speed proposal for the robot.

In order to make the robot drive along a smooth trajectory around larger obstacles, two schemes have been investigated. The first consists of exchanging the GoPoint behaviour with what we call a PathGoPoint behaviour. This uses sonar-data to construct a grid map used to compute a path around obstacles. In the other method, the vehicle controller detects when the sum of a behaviour fusion is close to zero. The controller will then start up another behaviour called Swirl. This behaviour will produce a steering angle, perpendicular to the object, to be added in the fusion process. Details on our strategy for smooth control can be found in [20].

8   Implementation

In this section we discuss the the hardware facilities of the robot, and the actual software implementation structure.

8.1   Hardware

The described system has been implemented on a Nomadic Technologies *Nomad 200* platform. It is a circular robot that is 125 cm high and with a diameter of 53 cm. There is an on-board 133 MHz Pentium computer with 128 MB RAM running Linux OS. The computer can communicate with the the outside world through a radio Ethernet link. Our robot is equipped with a wide range of sensors and actuators. Some of them are displayed in Figure 8.

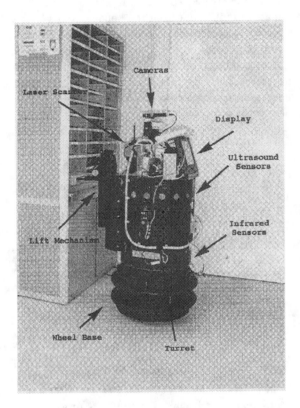

Fig. 8. Hardware Placement.

Sensors The sensors are both of proprioceptive and exteroceptive type. The proprioceptive sensors give the robot it's body awareness, while the exteroceptive sensors sense the world around it. We will describe all exteroceptive sensors in the following paragraphs, see also Figure 8.

Ultrasound Sensors One of the primary sensors on the platform is an omni-directional ultrasound sonar ring with 16 equiradial Polaroid elements. The sonars can measure distances from 15 cm up to 8 m and are placed approximately 0.8 m above the floor making it possible to detect obstacles at table height even at close range.

Infrared Sensors The infrared sensors are placed in a ring, similar to the sonar ring, but 38 cm above the floor. The measurement range is up to 40 cm. They can be used for last minute detection of obstacles that cannot be detected by the ultrasound sensors, e.g., a low coffee table.

Proximity Laser Scanner On top of the robot body we have mounted a SICK PLS-200 proximity laser scanner. It measures distances in a half-plane, approximately 95 cm above the floor. The sensor provides 361 readings resulting in an angular resolution of 0.5 degrees. The maximum measurement range is 100 m with an accuracy of 5 cm.

Odometry The odometry measures the rotation and translation of the wheels and the turret. The accuracy is high, but due to slippage on the floor etc. the measurements cannot be fully trusted.

Vision system A camera head, consisting of two relatively fixed colour cameras, is mounted on a pan-tilt unit on top of the robot. The cameras are connected to a Fujitsu image processing board.

Actuators The actuators of the robot enables it to locomote and to manipulate. A short description of the actuators are given below.

Wheel Base The wheel base consists of a three wheel synchro-drive, such that they always point in the same direction. The robot has a zero gyro-radius drive which enables it to turn on the spot. The maximum speed of the robot is 61 cm/s and the maximum turning speed is 60 degrees/s.

Turret The turret is mounted on the wheel base with a rotating joint. The joint can be rotated with a maximum turning speed of 90 degrees/s. All the sensors, except odometry, are mounted on the turret.

Lift mechanism A lift mechanism is mounted on the front of the robot. The lift has a gripper that can grip objects that are up to 15 cm wide and weight less than 9 kg. The lift can reach object from the floor and up to a height of 122 cm.

8.2   Software

The objective of the software implementation was to fulfil the following criteria:

- Easy integration of new behaviours
- Easy interfacing of new hardware devices
- Transparent performance on multiple platforms

- Efficient runtime performance
- Simple debugging

The implementation of the software architecture has been performed using an object-oriented framework. We have used C++ to code all parts of the architecture, except for some Nomad supplied code that is written in C. Separate parts of the system, e.g., Resources, Behaviours, Controllers etc., run as separate processes communicating using sockets. This makes it possible to simultaneously run the system on several different platforms. The programming in C++ is done in correspondence with the ANSI standard and the operating system specific parts are wrapped in the software package Adaptive Communication Environment (ACE), to promote portability [21]. The total system is at the moment (December 1998) of approximately 68000 lines of code, including comments.

Process management All the software components described in Section 4 are processes. These are connected to each other in a network using global and local sockets.

The *process manager* keeps track of all processes used in the architecture. It is similar to the function of the ORB in CORBA. Each new process has to inform the process manager of its name, host, and address. It maintains a database for this information as well as information about dependencies between processes.

If a process wants to establish communication with another process, the process manager will provide the necessary address. The process manager will also start processes when they are required.

The process manager has an *executor* daemon on each computer used in the system. On demand from the process manager, an executor will start or kill processes used in the system. The executor will also monitor the processes it has started and report unexpected terminations. The executor can be viewed as a tool for the process manager to reach over all the computers in the system. The executor will always work as a slave for the process manager, except for the instance when the communication to the process manager is broken. In such an event, the executor will kill all the processes it has spawned, and return to an initial state waiting for a process manager to leash it again. Thus, the executor does not have to be restarted.

ACE - Adaptive Communication Environment The adaptive communication environment is an object-oriented toolkit that implements strategic and tactical design patterns, to simplify the development of concurrent, event-driven communication software. It provides a rich category of classes and frameworks that perform communication tasks across a wide range of operating system platforms. It can for example handle event demultiplexing and handler dispatching, interprocess communication, shared memory management, message routing, dynamic configuration of network services, threading, and concurrent control. For more information, see [21].

9   Results

The system presented above has been evaluated in our laboratory, including the living room shown in Figure 1. The evaluation has included experiments on localisation, map building, and fully autonomous mission execution for tasks like mail delivery.

Localisation involving the sonar system is described in detail in [16,22]. In general, it is possible to localise the system with an accuracy of 5 cm within a room of the laboratory. Using the laser scanner, a continuous updating of ego-position can be achieved with an accuracy of 1-5 cm, depending on the environment. In certain situations it is only possible to detect the wall at the side of the robot, i.e., in a long corridor. During such situations the longitudinal uncertainty grows significantly until a feature like a door becomes visible. The laser results are reported in further detail in [17,18].

For evaluation of the fully integrated system, a mail delivery scenario has been defined. The robot receives the command "Robot Deliver Mail in *room*" from the speech interface, where *room* is one of several pre-defined rooms. The robot will then plan a path to the mail-slot, drive to the mail slot, pick-up mail and finally drive to the designated room and announce arrival of mail. During this mission it will avoid obstacles and re-localise, if the uncertainty in position grows beyond a certain bound. This task has been carried out more than 100 times. The system has a reliability of about 90%, where the major source of problems is due to loss of localisation, which can be provoked if the robot encounters a large number of obstacles, or drives a long distance without any cues for re-localisation. This might happen in a long corridor with no or few landmarks that allow re-localisation. The laser range based localisation is expected to solve this problem, once it has been fully integrated. Another source of error is the docking to pick-up mail. It is critical that the robot is positioned with an accuracy of a few centimetres, as it otherwise will miss the mail-tray. To achieve this the laser-based corner docking is used. Due to lack of tactile feedback on the fork-lift it is however, possible to miss the tray.

Recently the system was demonstrated at a trade fair, where it navigated for 8 hours, only interrupted by replacement of batteries, every hour. During this period the robot moved about in an area of 20 by 20 meters. The map of the demonstration site was acquired during an explore session and subsequently used for relocalisation, when the errors grow beyond a fixed threshold. The robot performed robustly during the demonstration period.

Overall, it is a robust and versatile system. The robot can with little effort be transferred to a new environment and be taught the layout of the plant using exploration behaviours. Once a topological map and landmarks for the involved rooms have been acquired, it can be sent on missions.

10   Summary

A fully operational robot system has been constructed. The system has been developed using a behaviour-based approach and integrated using a hybrid deliberative architecture. The system includes facilities for spoken and gesture based commanding of the robot. Once instructed to carry out a particular mission, it will generate a plan to complete the mission. The list of tasks to be accomplished is given to a state manager that configures the set of behaviours, and monitors the execution of individual tasks. In the event of errors, a simple recovery strategy is deployed. For localisation, the system utilises sonars and a laser scanner. A map of the environment is automatically constructed using an exploration strategy. The system has been implemented on a Nomad 200 robot from Nomadic Technologies Inc.

The system is capable of performing a variety of tasks like mail delivery and point-to-point based navigation in a natural environment, like a regular living room. Extensive experiments have demonstrated that the system has a very high degree of robustness.

The primary limitations of the system is in terms of robust localisation in the presence of few natural landmarks. In addition, the system has problems when it encounters low obstacles and they frequently will not be detected by the present set of sensors.

The system has been built by a set of 10 graduate students, that each have contributed methods within their particular area of expertise. The basic architecture has been carefully designed to allow for easy (and flexible) integration of new behaviours. Today a "plug-n-play" functionality is available for adding new behaviours, with a minimum need for behaviour designers to know the internal structure of the system.

Presently the system is being ported to a Nomadic Technologies XR4000 robot, that includes an on-board PUMA 560 manipulator. This robot has a richer set of sonars (just above the floor and at 80 cm height), and a better SICK scanner (LMS-200). These facilities will allow for more robust navigation. The new challenge will be integration of manipulation into the system, which introduces a number of interesting research questions in terms of coordination and non-holonomic path planning. The short-term goal is to extend the current scenario to include the ability to open and close doors and to ride in an elevator. The manipulation system also opens up new problems in grasping, using visual servoing, and force-torque sensing.

11   Acknowledgement

The service robot has been constructed by a team consisting of Magnus Andersson, Henrik I Christensen, Magnus Egerstedt, Martin Eriksson, Patric Jensfelt, Danica Kragic, Mattias Lindström, Anders Orebäck, Lars Pettersson, Hedvig Sidenbladh, Dennis Tell, and Olle Wijk. The entire team has been instrumental in this effort. Throughout the effort valuable guidance and advice was received from Ronald C. Arkin, Jan-Olof Eklundh, Bo Wahlberg, and Anders Lindquist.

The Centre for Autonomous Systems is sponsored by the Swedish Foundation for Strategic Research, without this support the effort would never have been possible.

References

1. H. P. Moravec, "Towards automatic visual obstacle avoidance," in *Proceedings of Int. Joint. Conf. on Artificial Intelligence in Cambridge,MA*, p. 584, 1977.

2. H. P. Moravec, "The Stanford cart and the CMU rover," *Proceedings IEEE*, vol. 71, pp. 872 – 884, 1983.

3. R. Brooks, "A hardware retargetable distributed layered architecture for mobile robot control," in *Proceedings of the IEEE International Conference on Robotics and Automation*, 1987.

4. R. C. Arkin, "Integrating behavioral, perceptual, and world knowledge in reactive navigation," in *Robotics and Autonomous Systems, Vol. 6, pp. 105-22*, 1990.

5. R. Simmons, "Structured control for autonomous robots," in *IEEE Transactions on Robotics and Automation*, 1994.

6. J. R. Firby, "Modularity issues in reactive planning," in *Third International Conference on AI Planning Systems*, (Menlo Park, CA), pp. 78–85, AAAI Press, 1996.

7. J. R. Firby, *Adaptive Execution in Complex Dynamic Worlds*. PhD thesis, Computer Science Dept., Yale University, 1989. TR-YALEU/CSD/RR #672.

8. M. Beetz, W. Burgard, A. B. Cremers, and D. Fox, "Active localization for service robot applications," in *Proceedings of the 5th International Symposium on Intelligent Robotic Systems '97*, 1997.

9. U. D. Hanebeck, C. Fischer, and G. Schmidt, "Roman: A mobile robotic assistant for indoor service applications," in *Proceedings of the International Conference on Intelligent Robots An Systems 1997*, 1997.

10. S. Thrun, M. Bennewitz, W. Burgard, A. Cremers, F. Delaert, D. Fox, D. Hahnel, C. Rosenberg, N. Roy, J. Schulte, and D. Schulz, "Minerva: A second generation museum tour-guide robot." CMU Tech Report (url://www.cs.cmu.edu/~thrun/papers).

11. I. Nourbakhsh, "The failures of a self reliant tour robot with no planner," tech. rep., Carnegie Mellon University, Robotics Institute, url: http://www.cs.cmu.edu/~illah/SAGE, 1999.

12. J. R. Firby, P. Prokopowicz, and M. J. Swain, *Artificial Intelligence and Mobile Robotics*, ch. The Animate Agent Architecture, pp. 243–275. Menlo Park, CA.: AAAI Press, 1998.

13. I. Nourbakhsh, *Artificial Intelligence and Mobile Robotics*, ch. Dervish: An Office Navigating Robot, pp. 73–90. Menlo Park, CA.: AAAI Press, 1998.

14. P. Agre and D. Chapman, "What are plans for?," *Robotics and Autonomous Systems*, vol. 6, pp. 17–34, 1990.

15. H. Asoh, S. Hayamizu, I. Hara, Y. Motomura, S. Akaho, and T. Matsui, "Sociall embedded leraring of office-conversant robot jijo-2," in *Int. Joint Conf. on Artificial Intell. 1997*, 1997.

16. O. Wijk and H. I. Christensen, "Extraction of natural landmarks and localization using sonars," in *Proceedings of the 6th International Symposium on Intelligent Robotic Systems '98*, 1998.

17. P. Jensfelt and H. I. Christensen, "Laser based position acquisition and tracking in an indoor environment," in *International Symposium on Robotics and Automation - ISRA'98*, (Saltillo, Coahuila, Mexico), December 1998.

18. P. Jensfelt and H. I. Christensen, "Laser based pose tracking," in *International Conference on Robotics and Automation 1999*, (Detroit, MI), May 1999.

19. H. Sidenbladh, D. Kragic, and H. I. Christensen, "A person following behaviour," in *IEEE International Conference on Robotics and Automation 1999*, (Detroit, MI), May 1999. (submitted).

20. M. Egerstedt, X. Hu, and A. Stotsky, "Control of a car-like robot using a dynamic model," in *IEEE International Conference on Robotics and Automation*, May 1998. Accepted for presentation.
21. D. C. Schmidt, "The adaptive communication environment: Object-oriented network programming components for developing client/server applications," in *11th and 12th Sun Users Group Conference*, 1994.
22. O. Wijk, P. Jensfelt, and H. I. Christensen, "Triangulation based fusion of ultrasonic sonar data," in *IEEE Conference on Robotics and Automation*, IEEE, May 1998.

Vision-Based Behavior Control of Autonomous Systems by Fuzzy Reasoning

Wei Li[1], Friedrich M. Wahl[2], Jiangzhong Z. Zhou[3], Hong Wang[1], and Kezhong Z. He[1]

[1] Department of Computer Science and Technology
Tsinghua University, Beijing 100084, P. R. China
[2] Institute for Robotics and Process Control
Technical University of Braunschweig
Hamburger Str. 267, D-38114 Braunschweig, Germany
[3] Institute for Computer Peripherals
Chinese Academic of Sciences, Wuhan 430050, P. R. China

Abstract. Vision-based motion control of an autonomous vehicle operating in real world requires fast image processing and robustness with respect to noisy sensor readings and with respect to varying illumination conditions. In order to improve vehicle navigation performance in out-door environments, this paper presents methods for recognizing landmarks based on fuzzy reasoning. Firstly, a fuzzy thresholding algorithm is proposed to segment roads and to extract white line marks on streets from images. Secondly, some special domain knowledge about edges on roads represented by fuzzy sets is integrated into the rule base of a fuzzy edge detector. Based on this, the fuzzy thresholding algorithm is adopted to recognize road edges. In addition, a method for path planning in sensor space is presented and a path following behavior based on a fuzzy rule base is defined to control vehicle motion. The proposed methods are applied to navigate the THMR-III autonomous vehicle in out-door environments. Some experimental driving maneuvers have been performed to prove their effectiveness and their robustness. ...

1 Introduction

The development of robust and efficient methods for recognizing landmarks in out-door environments is a key issue in vision based motion control of autonomous vehicles [11]. For example, extraction of white lines on streets is one of the most popular approaches for control of vehicle motion on streets. The basic algorithms for such applications are based on region oriented image segmentation or edge detection. In machine vision, different approaches have been proposed for region based image segmentation and edge detection, such as thresholding algorithms based on gray level statistics or based on the evaluation of an estimate of the local gradients or of the local second derivatives in [5]. These approaches, however, might be infeasible for vehicle navigation in out-door environments, since this application requires not only image processing in real-time, but also

Christensen et al. (Eds.): Sensor Based Intelligent Robots, LNAI 1724, pp. 311–325, 1999.
© Springer-Verlag Berlin Heidelberg 1999

robustness with respect to noise and with respect to illumination variations. For example, a thresholding algorithm has been implemented on the THMR-III vehicle [14]. In experiments, it has been noticed that thresholds largely depend on illumination conditions and on the properties of the environments, i.e., different thresholds should be determined to adaptively process different regions in the same picture to obtain a desired performance. Recently, some approaches based on computational intelligence have been proposed for image processing. In [1][2], Bezdek et al. develop a common framework for designing both traditional and nontraditional (learning model) edge detectors. In [6][9], some researchers present algorithms for edge extraction by the use of neural networks. Especially, edge detectors based on fuzzy reasoning show their robustness with respect to noise [12][13]. This paper presents some methods for vision based motion control of an autonomous vehicle by fuzzy inference [7][8]. Firstly, a fuzzy thresholding algorithm is proposed to segment roads and to extract white line marks on streets from images. Secondly, some special domain knowledge about edges on roads represented by fuzzy sets is integrated into the rule base of a fuzzy edge detector. Based on this, the fuzzy thresholding algorithm is adopted to recognize road edges. In order to control vehicle motion, we present a method for generating a path in sensor space and define a path following behavior based on a fuzzy rule base. Some experimental driving maneuvers have been performed to demonstrate the effectiveness and robustness of the proposed approaches [7][8]. This paper is organized as follows. In the next section we briefly introduce the vision system of the THMR-III vehicle. In section 3, we propose a new fuzzy segmentation algorithm to extract roads from images and directly generate paths based on image information. In section 4, we discuss recognition of white lines on roads by fuzzy reasoning and define a path following behavior to control vehicle motion. In section 5 we integrate special knowledge about roads represented by fuzzy sets into the rule base of a fuzzy edge detector and use the fuzzy thresholding algorithm to extract edges on streets. In section 6, we give some concluding remarks on the proposed methods.

Fig.1: Autonomous vehicle THMR-III

Fig.2: Vehicle's view environment 1

2   Vision System of the THMR-III Vehicle

The THMR-III autonomous system, as shown in Fig. 1, is equipped with a vision system, which consists of a high-level and a low-level sub-system. The high-level sub-system includes an image processing pipe, a CCD camera installed at the front top of the vehicle as well as a sun workstation; the low-level sub-system has been built up using a PC486 computer, a high-speed image processing board VIGP-2M and a CCD camera mounted at the front middle of the vehicle. The placements of the cameras have been selected according to navigation requirements discussed in [14]. The distance of the top camera to ground is 2m, while the distance of the lower camera to ground is about 1m. Image information from the top camera is mainly used to track far roads within regions from 8m to infinite. Image information from the lower camera is mainly used to avoid obstacles and to track close roads region from 4m to 12m. All images used in this paper are taken by the lower CCD camera (for example, see image in Fig. 2).

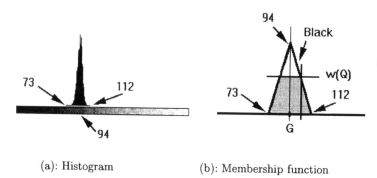

(a): Histogram   (b): Membership function

Fig.3: Generation of membership function based on histogram

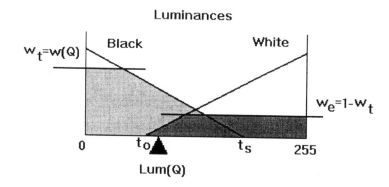

Fig.4: Fuzzy inference and defuzzification

3   Road Segmentation by Fuzzy Thresholding

One of the most important problems in vision is to identify subimages which represent objects. The partitioning of an image into regions is called segmentation [5]. For control of vehicle motion on streets, it is very important to partition an image into subimages so that a road can be extracted from the image. Since the interpretation of images of real scenes taken during vehicle motion are usually incomplete and ambiguous, it is very hard to reliably determine what is and what is not a border of a road in the images. Therefore, we recommend fuzzy logic to deal with such uncertainty.

Fuzzy logic inference is based on the theory of fuzzy sets, as introduced by Zadeh [18]. A fuzzy set A in a universe of discourse X is defined by its membership function $\mu_A(x)$. For each $x \in X$, there exists a value $\mu_A(x) \in [0, 1]$ representing the degree of membership of x in X. In a fuzzy logic expert system, membership functions assigned with linguistic variables are used to fuzzify physical quantities. A fuzzy rule base is used to refer fuzzy inputs. For example, a simple fuzzy inference rule relating input v to output u might be expressed in the condition-action form:

> IF v *is* W THEN u *is* Y

where W and Y are fuzzy values defined on the universes of v and u, respectively. The response of each fuzzy rule is weighted according to the degree of membership of its input conditions. The inference engine provides a set of control actions according to fuzzified inputs. Since the response actions are in a fuzzy sense, a defuzzification method is required to transform fuzzy response actions into a crisp output value. A widely used defuzzification method is the centroid method. In order to extract roads in images, as the one shown in Fig. 2, we use the following segmentation rules:

> IF *the gray level of a pixel is close to the gray level of the road*
> THEN *make it to black*
> ELSE *make it white*

We implement this fuzzy thresholding algorithm as follows: Firstly, we define a window that contains the road part in the image, as shown in Fig. 2. Then, we analyze the gray level histogram of the image within the window (in an initial phase, this can be done by interactive image manipulation; during the vehicle's movement phase, this information is updated continuously and automatically); such a histogram is shown in Fig. 3(a). Thirdly, we construct membership functions. In this paper, triangular or trapezoidal functions are used to define membership functions. For example, the "input-black" membership function shown in Fig. 3(b) are determined by the three points (73, 94, 112) of the histogram in Fig. 3(a). For the output membership functions, their parameters t_s and t_o are determined based on our experiments. Usually, we have $0 < t_s < 255$ and $t_o = 255 - t_s$. Subsequently, we calculate gray level differences between all pixels of the entire image and the gray level corresponding to the histogram's maximum by fuzzy reasoning. Fig. 4 shows the fuzzy rule base and the membership functions. By using centroid defuzzification, we compute the gray level of the pixel. The following simple algorithm is proposed to segment the image:

Step 1: Preprocess the image by a Gaussian smoothing algorithm;

Step 2: Calculate the histogram in the defined window;

Step 3: Construct a membership function;

Step 4.1: Fuzzify a pixel $Q(i)$ based on the defined membership function: $w_t = w(Q)$ and $w_b = 1 - w_t$;

Step 4.2: Classify this pixel into white or black based on w_t and w_b by the centroid defuzzification.

By using this algorithm, lanes can be segmented very clearly, as shown in Fig. 5(a). Next step, we generate paths based on image information for vehicle motion in two steps: Firstly, we decompose the image into several subimages with a height h; secondly, we calculate the centroid of the dark pixels, i.e., the first-order moment, in each subimage. As example, Fig. 5(b) shows a generated path based on image information. For further interpretation, we decompose the whole image into nine regions within some windows, as indicated in Fig. 5(b). According to a priori knowledge, the top regions are not considered any further. In section 5, we will present a path following behavior to control vehicle motion.

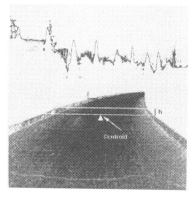

(a): Calculation of centroid   (b): A path for vehicle motion

Fig.5: Path planning based on image segmentation

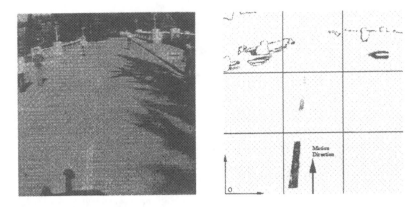

Fig.6: Vehicle's view in environment 2   Fig.7: White line for vehicle motion

4   White Line Mark Recognition

One of the most popular approaches for vehicle motion control on streets is to extract white line marks on streets, as the ones shown in Fig. 6. Often it is not easy to recognize them, since with increasing abrasion of the streets the markings become degraded; in addition, images usually are taken under varying illumination conditions. Our method for recognition of white lines is very similar to that for road segmentation discussed in section 3. Here, we use the following rules to extract white lines:

> IF *the gray level of a pixel is close to the gray level of a white line*
> THEN *make it black*
> ELSE make it white

In order to implement these rules, we decompose the whole image into nine regions and calculate the histograms of subimages within one window or within more windows containing white lines, as shown in Fig. 6-7. According to a priori knowledge, the left-top and right-top regions should not be considered. Based on the histogram, we construct a membership function. Fig. 7 shows the extracted white line markings which are clear and suitable for robot motion control. Fig. 8-9 show another example for extraction white line during vehicle motion. In [15]–[17], fuzzy logic-based behavior approaches are used to control mobile robot motion. Here, we define a path following behavior to track planned paths as, e.g., the one shown in Fig. 5(b) or white lines as the ones shown in Fig. 7 and Fig. 9.

5   Fuzzy-Logic-Based Behavior Control

In robot applications in the real world, a mobile robot should be able to operate in uncertain and dynamic environments. Behavior-control [3] and [4] shows

Fig.8: Vehicle's view in environment 3   Fig.9: White line for vehicle motion

potentialities for robot navigation in unstructured environments since it does not need building an exact world model and complex reasoning process. However, before behavior-based control is used to navigate a mobile robot in the real world perfectly, much effort should be made to solve problems with it, such as,

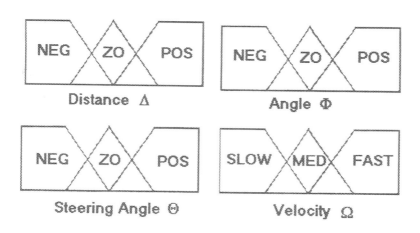

Fig.10: A fuzzy-logic-based *path following behavior*

the quantitative formulation of behavior, and the efficient coordination of conflicts and competition among multiple types of behavior, and so forth. In order to overcome these deficiencies, some fuzzy-logic-based behavior control schemes have been proposed in [15][16][17]. In this paper, we use fuzzy logic to implement *following edge behavior* to control vehicle motion. Fig. 7 shows a defined coordinate system in the image space, i.e., in the sensor space. The vehicle control system uses two inputs and two outputs. The first input signal is an angle $\Phi = \Phi_w - \Theta_a$, which is the difference between white line orientation Φ_w and vehicle motion direction Θ_a. In order to compute Φ, the vehicle motion direction is mapped into the sensor space. Φ has a negative value in both cases in Fig. 7 and Fig. 9. The second one is a deviation Δ between a vehicle's position and the white line on a horizontal axis. If the vehicle is located in the right side of the white line, Δ is defined as positive, as shown in Fig. 7; otherwise, Δ is defined as negative, as shown in Fig. 9. The outputs from the control scheme are the control signals of a vehicle velocity Ω and a steering direction Θ. The linguistic values, POS (positive), $ZERO$, NEG (negative) are defined for the angle Φ and for the deviation Δ. Similarly, POS (positive), $ZERO$, NEG (negative) are defined for the steering direction Θ, and $FAST$, MED (medium) and $SLOW$ are defined for the velocity Ω. Fig. 10 shows the defined membership functions for the inputs and the outputs. The rule base can be formulated based on our experience. The vehicle can be run fast if there is no position deviation between the vehicle and a white line and if angle Φ is close to zero. Therefore the fuzzy control scheme should output the control signals as follows: The steering direction Θ is close to zero and the vehicle's velocity Ω is fast. If the angle Φ is negative and the position deviation Δ is positive, the vehicle could keep its present direction and run at a medium speed, since the position deviation could be reduced during vehicle motion, as shown in Fig. 7. The vehicle should run at a low speed if both the angle Φ and the position deviation Δ are negative, as shown in Fig. 9. In this case, the fuzzy control scheme should generate a $Slow$ speed and an opposite steering angle Θ to reduce Φ and Δ. Some fuzzy rules from the rule base are listed below, in order to explain, at least in principle, how the path following behavior is realized:

Rule 1: IF (Δ is $ZERO$ and Φ is $ZERO$) THEN (Ω is $FAST$ and Θ is $ZERO$)
Rule 2: IF (Δ is NEG and Φ is POS) THEN (Ω is MED and Θ is NEG)
Rule 3: IF (Δ is NEG and Φ is NEG) THEN (Ω is $SLOW$ and Θ is POS)
Rule 4: ......

6   Road Edges Extraction by Fuzzy Reasoning

Extraction of road edges is another important issue for vehicle motion in unstructured roads. Basically, it is more difficult to detect road edges than to extract white line marks due to environment changes. For example, there are a lot of tree shadows on the street in Fig. 6 due to strong solar illumination.

(a): Neighbouring pixels of the Q

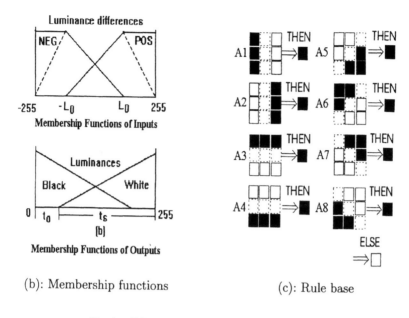

(b): Membership functions   (c): Rule base

Fig.11: Edge extraction by fuzzy reasoning

In this section, we discuss a new method for extraction of road edges based on fuzzy reasoning in three steps: Firstly, we detect edges in a given direction by a fuzzy edge detector; secondly, special knowledge about roads represented by fuzzy sets is integrated into the rule base of the fuzzy edge detector; thirdly, a fuzzy threshold algorithm is applied to segment road edges. In [12], the idea of detecting edges by fuzzy inference in order to classify every pixel in an image into white or black has been proposed. In order to decide if a pixel belongs to a border region or to a uniform region, luminance differences between the pixel and its neighboring ones are computed, as shown in Fig. 11(a). If a pixel meets one of the following conditions, it should be black, i.e., it belongs to a border region. Otherwise, the pixel should be white, i.e., it belongs to a uniform region. In [12], the membership functions of the inputs are defined by triangular functions with peaks at $Dif\_Lum(i) = L_0$ and $Dif\_Lum(i) = -L_0$, as shown by the dotted lines in Fig. 11(b). Using these types of membership functions, the maximum degree $\mu(L_0) = \mu(-L_0) = 1$. In the other word, the luminance difference $Dif\_Lum(i) = L_0$ (or $Dif\_Lum(i) = -L_0$) contributes the strongest weight to the white color (or the black color). In our study, we note that instead of luminance difference equal to L_0 (or $-L_0$), the luminance differences being larger than L_0 (or smaller than $-L_0$) do contribute the strongest weight to the white color (or the black color). Therefore, we replace the triangular membership functions by the trapezoidal membership functions, as shown by the solid line in Fig. 11(b). The value L_0 could be determined by the average value of all luminance differences. Fig. 11(b) also shows the membership functions of outputs. All rules shown in Fig. 11(c) are listed below:

Rule A1: IF $Dif\_Lum(1)$ is NEG and $Dif\_Lum(2)$ is NEG and $Dif\_Lum(3)$ is NEG and $Dif\_Lum(5)$ is POS and $Dif\_Lum(6)$ is POS and $Dif\_Lum(7)$ is POS THEN $Lum(Q)$ is BLACK

Rule A2: IF $Dif\_Lum(5)$ is NEG and $Dif\_Lum(6)$ is NEG and $Dif\_Lum(7)$ is NEG and $Dif\_Lum(1)$ is POS and $Dif\_Lum(2)$ is POS and $Dif\_Lum(3)$ is POS THEN $Lum(Q)$ is BLACK

Rule A3: IF $Dif\_Lum(1)$ is NEG and $Dif\_Lum(7)$ is NEG and $Dif\_Lum(8)$ is NEG and $Dif\_Lum(3)$ is POS and $Dif\_Lum(4)$ is POS and $Dif\_Lum(5)$ is POS THEN $Lum(Q)$ is BLACK

Rule A4: IF $Dif\_Lum(3)$ is NEG and $Dif\_Lum(4)$ is NEG and $Dif\_Lum(5)$ is NEG and $Dif\_Lum(1)$ is POS and $Dif\_Lum(7)$ is POS and $Dif\_Lum(8)$ is POS THEN $Lum(Q)$ is BLACK

Rule A5: IF $Dif\_Lum(4)$ is NEG and $Dif\_Lum(5)$ is NEG and $Dif\_Lum(6)$ is NEG and $Dif\_Lum(1)$ is POS and $Dif\_Lum(2)$ is POS and $Dif\_Lum(8)$ is POS THEN $Lum(Q)$ is BLACK

Rule A6: IF $Dif\_Lum(1)$ is NEG and $Dif\_Lum(2)$ is NEG and $Dif\_Lum(8)$ is NEG and $Dif\_Lum(4)$ is POS and $Dif\_Lum(5)$ is POS and $Dif\_Lum(6)$ is POS THEN $Lum(Q)$ is BLACK

Rule A7: IF $Dif\_Lum(6)$ is NEG and $Dif\_Lum(7)$ is NEG and $Dif\_Lum(8)$ is NEG and $Dif\_Lum(2)$ is POS and $Dif\_Lum(3)$ is POS and $Dif\_Lum(4)$ is POS THEN $Lum(Q)$ is BLACK

Rule A8: IF $Dif\_Lum(2)$ is NEG and $Dif\_Lum(3)$ is NEG and $Dif\_Lum(4)$ is NEG and $Dif\_Lum(6)$ is POS and $Dif\_Lum(7)$ is POS and $Dif\_Lum(8)$ is POS THEN $Lum(Q)$ is BLACK

where $Dif\_Lum(i)$ is the luminance differences between the pixel and its neighboring pixels, which could be a positive or negative value in $[-255, 255]$. Fig. 12 illustrates how to compute the black color strength of the pixel, w_t, using **Rule A8**. First, the luminance differences between the pixel with its neighboring pixels 1, 2, 4, 5, 6 and 8, are computed and are denoted by, $Lum\_Dif(1)$, $Lum\_Dif(2)$, $Lum\_Dif(4)$, $Lum\_Dif(5)$, $Lum\_Dif(6)$, and $Lum\_Dif(8)$. Then, $Lum\_Dif(1)$, $Lum\_Dif(2)$, and $Lum\_Dif(8)$ are fuzzified using the membership functions, NEG (negative),to get their membership degrees $\mu(1)$, $\mu(2)$, $\mu(8)$, respectively, shown in Fig. 12(a); while $Lum\_Dif(4)$, $Lum\_Dif(5)$, $Lum\_Dif(6)$ are fuzzified by the membership functions, POS (positive), $\mu(4)$, $\mu(5)$, $\mu(6)$, respectively, shown in Fig. 12(b). The black color strength of **Rule A8**, $w(1)$, is computed by $w(8) = (\mu(1) + \mu(2) + \mu(8) + \mu(4) + \mu(5) + \mu(6))/6$. Obviously, the more negative all $Lum\_Dif(1)$, $Lum\_Dif(2)$, and $Lum\_Dif(8)$ are and the more positive all $Lum\_Dif(4)$, $Lum\_Dif(5)$, $Lum\_Dif(6)$ are, the bigger the weight value, $w(1)$, of the black color is. For each rule, **Ai** (i=1,...8), the fuzzy reasoning outputs its correspond weight $w(i)$. Then, we can get w_t and w_e by $w_t = max\{w(i)\}$ and $w_e = 1 - w_t$. The last step, the luminance of the pixel, $Lum(Q)$, is computed by centroid defuzzification based on w_t and w_e.

Each rule produces a different weight based on edge direction. For example, **Rule A1** and **Rule A2** produce the maximal weight to detect any vertical edge; while **Rule A3** and **Rule A4** produce the maximal weight to detect any horizontal edge. Fig. 13(a) shows how the image is processed by all rules. The

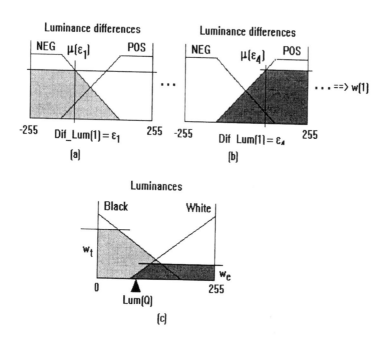

Fig.12: Image processing by fuzzy reasoning

images in Fig. 13(b) and 13(c) are generated by the vertical and horizontal rules, respectively. This characteristic can be used to strengthen road edges by activating the corresponding rules according to their orientations. For example, we only use **Rule A5** and **Rule A6** to detect edges in Fig. 6 since the right road edge in the image is orientated in the related direction. The image in Fig. 14 shows that the road edge is really stronger than others.

The second technique used in road edge extraction is to integrate some special knowledge represented by fuzzy sets into the rule base of the fuzzy edge detector. In doing this, we propose the following strategies:

IF *a pixel belongs to a border region and it is close to a feature*
THEN *make it black*
ELSE *make it white*
IF *a pixel belongs to a uniform region and it is close to a feature*
THEN *make it white*
ELSE *make it black*

Features in our context means road edge segments in an image. In order to build a membership function of a road edge, its initial points on the road edge have to be determined. Before the vehicle starts to move the membership function of road edges has to be adjusted to indicate the image region representing the road

(a): Alle edge detecion

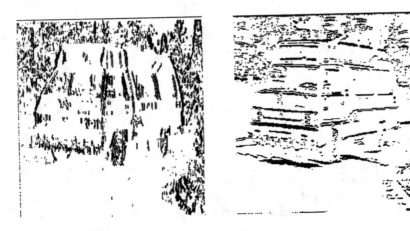

(b): Vertical edge detection   (c): Horizontal edge detection

Fig. 13: Edge detection by fuzzy reasoning

border; the region is defined by interactively pointing to several points of the road edges. For example, Fig. 15 shows the membership function for extraction of road edges. The parameter d_P is the shortest distance of a pixel P to the edge template T_s. On the basis of this membership function, a weight $w(d_P)$ is computed by the shortest distance. When the vehicle moves, it determines whether a pixel P in an image belongs to a road edge by the following algorithm:

Step 1: Compute luminance differences of the pixel P, $Lum\_Dif(i)$ $(i = 1, ..., 8)$ and its neighboring pixels;

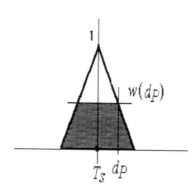

Fig.14: Road extraction step 1   Fig.15: Road membership function

Step 2: Based on defined membership functions, fuzzify the luminance differences $Lum\_Dif(i)$ $(i = 1, ..., 8)$ according to THEN-RULES to obtain $w(i)$ $(i = 1, ..., 8)$;

Step 3: Determine initial points of an edge template according to the last processed image.

Step 4: Compute the shortest distance d_P of the pixel P to the edge template and compute a weight $w(d_P)$ based on the defined membership function in Fig. 14;

Step 5: Compute $w_t = max(w(i), w(d_P))$ the strengths of THEN-RULES and ELSE-RULES by MIN-MAX inference;

Step 6: Classify the pixel under consideration into white or black by the centroid defuzzification.

Fig. 16 shows an image that has been processed by this algorithm. It is very clear, that the region where the road edge is located is strengthened. The last step is to segment road edges by the fuzzy thresholding algorithm as follows: Firstly, we select all pixels whose weight w_t is greater than $alpha < l$. Secondly, we calculate their mean value ϵ and standard deviation σ. Thirdly, we use $\eta = \epsilon - \sigma$ as a threshold to binarize the image, i.e., if the gray level of a pixel is greater than η, this pixel is classified as black, and otherwise this pixel is classified as white. Fig. 17 shows a right road edge which has been extracted from the environment in Fig. 6. It can be seen that shades of trees and other noise have been removed, i.e., the extracted edge can be used for vehicle motion control. In order to extract road edges on both sides in Fig. 2, we activate **Rule A5**, **Rule A6**, **Rule A7** and **Rule A8**. Fig. 18 shows the extraction of road edges in Fig. 2.

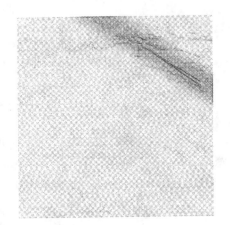

Fig.16: Road extraction step 2   Fig.17: Extracted road edge

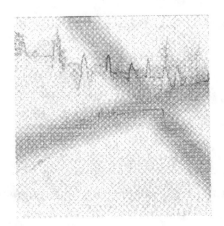

(a): Road extraction   3ex (b): Extracted road edge

Fig.18: Road edge detection by fuzzy reasoning

7 Conclusion

This paper presents new methods for recognizing landmarks by fuzzy thresholding and fuzzy edge detection. Based on fuzzy image segmentation, we generate
paths in a sensor space and define the path following behavior to control vehicle
motion in out-door environments. One of the advantages of representing road
edges by using fuzzy sets, is that it is not necessary to acquire exact knowledge
about roads. The proposed methods are robust with respect to noise and to
illumination conditions.

References

1. Bezdek J. C., Chandrasekhar R. and Attikiouzel Y., A geometric approach to edge detection, IEEE Transactions on Fuzzy Systems, **6**(1) (1998) 52–75
2. Bezdek J. C. and Shirvaikar M., Edge detection using the fuzzy control paradigm, Proc. of 2nd Euro. Congress on Intelligent Techniques and Soft Computing, Aachen, Germany, (1994) 1–12
3. Brooks AR. A. A., A robust layered control system for a mobile robot, IEEE J. Robot. Automation **RA-2**, (1986), 14–23
4. Arkin R. C. and Murphy R. R., Autonomous Navigation in a Manufacturing Environment, IEEE Trans. Robot. Automation **RA-6**, (1990), 445–454 Congress on Intelligent Techniques and Soft Computing, Aachen, Germany, (1994) 1–12
5. Jain A. K., Fundamentals of Digital Image Processing, Prentice-Hall, Inc., Englewood Cliffs, NJ, 1989
6. R. Lepage and D. Poussart, Multi-Resolution Edge Detector, Proc. IJCNN, 1992
7. Li W. et al., Detection of road edge for an autonomous mobile vehicle by fuzzy logic inference, Proc. of International Conference on Neural Information Processing, (1995), 413–416
8. Li W., Lu G. T., and Wang Y. Q., Recognizing white line markings for vision-guided vehicle navigation by fuzzy reasoning, Pattern Recognition Letters, **18**(8) (1997) 771–780
9. Moura L. and Martin F., Edge detection through cooperation and competition, Proc. of IJCNN, 1991
10. Pedrycz W., Fuzzy control and fuzzy systems, Research Studies Press Ltd., Second edition, 1993
11. Regensburger U. and Graefe V., Visual recognition of obstacles on roads, Proc. of IEEE/RSJ International Conference On Intelligent Robots and Systems, (1994) 980–987
12. Russo F. and Ramponi G., Edge extraction by FIRE operators, Proc. of the Third IEEE International Conference on Fuzzy Systems, (1994) 249–253
13. Tao C. W., Thompson W. E., and Taur J. S., A fuzzy If -Then approach to edge detection, Proc. of 2rd IEEE International Conference on Fuzzy Systems, (1993) 1356–1361
14. Yao X. H., The research of multiple-sensor based local planning and navigation for the autonomous land vehicle, The Master Thesis, Department of Computer Science, Tsinghua University, 1994
15. Pin F. and Watanabe Y., Driving a car using reflexive behavior, Proc. of the 2nd IEEE International Conference on Fuzzy Systems, (1993) 1425–1430
16. Ruspini E. H., Fuzzy logic-based planning and reactive control of autonomous mobile robots, Proc. of the 1995 International Conference on Fuzzy Systems, (1995) 1071–1076
17. Li W., Fuzzy-logic-based reactive behavior control of an autonomous mobile system in unknown environments, Engineering Application of Artificial Intelligence, **7**(5) (1994), 521-531
18. Zadeh L. A., Fuzzy sets, Information and Control, **8** (1965), 338–353

Author Index

Lecture Notes in Artificial Intelligence (LNAI)

Lecture Notes in Computer Science